Advances in Industrial Engineering, 16

Quality through Engineering Design

Advances in Industrial Engineering

Vol. 1. Performance Criteria and Incentive Systems (S. Globerson)
Vol. 2. Techniques for Multiobjective Decision Making in Systems Management (F. Szidarovszky, M. E. Gershon and L. Duckstein)
Vol. 3. Analysis and Design of Hierarchical Control Systems: With Special Reference to Steel Plant Operations (T. J. Williams, ed.)
Vol. 4. Productivity Improvement: Employee Involvement and Gainsharing Plans (E. M. Dar-El)
Vol. 5. Manufacturing Research: Organizational Issues (A. Gerstenfeld, H. J. Bullinger and H. J. Warnecke, eds.)
Vol. 6. Innovation: A Challenge to the Engineer (K. Holt)
Vol. 7. Manufacturing Research Perspectives: U.S.A.–Japan (A. Gerstenfeld, C. Ganz and T. Sata, eds.)
Vol. 8. Applications of Fuzzy Set Methodologies in Industrial Engineering (G. W. Evans, W. Karwowski and M. R. Wilhelm, eds.)
Vol. 9. People and Product Management in Manufacturing (J. A. Edosomwan, ed.)
Vol. 10. Integrative Management, Innovation and New Venturing: A Guide to Sustained Profitability (M. Terkel)
Vol. 11. Surface Modeling for CAD/CAM (B. K. Choi)
Vol. 12. Planning and Managing Industrial Logistics Systems (E. R. Sims, Jr.)
Vol. 13. Computer-Aided Process Planning (H. P. Wang and J. Li)
Vol. 14. Productivity Measurement and Management at the Company Level: The Japanese Experience (K. Kurosawa)
Vol. 15. Industrial Applications of Knowledge-Based Diagnosis (G. Guida and A. Stefanini, eds.)
Vol. 16. Quality through Engineering Design (W. Kuo, ed.)

Advances in Industrial Engineering, 16

Quality through Engineering Design

Edited by

Way Kuo

Department of Industrial &
Manufacturing Systems Engineering
Iowa State University
Ames, IA, U.S.A.

Managing Editor

Marcia Martens Pierson

Department of Industrial &
Manufacturing Systems Engineering
Iowa State University
Ames, IA, U.S.A.

ELSEVIER
Amsterdam–London–New York–Tokyo 1993

ELSEVIER SCIENCE PUBLISHERS B.V.
Sara Burgerhartstraat 25
P.O. Box 211, 1000 AE Amsterdam, The Netherlands

ISBN: 0 444 89667 8

This book is printed on acid-free paper.

Printed in The Netherlands

Preface

Robert Lundegard of the National Institute of Standards and Technology (NIST), Kimiko Bowman of the Oak Ridge National Laboratory (ORNL), and P. K. Pal of the Indian Statistical Institute initiated the idea for the Quality through Engineering Design Golden Conference planned for Bangalore, India, January 11-14, 1993. The original conference planning was also strongly endorsed and supported by P. L. Heydemann of the U.S. Embassy in Delhi, and Junichi Noguchi of the Union of Japanese Scientists and Engineers. With these endorsements, we were able to obtain key financial support from various organizations.

BACKGROUND

Competitiveness in the world market is an essential element of economic development. Focusing on advanced materials and parts manufacture, this conference intends to foster understanding and collaboration among scientists and engineers in the fields of conceptual approaches to engineering design, computational approaches in design, statistical design and analysis, reliability, and quality control.

In order to disseminate the state-of-the-art information on engineering design in the context of quality, we have invited twenty world-class scholars to present their views and to share their visions with each other. Also included in the program are fourteen contributed presentations and a number of poster papers solicited and reviewed from over one hundred submittals from many countries in response to the call for papers. In all aspects, we emphasize originality, applicability, quality, and relevance to the theme of quality through engineering design. Indeed, the conference addresses, perhaps for the first time, the timely issues of quality and engineering design in a formal way.

EDITING

All the invited and contributed papers have been extensively reviewed by specialists of an editorial review board who provided technical suggestions for authors of accepted papers. In addition, Marcia Martens Pierson of the Engineering Editorial Services Office at Iowa State University (ISU) carefully edited the papers and the staff at its Technical Illustration Office formatted the papers and prepared camera-ready copy. Through an iterative process and careful design, the end result gives birth to the final version of this book.

GOAL

In addition to the quantitative models presented in this book, applications, strategic considerations, and validated managerial practices are also considered. It is my hope that readers will find this book useful. Also this book illustrates the great value of developing communication channels between engineers and statisticians and between academic theoreticians and industrial practitioners and of differentiating future design needs from the present design practice. Although the conference is entitled the "Indo-U.S.-Japan Conference on Quality through Engineering Design," the intent is to present problems and solutions across the boundaries of nations and cultures. Indeed, it is this quality that is essential to expanding world economics and enhancing human living standards.

USAGE

Engineering faculty and graduate students will find this book useful as a supplementary text for graduate courses in concurrent engineering and engineering design where they learn the stochastic nature of design and the essential elements of design for quality and reliability. For statisticians, the book provides information on applying quality and reliability methodologies to engineering problems that will help them collaborate with engineers and also provides challenging problems in robust design for graduate students in statistics. Graduate students in management will find the strategic considerations of quality useful in today's business because it is quality that guarantees success; this book provides first-hand information in that respect.

This book can also be used as a reference book for engineering design practitioners who put emphasis on issues related to quality design. They will benefit from modern statistical techniques and tutorials presented in the book and from the many successful, real applications of quality and reliability techniques to various design projects.

ACKNOWLEDGMENTS

Both the Indo-U.S.-Japan Conference and publication of this book are supported by the National Institute of Standards and Technology (NIST), National Science Foundation (NSF), and Army Research Office of the United States of America; and the Indian Statistical Institute (ISI) of India. In addition, the Union of Japanese Scientists and Engineers of Japan has cosponsored the conference, and the College of Engineering at ISU and Collins Avionics and Communications Division of Rockwell International have committed funds to support publication costs. In particular, we would like to acknowledge Thom Hodgson of the Division of Design and Manufacturing Systems and O. Shinaishin of the Division of International Programs, both with NSF; David T. Kao and George Burnet of the College of Engineering at ISU; and Keith Jessen of Rockwell International.

Many have contributed their time and knowledge for publishing this book. In particular, Robert Lundegard of NIST and Kimiko Bowman of ORNL, who have planned all activities and participated in every detail from the beginning, deserve special recognition. Steve Carpenter of NIST, and B. L. S. Prakash Rao, M. A. Gopinath, and J. K. Ghosh of ISI are acknowledged for their support and promotion of activities involved in organizing the conference and publishing the book.

On behalf of the Advisory Board Members, I want to thank the invited speakers whose original papers make this book a truly valuable and quality document in engineering design. I admire their dedication and sincerity in contributing to this historical activity without requesting any nonacademic compensation. We are also very grateful for the contributed authors who enlighten the theme of design and substantiate the breadth of this subject through their practice.

We are indebted to the following individuals for their technical evaluation and constructive comments made on the selected papers: Robert Lundegard of NIST, Kimiko Bowman of ORNL; John Jackman, Roger Berger, Lin-Lin Chen, and Richard Linn of the Industrial and Manufacturing Systems Engineering Department (IMSE), Steve Vardeman of IMSE/Statistics, Pat Patterson of the Materials Science and Engineering Department, all with ISU; and Raghu Kacker of NIST.

I would like to thank Marcia Martens Pierson and her associates Kurt Plagge, David Sauke, Carol Gostele, and Maxine Bogue in the ISU Engineering Editorial Services Office and Kary Chien and Tom Lynch of the IMSE Department and Kurt James for their first-class job in preparing this book. Furthermore, both Donna Cerka of IMSE at ISU and Kaye Wade of the Computing and Applied Mathematics Laboratory at NIST are acknowledged for their exceptional administrative support. Such dedicated supporting staff have made this undertaking possible.

From conception to final product, it is the teamwork that makes *Quality through Engineering Design* a quality design and hence a quality product. To those mentioned above and those who have worked behind the scenes, I extend my grateful appreciation.

Contents

Introduction

Competitiveness in international markets is now recognized as an essential element of national economic development. Quality of products, as perceived by the customers, is the most important factor in facing this competition. Achieving competitive levels of quality requires leadership from the top, good management practices, effective and efficient operating systems, and use of appropriate state-of-the-art engineering design approaches.

This book focuses on quality-related engineering design. The application areas of particular interest are advanced materials formulation and processing as well as design and manufacture of electro-mechanical devices. Among the many considerations for designing products that are robust to manufacturing variations and customer use are concurrent engineering, computer-aided simulation, and physical experimentation. Combining these areas with a focus on quality is a highly complex task toward which engineers, computer scientists, statisticians, and operation research analysts must work. More extensive computer simulation and physical experimentation are required to achieve quality goals and reduce the risk of poor quality to levels that will justify the necessary capital investment in tooling and market development. It is hoped that the results of such research will add greater power to the now widely accepted computer-aided design (CAD) systems.

PURPOSE

The purpose of this book is to encourage the exchange of information on computer-based and statistical approaches to the engineering design of quality products and processes. Computer-aided design and manufacturing (CAD/CAM) approaches have been found to be very effective in the design and manufacture of electro-mechanical devices. Statistical approaches have been found to be effective in process modeling and optimization. This book addresses the question of building quality into products and processes through design. Conceptual frameworks and the use of computer-based approaches and experimentation are featured considerations. Case histories are given an explicit emphasis. The intent is to promote a better understanding and collaboration between the world's leading scientists and engineers in these fields and at the same time benefit manufacturing industries.

CONCURRENT ENGINEERING

Concurrent engineering, also known as simultaneous design, is being recognized as the most effective and efficient approach to engineering design.

Because concurrent engineering is a dramatic departure from past practice, there is an immediate need for a new design environment and technology requiring extensive interdisciplinary cooperation and integration of such diverse fields as engineering, manufacturing, material management, and marketing.

TRADITIONAL PRODUCTION TECHNOLOGY

Prior to the Industrial Revolution, individual craftsmen typically directed the production of an item needed to perform a certain function all the way from the basic recognition of need for the item through its initial conceptualization, design, and manufacture. In the process, single individuals usually considered all the problems of how the product should be configured, what materials should be used, and how they should be worked. Then they produced them, often one item at a time, by using simple hand tools and a fair amount of customized fitting or mating of parts. Depending on the skills of the craftsmen, the result was often a high-quality item capable of fulfilling its intended function over a long period of time. In the meantime, such items were rather expensive and were not widely available.

The Industrial Revolution, however, introduced mass production with standardization and reliance on machine tools that were operated by a widely available, but relatively unskilled labor force. Lately, increasingly complex and yet inexpensive products have resulted from the use of practices instituted at that time. On the other hand, it is now well recognized that a number of existing practices inimical to today's manufacturing environment of high product mix with low production volume find their roots in the very methods that made the Industrial Revolution successful. For example, increasing job specialization on the one hand and the separation of product designs, production, distribution, and field maintenance functions on the other have led to myopic engineering decisions that have a negative impact on ultimate product quality and cost.

RECENT DEVELOPMENTS

Recent efforts toward the integration of product and process design, involving design for manufacturing, design for assembly, design for quality and reliability, design for automation, and the like, attempt to address some of the problems spawned by job specialization and separation. The new concept of integrating engineering design with system manufacturing yields such significant advantages as reducing market lead time, increasing product quality and reliability, and reducing life cycle cost.

Other examples of modern manufacturing problems stemming from methods of the Industrial Revolution might include the negative effects of outdated standards and mass production techniques and principles. Application of these principles to the manufacture of complex products tends to create large, costly inventories and an inability to respond to rapid market changes. These problems are, to some extent, addressed by recent trends toward flexible automation, just-in-time production, new scheduling practices, increasingly precise machine tools, and systematic allocation of critical tolerances, as well as the relaxation of noncritical tolerances.

Despite the fact that many of the trends just mentioned are important, experts believe that they lack an over-arching rational framework to guide their implementation and that they do not always take full advantage of modern managerial and technological capabilities. New capabilities in measurement technology, information processing, automation, precision engineering, and machine control and statistical quality management demand a rethinking of the entire product engineering process. This rethinking should also result in new paradigms for manufacturing organizations.

Bear in mind that manufacturing industries use many types of planning activities, descriptive documents, information resources, expertise, and strategies to model the manufacturing process for a product. A key issue in future manufacturing will be the use of human creativity to manufacture things designed to fit people. Too often people are expected to adjust to the design.

ENGINEERING DESIGN

Engineering design continuously plays an important role in manufacturing because of the shorter product lives observed by the users. Many have recently discussed design theory and philosophy and have extended it to investigating design methodologies. Since there are many technology designers, one would believe that studying design theory could improve and enhance the design capability. However, Yoshikawa [1] pointed out that fewer commonalities are found among the designers of various technologies because of lack of communication. In fact, the necessity of studying design theory has been controversial. Many design theoreticians emphasize creativity, intelligent CAD/CAM, AI, design process/protocols, information processing, and others in a deterministic context. It is a pity, in my opinion, that quality and reliability are overlooked by these theoreticians even though quality and reliability are among the key factors to be evaluated by the end-users of the designed products.

QUALITY ENGINEERING

Although statistical methods, in particular, experimental design, have existed and been applied to many disciplines for years (including Box's evolutionary operation [2]), classical manufacturing and engineering approaches were too often viewed as deterministic until recently when the quality engineering approach proliferated because of Taguchi's effort in promoting experimental design. The utmost goal for manufactured products should be to enhance human living standards, which, however, is not achievable through cost-accounting principles. Instead, it is quality that warrants profit because of world competitiveness and short product life cycle. Such a quality warranty can be provided, according to Taguchi, by using experimental design for products or processes so that they are robust to environmental conditions and component variations and so that they minimize variation around a target value.

QUALITY IN SOFTWARE DEVELOPMENT—AN EXAMPLE

Poor quality software that results in damage to the user and possible lawsuits is too costly for any business to accept and must be avoided by prevention instead of repair. According to Deming [3], business drifts to the lowest bidder without adequate measures of quality, and low quality and high cost are the inevitable result. The same argument applies equally to software development. All these quality concerns have to be embedded in the design stage.

A distinction is usually drawn between the product and the process used to manufacture the product. Similar distinctions can be drawn between the product of a piece of software, that is, the output, and the software itself, which is analogous to the factory. With proper justification, modeling quality software development is no different from modeling the manufacturing plant.

A typical quality programming process can be divided into the following stages:

- modeling
- requirement specification
- concurrent software design and test design
- test and integration
- software acceptance

At each stage, quality control and design are an integral part of the software development activities. Furthermore, because of the dynamic nature of software usage, reliability testing and life-cycle cost analysis should be considered before and after delivery.

Once the life-cycle cost analysis is completed, a strategy can be defined either to minimize the cost subject to reliability and quality requirements or to maximize the reliability and quality standards subject to a restricted cost plan. There may not be a unified solution for all products and manufacturing processes, but there certainly exists an optimal effort to achieve the objectives. According to Taguchi and Phadke [4], the main technological reason for Japan's success in producing high-quality products at low cost is the emphasis on optimizing both the product and process designs for manufacturability, quality, and reliability as well as the availability of efficient experimentation and simulation methods for performing optimization. A complete software life-cycle analysis based on the concept of simultaneous design is suggested in Fig. 1.

TECHNICAL SCOPE

The book features successful applications of new methods for the design of products and processes and also focuses on needs and opportunities for improved approaches beyond the demonstrated state-of-the-art. Engineers can gain insight into the requirements that must be faced in the design of world-class products. Invited papers and carefully selected contributed papers display

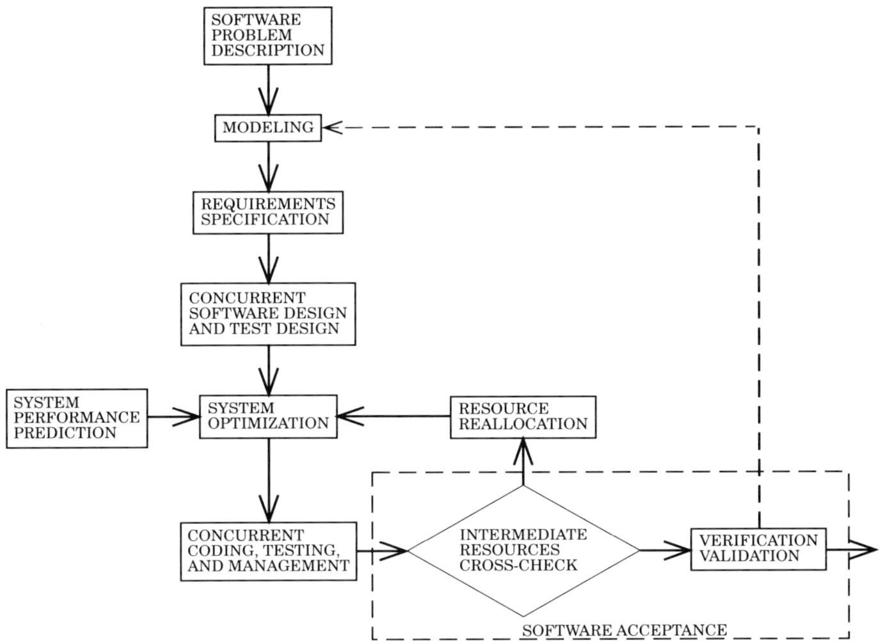

Fig. 1. A process for reliability-related quality programming.

the breadth of successful or promising approaches involving the engineering, statistical, computing, and systems or operations research disciplines. Each author is a leader in a particular technical community that has recognized and specialized strength in statistical analysis, experimental design, engineering design, computer simulation, and material technology. The book is organized in eight major topical areas.

1. **Conceptual and Methodological Design**. Here we lay the foundations for conceptual design and then discuss design theory from a scientific point of view to enable researchers to evaluate design performance systematically. Approaches are presented in which quality is formally linked to design at the conceptual design stage and in which variations in design and measures of design performance are recognized as well. Taguchi discusses how design tasks are performed at different sites in the United States and Japan and then compares robust technology developments in both countries. A review and report on some recent advances are made in on-line quality control methods, statistical process control, and automatic process control. Reviews of the fundamentals of the axiomatic approach and the implications of the design axioms on quality control are also discussed. Finally, one paper highlights the use of discrete optimization to build quality into the design of circuits and networks.

2. **Management through Strategic Considerations of Design.** As illustrated in the "Quality in software development—an example" section, product and process should be designed strategically, including not only the tools and methodologies employed but also the design philosophy carried out by the designer and the associated life-cycle cost along with its optimization. Considering total quality control and management, we present in this chapter many case studies ranging from systematic fool proofing techniques to off-highway dump truck and spacecraft mechanics practiced in a number of industries in Japan, Canada, and India.

3. **Robust Design and Data Analysis.** A design is called robust if the product or process performs consistently on target and hence is insensitive to factors that are difficult to control. According to Taguchi's philosophy on quality engineering, three stages are considered in the development of a product or process: system design, parameter design, and tolerance design. Robust design in this chapter includes a generalization of the blocking concept in the design of experiments through stratified replications, suggestions of statistical methods for the target mean problem and variance components estimation, statistical properties of Hadamard matrices of Order 16, application of experimental design to industrial problems, multivariate generalization of Taguchi's signal-to-noise ratio for industrial experiments, and exploratory data analysis for the characterization of high-performance turning machines. This chapter introduces many new methods and techniques, including block plots to examine the interaction among various factors.

4. **Statistical Approaches and Reliability.** Applications of statistical approaches and reliability are introduced in the design process. Controlling statistical uncertainties is the key to producing high-quality products. Both experimental design and a tutorial on accelerated testing are provided along with an introduction of time-dependent availability for modeling performance of flexible manufacturing systems and electronic parts. Reliability techniques are implemented in design-engineering problems including those of noncoherent systems.

5. **Tolerances and Discrete Parts Design.** Tolerance design and design for assembling parts are discussed. We intend to find ways for hardware to quantify design tolerance and specify design parameters. Excellent case studies are discussed to illustrate both the feasibility and shortfalls of the present design methodology. First, issues in assessing conformity of dimensions of manufactured parts to design tolerance are investigated. Dimensional tolerancing is a language used to communicate product quality specifications. The weaknesses of the language of tolerances is discussed and suggestions for improving research and standards are provided. Readers can find very original work related to tolerances and discrete sampling with coordinate measurements in this chapter.

6. **Finite-Element Method for Design**. The finite-element method is a widely applied tool used in solving engineering problems. In this chapter, the robust-design concept is formulated in the context of the finite-element method to ensure response on target and to render the design insensitive to variation. Because of its versatility and adaptability, the finite-element method is also illustrated in analyzing performance of multipad disc brakes and tilting-pad journal bearings. All these applications permit a wider extension of CAD for robust design.

7. **Computing in Design.** Issues related to computation, system modeling, and system simulation are presented. This chapter discusses research on optimal sampling procedures, executable CAD/CAM software, a nonlinear optimization technique, polynomial regression, computer simulation, and knowledge-based integrated system approaches for engineering design. Optimal sampling procedures and case studies for power transmission systems, dry cells, and optic fibers are discussed as well. It is shown in this chapter that engineering design is truly a multifaceted subject that requires various computation techniques.

8. **Materials Processing and Metrological Design.** Since modern products influence large manufacturing markets, we address quality and reliability from a life-cycle approach that can best provide a global view of design. Materials processing plays a very important role in this aspect of design. The subprocesses involved in the manufacture for advanced ceramic materials are discussed by highlighting major parameters that affect properties of the ceramics. A case for the application of statistics in the development of robust processes and its role in experimentation is also described. At Japan's National Research Laboratory of Metrology, a computer-aided measurement and process design and control system (CAMPS) is introduced for plastic and ceramic injection molding processes as a system applied for securing production quality prior to starting mass production. It emphasizes the important role of product design and production process design.

Quality in engineering design is thus addressed in this book in a formal way; that is, the book not only sets forth the details on a philosophical basis but also sets up a science-based foundation for design considerations and applications in manufacturing and software engineering.

REFERENCES

1 Yoshikawa, H., Design philosophy: The state of the art, in: Proc. 39th CIRP General Assembly, Annals of the CIRP 38, no. 2, 1989, pp. 579-586.

2 Box, G. E. P., Appl. Stat., 6, 1957.

3 Deming, W. E., Quality, Productivity and Competitive Position, M.I.T. Center for Advanced Engineering Study, Cambridge, Mass., 1982.

4 Taguchi, G., Phadke, M. S., Quality Engineering Through Design Optimization, International QC Forum II, No. II, 1985, pp. 27-48.

Chapter 1
Conceptual and Methodological Design

Here we lay the foundations for conceptual design and then discuss design theory from a scientific point of view to enable researchers to evaluate design performance systematically. Approaches are presented in which quality is formally linked to design at the conceptual design stage and in which variations in design and measures of design performance are recognized as well.

Taguchi discusses how design tasks are performed at different sites in the United States and Japan and then compares robust technology developments in both countries. A review and report on some recent advances are made in on-line quality control methods, statistical process control, and automatic process control. Reviews of the fundamentals of the axiomatic approach and the implications of the design axioms on quality control are also discussed. Finally, one paper highlights the use of discrete optimization to build quality into the design of circuits and networks.

Quality through Engineering Design - W. Kuo (Editor)

Robust Technology Development

G. Taguchi

2-12-15 Higashi-Nakano
Nakano-Ku Tokyo 164 Japan

Abstract

In order to improve the quality and cost of any product, we need a good product design and good manufacturing processes. Drawings and specifications are results of design activity. How to develop good specifications and drawings quickly for engineering research work is the most important strategy for any company in meeting world competition.

What Japanese engineers have been doing is to improve product design and processes mostly at the actual design stage after product planning has occurred, while United States (American) engineers at research and development (R&D) laboratories improve them before product planning. Results of basic technology development at the R&D stage are difficult to repeat or reproduce in downstream conditions, large-scale manufacturing conditions, and various customer's conditions because those conditions are quite different from the ones in the laboratory. Since Japanese engineers do most of their research work at the actual design stage on actual products and processes rather than at the R&D stages, they naturally have better repeatability and reproducibility than researchers working with small-scale laboratory experiments. However, the Japanese way requires a high budget, many engineers, and overtime. Currently Japanese companies are suffering from a shortage of engineers.

The best way to do engineering work is in laboratories as is done in the United States. However, we have to find a method to improve the repeatability of the technology developed at R&D laboratories in downstream conditions. That is the subject of this paper.

QUALITY AT DIFFERENT STAGES

In quality engineering, improving the quality of shipped products requires starting with the most upstream stage. For this purpose, research is divided into the following stages.

Downstream quality

In the case of a car engine, for example, downstream (user/consumer) quality includes such concerns of the customer as fuel consumption, noise, vibration, functional failure rate, harmful contents in exhaust gas, etc. These downstream quality items are important for management.

Midstream quality

Midstream quality involves the manufacturing arena and includes such items as dimensions in the drawing or specifications. These quality concerns are taken into account in process or product control and are measurable by gauges or measuring devices. Midstream quality is important on the production floor and at trading.

Upstream quality

Upstream quality includes concerns relating to product development, such as smaller-the-better, larger-the-better, nominal-the-best, or operating window. In quality engineering, using nondynamic signal-to-noise (S/N) ratios such as dimensions or other quality characteristics obtained under noise conditions is often recommended. Nondynamic characteristics, as substitutes for dynamic characteristics that will be explained later, are often applied in product development as the next best approach to dynamic characteristics.

Furthermost upstream quality

Furthermost upstream quality pertains to technology development and is the measure of how close an engineering basic function is to the ideal performance. Such a measure is called the dynamic characteristic. The ability to machine to any target value for engine shape and transformability of fuel to energy belong to this category. It is important in quality engineering to study the robustness of the generic function of technology.

The objective of this paper is to explain dynamic characteristics through two examples with the intent of helping change the style of development in Japan from a design-oriented one to a technology-oriented one. Development work in the United States and Europe is done within research and development (R&D) laboratories, where product development and manufacturing are more technology oriented. In Japan, most design development starts after product planning. As a result, product design and manufacturing engineering divisions in Japan are large, while in the United States and Europe smaller divisions attempt to accomplish the same tasks. One reason is that research projects in Japan are product-to-product in nature and obviously do not fit well with the new era of multivariety and small quantity production.

In the United States and Europe, however, technology development is not developing a particular product design or a process design for an already planned product but rather developing the technology most applicable to a wide range of products including future products. Technology development has the following features:

- precedency before product planning
- flexibility
- repeatability

It is important to develop a robust technology in such a way that is easily transferable to various product designs and manufacturing processes. Therefore, no particular product is targeted in technology development but flexibility is required. This concept is applicable to manufacturing areas also. Instead of trying to manufacture products to meet the given drawing, a flexible technology should be established so that the technology may be applied for various drawings within a certain range. Precedency before product design and flexibility are therefore required. Japanese executives have finally realized this necessity and have started to invest budget for fundamental and basic technology development. It is a good trend, but there is a problem.

In the United States and Europe where much more R&D work is conducted, the problem is repeatability at downstream conditions, one of the above requirements. R&D work has been hindered by the fact that the technology developed in a laboratory cannot be reproduced in large-scale production or in the market. In Japan, however, actual products are developed in design or manufacturing areas, and fewer problems in repeatability occur compared with the United States and Europe. In order to improve repeatability, the concept of conducting product design and manufacturing techniques at the same time (called concurrent engineering or simultaneous engineering) through the use of the S/N ratio for nondynamic characteristics has been brought out. However, the problem with regard to research efficiency still exists. The nondynamic characteristic-type S/N ratio (or operating window) approach belongs to the next-to-the best approach; the best approach is to use dynamic characteristic-type S/N ratios. Using the dynamic characteristic approach creates individual signals for the design of individual products and is followed by the evaluation of the technology. The author has been increasingly advocating this direction and is gaining more alliances. The following example shows why nondynamic characteristics (or an operating window) is the next-to-the best approach.

LARGER-THE-BETTER CHARACTERISTIC, THE CASE OF STRENGTH

Company S in Taiwan manufactures sticking paper (one side glued) with a certain sticking strength. Customers require a specific strength for a specific area of paper, and the company suffered market problems when they did not meet the specification.

As shown in Fig. 1, paper is glued over a required width and length, and the unglued part is pulled with 1 kgf force and is required to hold at least 10 sec. Evaluation by such a test is called specification. Specification, provided to prevent quality problems in the market, is also called a quality characteristic. Quality characteristics are given by the customer and sometimes provided beforehand by the manufacturer. However, evaluation through specification is an inefficient method and should not be used in technology development. This is a very important point for quality engineering. Specifications are useful for daily control during the manufacturing stage or for design review, but both are for inspection and not for improvement. They are useful only for inspection after the improvement to see if the quality meets the customer's requirement, i.e., midstream quality.

Nondynamic characteristics are more technically oriented than inspection but not overly so. The S/N ratio of a larger-the-better characteristic is to measure the tensile strength under various noise conditions (e.g., humidity, deterioration, or ability of objects to stick) and treats the outcome as the larger-the-better where the mean is maximized and variability is minimized. Although it is not absolutely objectionable, it is not recommended from a quality engineering viewpoint for the following reason.

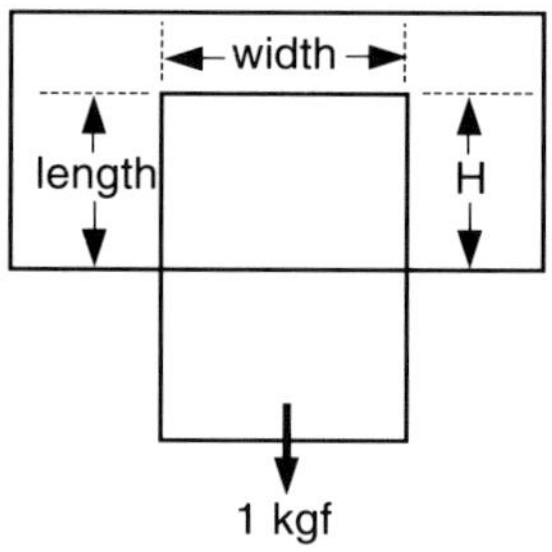

Fig. 1. Pulling strength test.

By testing with a certain method, we get the average of a certain result. Figure 1 shows that the test result indicates the average strength of a certain area determined by a given width and length. We do not know whether the paper sticks uniformly across the whole area or if the upper half has 1.5 times the average and the lower half is 0.5 times the average.

The author asked the people from Company S if it were true that the function of the glued paper is to stick with uniform strength at any point in the area. They agreed it was. In using a S/N ratio for dynamic characteristics, every point has uniform strength, which means that strength is proportional to the area. Determining the strength per unit area in order to meet the customer's requirement is called a tuning or adjusting activity in quality engineering. For such purposes, we need to determine what kind of test must be conducted and how to improve the quality.

Measuring the object to check if the objective is met is nothing but inspection. Many testing methods for dynamic characteristics are possible to determine whether strength is proportional to area. Different methods and measurement technology create a difference in technology level. The author inquired of the company whether a continuous peel-off tester was available. When the answer was yes, the following testing method was suggested. After considering the future market, we set the following levels as the signal factors for width on test pieces:

$M_1 = 5$ mm
$M_2 = 15$ mm
$M_3 = 50$ mm

When the width of a test piece is equal to zero, the peel-off strength must be equal to zero and the peel-off strength must also be proportional to the width.

After a test piece is stuck as shown in Fig. 1, the lower bottom is peeled off vertically from the stuck object. Peel-off strength is measured continuously to obtain the results shown in Fig. 2.

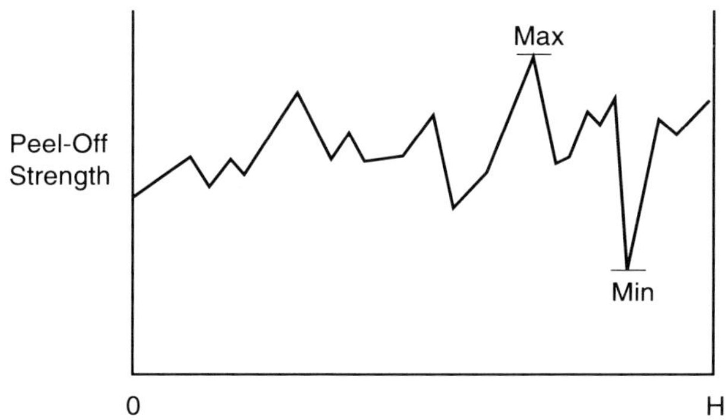

Fig. 2. Peel-off strength.

The maximum and minimum peel-off strengths are obtained, and such data are collected under the following noise factor combinations:

K: Sticking Object — K_1 = Cloth
K_2 = Plastic board

P: Deterioration — P_1 = After 5 minutes
P_2 = After 24 hours
P_3 = After deterioration test

Thus the total number of noise factor combinations is 2 x 3 x 2 = 12 as shown in Table 1. This set of data will provide the decomposition shown in Table 2.

Table 1. Data for dynamic characteristic.

		K_1			K_2		
		P_1	P_2	P_3	P_1	P_2	P_3
M_1	Min.	y_{11}	y_{21}	y_{31}	y_{41}	y_{51}	y_{61}
	Max.	y_{12}	y_{22}	y_{32}	y_{42}	y_{52}	y_{62}
M_2	Min.	y_{13}	y_{23}	y_{33}	y_{43}	y_{53}	y_{63}
	Max.	y_{14}	y_{24}	y_{34}	y_{44}	y_{54}	y_{64}
M_3	Min.	y_{15}	y_{25}	y_{35}	y_{45}	y_{55}	y_{65}
	Max.	y_{16}	y_{26}	y_{36}	y_{46}	y_{56}	y_{66}

Table 2. Decomposition of variation.

Source	f	S	V	
β	1	S_β		
$\beta \times N$	11	$S_{\beta \times N}$	$V_{\beta \times N}$	$\}V_N$
e	24	S_e	V_e	
T	36	S_T		

If such a data collection is too costly or time consuming, noise factors K and L can be compounded. In this example, all data in Table 1 are used.

The total sum of squares S_T is decomposed into the variation because of the proportional term S_β, variation of the proportional constant (change of sensitivity) $S_{N\times\beta}$, and residual variation S_e. Calculations are made as follows (f denotes the degrees of freedom):

$L_1, L_2, \ldots, L_{12}$ are linear forms when $M_1 = 5$, $M_2 = 15$, and $M_3 = 50$.

$$S_T = y_{11}^2 + y_{12}^2 + \ldots + y_{66}^2 \qquad (f = 36) \tag{1}$$

$$S_\beta = \frac{(L_1 + L_2 + \ldots + L_{12})^2}{12r} \qquad (f = 1) \tag{2}$$

$$S_{N\times\beta} = \frac{L_1^2 + L_2^2 + \ldots + L_{12}^2}{r} - S_\beta \qquad (f = 11) \tag{3}$$

$$S_e = S_T - S_\beta - S_{N\times\beta} \qquad (f = 24) \tag{4}$$

where

$$\left.\begin{aligned} L_1 &= M_1 y_{11} + M_2 y_{13} + M_3 y_{15} \\ L_2 &= M_1 y_{12} + M_2 y_{14} + M_3 y_{16} \\ &\vdots \\ L_{11} &= M_1 y_{61} + M_2 y_{63} + M_3 y_{65} \\ L_{12} &= M_1 y_{62} + M_2 y_{64} + M_3 y_{66} \end{aligned}\right\} \tag{5}$$

magnitude of input r is given by

$$\begin{aligned} r &= M_1^2 + M_2^2 + M_3^2 \\ &= 5^2 + 15^2 + 50^2 \\ &= 2{,}750 \end{aligned} \tag{6}$$

Linear equation L divided by the magnitude of input r is the peel-off strength per unit area β.

$$\beta = \frac{L_1 + L_2 + \ \dots \ + L_{12}}{12r} \tag{7}$$

is the average proportional constant.

The following value

$$S = 20 \log \beta \tag{8}$$

or

$$S = 10 \log \frac{1}{12r}\left(S_\beta - V_e\right) \tag{9}$$

is called sensitivity. The error variance is given by

$$V_e = \frac{1}{24} \times S_e \tag{10}$$

Although either Eq. (8) or (9) may be used as sensitivity, Eq. (9) is usually used. Equation (8) is used to adjust the average strength or for tuning.

For the dynamic characteristic S/N ratio or the most important index to improve the average sticking strength in this example, neither $S_{N\text{x}\beta}$ or S_e are desirable. Therefore, the S/N ratio is calculated by

$$\eta = 10 \log \frac{\frac{1}{12r}\left(S_\beta - V_e\right)}{V_N} \tag{11}$$

where

$$V_N = \frac{S_{N \times \beta} + S_e}{35} \tag{12}$$

It is important in improving the S/N ratio and also for tuning purposes to discover the factor that does not affect the S/N ratio much but significantly affects sensitivity. For such purposes, experiments should be designed by using orthogonal arrays that allow control factors such as glue temperature, binder type, or method of binder application to be varied.

THE CASE OF BIMETALS

A bimetal consists of two different metal plates that are stuck together and perform an off/on function by using warping due to temperature change. Since this is a digital function, the operating window has been used for improving product function.

In developing a switch under a sudden electric current change or with an excessive current, the operating window (which is a nondynamic characteristic) is used, and tests are made by the following method. Since electric current is the input signal, it is used as a measuring characteristic; and the stability of the on/off function is investigated. A two-level noise factor is set as

N_1 = negative-side worst condition

N_2 = positive-side worst condition

N_1 is the condition of noise factors, such as humidity, component variability and before-and-after deterioration, to perform from the on-to-off situation at a lower current. The opposite condition is N_2. Current is varied at a certain interval (using the up and down method), and the threshold current that changes the switch from off to on, x, and from on to off, y, is measured. Results are shown in Table 3.

Table 3. Threshold values.

	x	y
N_1	x_1	y_1
N_2	x_2	y_2
Total	X	Y

The magnitude of the operating window h and the two types of sensitivity S_m and S_{x-y} are calculated by the following equations:

$$\eta = 10\log\frac{\frac{1}{4}(S_m - V_e)}{V_e} \quad (13)$$

$$S = 10\log\frac{1}{4}(S_m - V_e) \quad (14)$$

$$S_{x-y} = \frac{1}{2}(Y - X)^2 \quad (15)$$

$$S_m = \frac{(X+Y)^2}{4} \quad (16)$$

$$V_e = \frac{\left(x_1^2 + x_2^2 + y_1^2 + y_2^2 - \dfrac{X^2 + Y^2}{2} \right)}{2} \tag{17}$$

The factor that does not affect the S/N ratio but affects $x–y$ or $S_{x–y}$ is used for tuning hysteresis. The factor affecting S_m is used to set the center current between on and off.

Although the method described here is generally recommended as one of the operating window approaches, the use of dynamic characteristics is more effective. Using the dynamic characteristic, one can investigate the relationship between current and displacement of the tip to see how the relationship follows Hooke's Law.

The input signal M is current and the output y is displacement. On/off data are not measured. The compounded noise factor is set the same way. The amount of displacement y is measured when the signal factor is changed at three levels: M_1, M_2, and M_3, and the noise factor is changed at N_1 and N_2, where N_1 and N_2 are the compounded noise conditions to make the displacement minimum and maximum, respectively. The data will be such as that shown in Table 4.

Table 4. Displacement.

	M_1	M_2	M_3
N_1	y_{11}	y_{12}	y_{13}
N_2	y_{21}	y_{22}	y_{23}

The dynamic characteristic S/N ratios of the proportional equation h and sensitivity S are calculated as follows:

$$\eta = 10 \log \frac{\frac{1}{r}\left(S_\beta - V_e\right)}{V_N} \tag{18}$$

$$S = 10 \log \frac{1}{r}\left(S_\beta - V_e\right) \tag{19}$$

$$S_T = y_{11}^2 + y_{12}^2 + \; \dots \; + y_{23}^2 \tag{20}$$

$$S_\beta = \frac{\left(L_1 + L_2\right)^2}{2r} \tag{21}$$

$$V_N = \frac{S_T - S_\beta}{5} \tag{22}$$

$$V_e = \left(S_T - \frac{L_1^2 + L_2^2}{\frac{r}{2}} \right) \times \frac{1}{4} \tag{23}$$

$$\left.\begin{aligned} L_1 &= M_1\, y_{11} + M_2\, y_{12} + M_3\, y_{13} \\ L_2 &= M_1\, y_{21} + M_2\, y_{22} + M_3\, y_{23} \end{aligned}\right\}$$

$$r = 2\left(M_1^2 + M_2^2 + M_3^2\right) \tag{24}$$

These calculations are made for the S/N ratio of the basic proportional equation. The control factor affecting sensitivity S is used for tuning, where the positions of on and off are set according to the customer's requirement. The tuning level varies depending on the extent to which the S/N ratio can be improved. Note that the calculation of S/N itself is not the important thing. The important thing is the thought process involved in coming up with the appropriate dynamic characteristic.

CONCLUSIONS

Two examples have been used to explain the dynamic characteristic S/N ratio that is more desirable than the nondynamic type in terms of efficiency because it fits basic technology better. Since engineers tend to use some basic function that essentially uses input/output relationships to achieve the objective function, all functions can be converted into dynamic characteristics.

The problem with R&D in the United States and Europe is that the results of research are not repeatable in downstream conditions. In order to conduct research with repeatability, implementing the dynamic characteristic to achieve the most robust technology must be considered as the core of R&D, and the difference between the results of the laboratory and downstream conditions can be corrected easily by tuning only. The repeatability of dynamic characteristics is great because a dynamic S/N ratio is an index of the certainty of functions, not of the output itself. By maximizing a dynamic S/N ratio, the center value of the output (such as sensitivity or functioning point) can be easily adjusted downstream afterwards. It is also important that such factors for adjustment be discovered at the R&D stage. And this can usually be determined through the same experiment.

In quality engineering, we must shift from product design to technology development, from Japanese style design work to American and European style R&D. However, efficiency requires the use of dynamic characteristic S/N ratios instead of the Western-style response analysis that studies the effects of all variables in order to understand the selected system.

Quality through Engineering Design - W. Kuo (Editor)

Some Aspects of Statistical Quality Control Methods

C. R. Rao[a], M. S. Srivastava[b] and Yanhong Wu[b]

[a]Statistics Department, Pennsylvania State University, University Park, PA 16802, U.S.A.

[b]Statistics Department, University of Toronto, Toronto, Ontario, M5S 1A1, Canada

Abstract

This paper reviews the statistical basis of the on-line quality control methods, statistical process control and automatic process control, and introduces some recent advances made in these two areas.

INTRODUCTION

Increasingly, statistical concepts and methods are being used in the production of commodities conforming to specification while minimizing operational costs. Some of the important statistical methods currently being used are statistical process control (SPC) for monitoring a production process and detecting changes, automatic process control (APC) for making adjustments in a process to control expected deviations, lot inspection (LI) for assessing the quality of manufactured goods, design of experiments (DE) for quality improvement, and estimating the latent quality of inputs (LQI). There is extensive literature on all these methods, except perhaps for LQI, an example of which is available in Khatri and Rao [1]. In this paper, we concentrate on the on-line quality control methods, SPC and APC, and discuss certain improvements and clarifications made in the last few years. We refer to recent papers by Box and Kramer [2] and Vander Wiel et al. [3], which contain an extensive review of these topics and provide a suitable framework for further discussion.

STATISTICAL PROCESS CONTROL (SPC)

Control charts

SPC is a general term used for monitoring a production process. The statistical methodology developed for this purpose is intended to detect, as early as possible, deviations from a specified probability model for the quality of successive items produced under a manufacturing process. More specifically, the interest is centered round the specified average value and the variance of the quality of goods. Another goal is to devise techniques that are simple enough for the technicians in charge of the production process to implement in a routine manner. Four well-known procedures for this purpose are based on the assumption that successive observations on the quality of goods are i.i.d. random variables from a given distribution.

Shewhart $\overline{x}$ and s control charts

Using Shewhart $\overline{x}$ and s control charts is the most classical but still very useful procedure. In the $\overline{x}$ chart, measurement of production of an individual item or the average of a chosen number (e.g., n) of successive items are plotted at fixed intervals of time. If a point falls outside the $\mu_0 \pm k\sigma$ limits, where μ_0 is the target value, σ^2 is the variation of the process, and k is a suitably chosen value (usually $3/\sqrt{n}$), deviation from the target value is suspected and an adjustment is made to correct the process. Similarly, a chart is maintained for some measure of variation, such as the standard deviation of the measurements in samples of a chosen size, to detect changes in the variation of the process.

Since these charts do not make use of the information supplied by all the previous observations, they may not be efficient for detecting a change in the process. For example, the delay detection time is rather long for small shifts. Several modifications were made to improve efficiency. The simplest are the run rules suggested by Dudding and Jennett [4] and Rao [5, p. 511], where warning lines are placed at $\mu_0 \pm k_0\sigma$, $k_0 < k$, and a change is declared when (1) a point falls outside $\mu_0 \pm k\sigma$; (2) two consecutive points fall outside $\mu_0 + k_0\sigma$ or $\mu_0 - k_0\sigma$; (3) three consecutive points are such that the first and third are above $\mu_0 + k_0\sigma$ and the second is between μ_0 and $\mu_0 + k_0\sigma$, or the first and third are below $\mu_0 - k_0\sigma$ and the second is between μ_0 and $\mu_0 - k_0\sigma$; or (4) five successive points are above or below μ_0 and so on. These rules considerably improve the performance of Shewhart's control chart and are likely to be nearly as efficient as the following more elaborate charts (see Champ and Woodall [6] for partial results). Further studies are being made to compare the relative performances of the various charts.

CUSUM chart

The CUSUM (cumulative sum chart) was introduced by Page [7]. To detect an upward shift, an alarm goes off at

$$N = \min\{n \geq 1 : S_n \geq h\}; \ S_n = \max\left[S_{n-1} + (x_n - k),\ 0\right]$$

where $S_0 = 0$, k and h are the reference value and control limit, respectively, and $x_1, x_2, \ldots$ are independent observations. In the normal case, k is chosen to be $\delta/2$, where $\delta = \mu_1 - \mu_0$ is the smallest change in shift that we wish to detect. The quantity h is chosen such that the average run length (ARL_0), i.e., the average number of observations between false alarms, has a specified value. If we want to detect shifts on either side, we can perform another CUSUM based on $-x_i$ with properly chosen reference and control values.

Shiryayev-Roberts procedure

Roberts [8] and Shiryayev [9] considered their procedure as a limit of a sequence of Bayes procedures given by Girshick and Rubin [10]. In the normal case with variance one, the Shiryayev-Roberts process can be written as

$$R_n = (1 + R_{n-1}) \exp[\delta(x_n - \delta/2)]$$

where $\delta = \mu_1 - \mu_0$ denotes the shift amount. An alarm goes off at $N = \min(n \geq 1: R_n \geq T)$ with a properly chosen T. A two-sided procedure can also be developed by mixing two processes that are used to detect upward and downward shifts.

EWMA procedure

The EWMA (exponentially weighted moving average) procedure was proposed by Roberts [11]. This procedure has recently received considerable attention in the quality control literature. Let $y_n = (1-\beta)y_{n-1} + \beta x_n, y_0 = 0$, where β is a smoothing parameter. An alarm goes off at $N = \min(n \geq 1: |y_n| \geq \sigma_\infty b)$, where $\sigma_\infty^2 = \lim Var(y_n) = \beta/(2-\beta)$ is the stable variance of the EWMA process y_n.

Operating characteristics of control charts

In this section, we compare the performances of the four control charts by accurately evaluating their operating characteristics. One of the main factors that is considered in choosing a control chart is the average delay detection time under the change point model. Another is the rate of false alarm. When no prior information is available for the change point, one usually fixes the average in control run length $E_\infty N = ARL_0$ at some specified level and chooses a procedure that minimizes the average delay time, where $E_k[\bullet]$ denotes the mean when the change is at time k.

Measuring the average delay time is a complicated problem since the delay detection time depends on the location of the change. Traditionally, the average out-of-control run length $ARL_1 = E_1N$, which is the average number of observations required to stop the process when the change occurs on the first observation, is the main measure for evaluating the efficiency of a procedure.

For the purpose of designing a control chart, an accurate evaluation of ARL_0 and ARL_1 is therefore crucial. For the Shewhart procedure as well as the procedure with warning lines, ARL_0 can be directly calculated for given control limits and warning lines based on the normal distribution. The evaluation for the other three procedures is complicated. For the CUSUM chart, the current method used in the literature is the numerical Markov chain method proposed by Brook and Evans [12]. This method significantly improves the traditional approximation based on Wald's formula. However, it has shortcomings because it is computationally very complex. The same method is also used recently by Lucas and Saccucci [13] for the EWMA procedure. In the following discussion, we introduce an alternative approximation based on the boundary correction technique. The basic idea is to correct the control limit by adding the mean of the overshoot at the crossing time N. The approximation is very simple and is also quite satisfactory in accuracy. We shall only give the results and refer to Siegmund [14], Pollak [15], and Wu [16] for details and related techniques.

For the CUSUM procedure, in the one-sided shift case,

$$ARL_0 \approx \frac{e^{\delta(d+2\rho)} - 1 - \delta(d+2\rho)}{\delta^2/2}$$

This is obtained simply by changing d to d + 2ρ in the continuous time formula for ARL_0. For the normal distribution, $\rho \approx 0.583$. Wu [16] has compared this approximation with the numerical values given by Van Dobben de Bruyn [17] and found that the approximation is quite satisfactory for $\delta \leq 1.2$ and the relative error is within 2%. Thus, for relatively small reference values, this approximation is good enough for practical use. The formula for the two-sided shift case can be similarly obtained by using the formula given by Van Dobben de Bruyn [17].

Approximations for ARL_0 and ARL_1 for the Shiryayev-Roberts (S-R) procedure were given by Pollak [15] and Wu [16]. The approximate formula for ARL_0 given by Wu [16] is simpler,

$$ARL_0 \approx T\exp[\delta\rho]$$

where $\rho = 0.583$ as before. Wu [16] also checked the accuracy of this approximation for T = 100, 300, and 500 with the simulated results. The comparison shows that for $\delta \leq 2.0$, the relative error is within 2%. The Shiryayev-Roberts procedure in quality control is not widely used although Pollak and Siegmund [18] and Srivastava and Wu [19] have shown that this procedure is as good as the CUSUM and better than EWMA.

Finally, we discuss the EWMA procedure. For a given shift value of δ that we want to detect, the EWMA procedure is difficult to implement since we have to find optimum values of two parameters, namely, the weight parameter ß and the control limit b, where ß and b are chosen such that for a fixed ARL_0 the average delay time achieves its minimum value for the given δ. A numerical search for the optimal value of ß has been carried out by Lucas and Saccucci [13] by using the Markov chain method. The search procedure is obviously very cumbersome and time consuming. Srivastava and Wu [19] considered the optimal design problem in the continuous time version of the EWMA procedure, where an obtained approximate solutions for ß and b when ARL_0 is large. These results show that the optimal value of ß increases with δ and decreases with ARL_0. This is in agreement with the numerical results of Lucas and Saccucci [13]. More specifically, the optimal ß is approximately equal to

$$\beta^* \approx 0.5117\delta^2 / \ln\left\{0.4083\delta^2 T\left[2\ln\left(0,4083\delta^2 T\right)\right]^{1/2}\right\}$$

when T = ARL_0 is relatively large. Also, for sufficiently large ARL_0, $\sigma_\infty b \approx 0.715\delta$. Thus, roughly speaking, the control limit for the EWMA process y_n

should be set at about 0.715δ if we want to detect a shift amount of δ. The numerical results of Lucas and Saccucci [13] show that the optimal ß is quite satisfactory, but the approximation for the control limit is not very accurate. Finally, we should point out that although the theoretical study shows that the EWMA procedure is less efficient than the CUSUM and S-R procedures (about 82% when ARL_0 is large), it is almost as good as the other two procedures for moderate ARL_0's.

Some general comments

(i) ARL_1 as the only measure for the average delay time might be misleading. An example is the fast initial response (FIR) technique discussed in Lucas and Croiser [20]. For instance, for the CUSUM procedure, instead of starting the procedure at 0, we may set $S_0 = s_0 > 0$. While it does not affect ARL_0 too much, it reduces ARL_1 significantly. The proposed value for $s_0 = h/2$ where h is the control limit. However, minimizing only ARL_1 is unreasonable because it only takes care of the case when the change occurs in the beginning. When the location of the change point is unknown, more reasonable measures for the average delay time should be considered. Two of them are the conditional stationary average delay time (CSADT) and the unconditional stationary average delay time (SADT) considered by Taylor [21] and Shiryayev [9]. The former measures the average delay time when there is no false alarm. The latter, however, measures the average delay time when many false alarms occur before the change point. The FIR technique reduces ARL_1 but increases SADT and CSADT. An appropriate application of FIR is to set the starting point at a certain state to minimize the difference between ARL_1 and CSADT or SADT.

(ii) The economic design of control charts depends on the information about the change point. When the prior distribution is available for the change point, a decision approach may be used to obtain an optimal Bayes control procedure. Duncan [22] has considered the economic design of the Shewhart procedure. However, this approach also needs the estimation of the false alarm cost and the loss rate for the delay detection time.

(iii) We may consider varying the sampling percentage or frequency in order to reduce the average delay time while keeping the sampling rate the same as under the fixed sampling plan. A simple dynamic sampling plan is of the following form. We set a switching limit that is within the control limit. When the detecting process is within the switching limit, we use a lower sampling rate. When the detecting process is outside the switching limit, a higher sampling rate is used. An alarm goes off when the detecting process crosses the control limit. This plan is the so-called two-rate sampling plan. The switching limit and the control limit should be designed to satisfy the requirements on ARL_0 and the average sampling rate when the process is in control. This can be achieved in two ways. In the lot or batch production case, we may consider changing the sampling percentage in each batch. In the one-at-a-time production case, the only way is to change the sampling frequency or sampling interval. Girshick and Rubin [10] seem to be the first to discuss this problem from a Bayesian point of view. Interest in the above dynamic sampling plan has been revived recently. In the one-at-a-time case, Reynolds, Amin, and Arnold [23] considered a two-sampling interval plan in the CUSUM procedure by using the Markov chain approach, which is shown to be significantly better than the fixed sampling plan. However, an obvious shortcoming of this approach is the computing time. Readers who are more interested in theoretical research are referred to Assaf [24] and Wu and Srivastava [25].

(iv) Most of the SPC techniques developed for monitoring a production process relate to one quality characteristic. Recently, attempts are being made to introduce multivariate control charts, which seem to provide additional information in detecting changes. Further theoretical work on these charts is necessary, and the feasibility of using them in a production process has to be explored. See Jackson [26] for recent literature on the subject.

(v) Much of the literature on control charts deals with detecting changes from a process characterized by i.i.d. variables. It would be worthwhile to work on designs of control charts when the underlying distribution is a specified stochastic process (perhaps stable with mild correlations between successive observations). It may also be necessary to design control charts intended to detect certain types of departures from a specified process. A recent contribution by Box and Ramirez [27] in this connection is of some interest.

(vi) In our discussion, we have given more attention to detecting shifts in the mean value of the process. Similar remarks apply to charts designed to detect departures in the variation of the process.

AUTOMATIC PROCESS CONTROL (APC)

General theory

APC is based on stochastic control theory developed by quality control engineers, where past observations are used to predict future observations and appropriate action is taken through what are called control variables to minimize deviations from the target value. SPC was developed in isolation from APC as a monitoring device to detect changes in a given process, and it was believed that these two were alternative techniques for achieving quality control. The issues associated with APC and SPC are clarified in recent papers by Box and Kramer [2], Vander Wiel et al. [3], and Harris [28]. These papers explore the rationales for process monitoring using SPC and for process adjustment using APC and also stress the need for integrating these two techniques in quality control work.

In the original control theory, a stochastic model of the type

$$y_t = f(u_{t-1}, u_{t-2}, \ldots) + \xi_t \tag{1}$$

is used for the observation at time t, where $u_{t-1}, u_{t-2}, \ldots$ are the values of a control (input) variable used in the past and ξ_t is a stochastic process indicating normal variations in a production process (ξ_t may reflect random changes in environmental variables, gradual wearing of a machine, and so on). Considering a suitable time series model for ξ_t, a minimum variance prediction is made of ξ_{t+1}, the value of the ξ-process at time t + 1, which we represent by ξ_{t+1}. If the value chosen for the control variable at time t is u_t, then the predicted value of y_{t+1} is

$$\hat{y}_{t+1} = f(u_t, u_{t-1}, \ldots) + \xi_{t+1}. \tag{2}$$

In order to minimize the variability in production, u_t is chosen such that the right-hand side of Eq. (2) is zero.

The solution so obtained does not depend on cost considerations. The first steps in this direction were taken by Bather [29] and Box and Jenkins [30]. The problem can be formulated in a very general setup by considering a quadratic loss function of the form

$$L(y) = ay^2 \tag{3}$$

where y denotes deviation from the target value and a is a suitable constant, and by then introducing inspection cost C_I for a single item and adjustment cost C_A. With a given cost structure specifying the values of a, C_I, and C_A, we wish to minimize the expected loss by making an appropriate choice of (1) the interval at which the items are to be inspected, (2) the decision rule to make adjustment or not, and (3) the amount of adjustment to be made. A solution to the problem under a general set up as in Eq. (1) for any time series ξ_t may be difficult. We present the solutions obtained under certain simple models for ξ_t.

As a datum of the problem, the values of a, C_I, and C_A have to be specified. Taguchi [31] chooses a = S/Δ^2 where S is the cost of scrapping or reworking a product with a deviation larger than the tolerance limit Δ that is specified.

Suppose that the observations are taken at intervals of m time points, which may be denoted by $y_m, y_{2m}, \ldots, y_{nm}, \ldots$ and our decision rule is to make an adjustment at time

$$N = \min\left(n \geq 1 : \left| \hat{y}_{(n+1)m} \right| > d\right) \tag{4}$$

in such a way that the quality characteristic process is new again, where $\hat{y}_{(n+1)m}$ is the predicted value of the deviation from the target value at time (n + 1)m, based on $y_m, \ldots, y_{nm}$. For given choices of m and d and an appropriate time series model for ξ_t, the problem reduces to evaluating the expected loss, which is a function of m and d, and finding the optimum m and d for which the expected loss is a minimum. In general, the expected loss function per item is of the form

$$C(m,d) = \frac{C_1}{m} + \frac{C_A}{mE(N)} + \frac{a}{mE(N)} E\left[\sum_{n=1}^{Nm} y_n^2\right] \tag{5}$$

We provide some details in the following sections for particular cases.

Taguchi's solution

Taguchi [31] assumed that (1) Y_n is uniformly distributed in (–d,d) when the process is in control, (2) the time when the process is out of control is approximately half of the last inspection interval, i.e., (m + 1)/2, (3) during the period the process is out of control, the deviation is approximated by d, and (4) the forecasted average adjusting interval u and control limit d have the approximate relationship

$$u = d^2\left(u_0 / d_0^2\right) \tag{6}$$

where u_0 and d_0 are the corresponding current values. With these assumptions and the quadratic loss function Eq. (3), Taguchi gives an approximate evaluation of average loss

$$C(m,d) = \frac{C_A}{u} + \frac{C_I}{m} + a\left[\frac{d^2}{3} + \left(\frac{m+1}{2} + l\right)\frac{d^2}{u}\right] \tag{7}$$

where there is a lag ℓ items produced from the time of measurement to the time of adjustment and u is the same as in Eq. (6). He obtains the optimum values of d and m as

$$d^* = \left(\frac{3C_A}{a} \cdot \frac{d_0^2}{u_0}\right)^{1/4}, \quad m^* = \left(\frac{2u_0 C_I}{a d_0^2}\right)^{1/2} \tag{8}$$

The exact nature of the stochastic model under which Taguchi's results are valid is not clear from his writings. See Taguchi, Elsayed, and Hsiang [32] for further details.

Random walk model

Srivastava and Wu [33] computed the expected loss function under the random walk model in the form

$$C(d,m) = \frac{C_A}{mE\left(z_N^2\right)} + \frac{C_I}{m} + \frac{a\sigma^2}{6}\left[\frac{mE\left(z_N^4\right)}{E\left(z_N^2\right)} + \frac{1}{2}\right] \tag{9}$$

Where $z_N = y_{mN} / \sigma\sqrt{m}$. Box and Jenkins [30] obtained an approximation for $E\left(z_N^2\right)$ and $E\left(z_N^4\right)$. Another approximation is given by Srivastava and Wu [33]. Simulation results show that the approximation given by Srivastava and Wu are much closer to the true value than those given by Box and Jenkins. The optimum values of m and d under the random walk model obtained by Srivastava and Wu are

$$m^* \approx \left(\frac{6C_I}{a\sigma^2}\right)^{1/2}, \quad d^* \approx \sigma\left[\left(\frac{6C_A}{a\sigma^2}\right)^{1/4} - 0.583\left(\frac{6C_I}{a\sigma^2}\right)^{1/4}\right] \tag{10}$$

Box and Jenkins' model

Box and Jenkins [30] consider the IMA (0,1,1) model for the quality characteristic z_t, which is more general than the random walk model. When the target value is assumed to be zero, $z_0 = 0$, $z_1 = a_1$, and

$$z_n = a_n + \gamma \sum_{j=1}^{n-1} a_j, \quad 0 \leq \gamma \leq 1, \quad n \geq 2$$

where the sequence (a_i) is i.i.d. normal random variables with mean zero and variance σ^2.

Consider the case $C_I = 0$ and the sampling interval m is unity. Then Eq. (5) becomes

$$C(d) \frac{C_A}{E(N)} + \frac{aE\sum_{1}^{N} z_i^2}{E(N)} \tag{11}$$

where N is the number of samples before an adjustment. The process z_n can also be written as

$$z_n = a_n + \hat{z}_n \tag{12}$$

where $\hat{z}_n$ is the predicted value of z_n based on the observations $z_1, \ldots, z_{n-1}$. Then, we have the expression

$$\hat{z}_n = E\left(z_n \middle| z_1, \ldots, z_{n-1} \middle| \right) = \gamma \sum_{i=1}^{n-1} a_i$$

An adjustment is made as soon as

$$N = \inf\left[n \geq 1 : \left|\hat{z}_{n+1}\right| \geq d\right] = \inf\left[n \geq 1 = \gamma \left|\sum_{i=1}^{n} a_i\right| \geq d\right] \tag{13}$$

After adjustment, we see that the quality characteristic process is renewed. Obviously $\gamma \Sigma a_i$ is a random walk with variance $(\gamma\sigma)^2$. Thus the optimal control limit obtained from the random walk model could be used with σ^2 replaced by $(\gamma\sigma)^2$.

Box and Kramer [2] also consider the case of general m and obtain the optimum values of m and d by simulation. Srivastava and Wu [34] give a solution similar to that for random walk.

Some comments about the methods

We should point out that the equally spaced inspection plan is obviously not the optimal sampling plan. As in SPC, the dynamic sampling plan can also be considered, see Bather [35] and Taylor [36]. However, it would be difficult to implement it in practical situations.

The formulas for the control parameters are derived under the assumption that the adjustment cost is larger than the inspection cost, $C_A > C_1$. When $C_1 > C_A$, different approaches are needed as discussed in Srivastava and Wu [37].

In many situations, measurement errors may also exist. Bather [29] considered the case when the measurement errors were i.i.d. and gave a procedure which is similar to the EWMA procedure but with time-dependent weight factors. For more recent work on this aspect we refer to Taguchi, Elsayed, and Hsiang [32] and Srivastava and Wu [38].

Some care is needed in choosing an appropriate model for use in APC. There is need to develop suitable statistical methodology to test whether a particular model fits the data and to choose between alternative models.

REFERENCES

1 C. G. Khatri and C. R. Rao, J. Comb. Inf. Syst. Sci., 16 (1991) 137–154.
2 G. E. P. Box and T. Kramer, Technometrics, 34 (1992) 251–285.
3 S. A. Vander Wiel, W. T. Tucker, F. W. Faltin and N. Doganaksoy, Technometrics, 34 (1992) 286–297.
4 B. P. Dudding and W. J. Jennett, Quality Control Charts, British Standard 600R, British Standards Institution, London, 1942.
5 C. R. Rao, Linear Statistical Inference and its Applications, 2nd ed., John Wiley and Sons, Ltd, New York, 1973.
6 C. W. Champ and W. H. Woodall, Technometrics, 29 (1987) 393–399.
7 E. S. Page, Biometrika, 41 (1954) 100–114.
8 S. W. Roberts, Technometrics, 8 (1966) 411–430.
9 A. N. Shiryayev, Theor. Probab. Its Appl., 13 (1963) 22–46.
10 M. A. Girshick and H. M. Rubin, Ann. Math. Stat., 23 (1952) 114–125.
11 S. W. Roberts, Technometrics, 1 (1959) 239–250.
12 D. Brook and D. A. Evans, Biometrika, 59 (1972) 539–647.
13 J. M. Lucas and M. Saccucci, Technometrics, 32 (1990) 1–30.
14 D. Siegmund, Sequential Analysis: Tests and Confidence Intervals, Springer-Verlag, Berlin, 1985.
15 M. Pollak, Ann. Stat., 15 (1987) 749–779.
16 Y. Wu, Some contributions to on-line quality control, Ph.D. diss., University of Toronto, Toronto, Ontario, Canada, 1991.
17 C. S. Van Dobben de Bruyn, Cumulative Sum Tests, Griffin, London, 1968.
18 M. Pollak and D. Siegmund, Biometrika, 72 (1985) 267–280.
19 M. S. Srivastava and Y. Wu, Comparison of EWMA, CUSUM and Shiryayev-Roberts procedures for detecting a shift in the mean, Ann. Stat., in press.

20 J. M. Lucas and R. B. Croiser, Technometrics, 24 (1982) 199–206.
21 H. M. Taylor, Technometrics, 10 (1968) 479–488.
22 A. J. Duncan, J. Am. Stat. Assoc., 51 (1956).
23 M. R. Reynolds, Jr., R. M. Amin and J. C. Arnold, Technometrics, 32 (1990) 371–396.
24 D. Assaf, Ann. Stat., 16 (1988) 236–253.
25 Y. Wu and M. S. Srivastava, On the dynamic sampling plans in on-line control charts, manuscript, 1992.
26 J. E. Jackson, Commun. Stat.-Theor. Meth., 14 (1985) 2657–2688.
27 G. E. P. Box and J. Ramirez, Qual. Reliab. Eng. Int., 8 (1992) 17–27.
28 T. J. Harris, Technometrics, 34 (1992) 298–306.
29 J. Bather, J. Roy. Stat. Soc., B 25 (1963) 49–70.
30 G. E. P. Box and G. M. Jenkins, Bull. Int. Stat. Inst., 34 (1963) 943–974.
31 G. Taguchi, Commun. Stat.-Theor. Meth., 14 (1985) 2785–2801.
32 G. Taguchi, E. A. Elsayed and T. Hsiang, Quality Engineering in Production Systems, McGraw Hill, New York, 1989.
33 M. S. Srivastava and Y. Wu, An improved version of Taguchi's on-line control procedure, J. Stat. Plan. Inf., in press.
34 M. S. Srivastava and Y. Wu, On-line quality control procedures based on random walk model and integrated moving average model of order (0,1,1), Technical Report 9303, Department of Statistics, University of Toronto, Toronto, Ontario, Canada, 1992.
35 J. A. Bather, Math. Oper. Res., 3 (1976) 209–224.
36 H. M. Taylor, Ann. Math. Stat., 36 (1965) 1677–1694.
37 M. S. Srivastava and Y. Wu, Comm. in Stat., (A) 20 (1991) 2149–2168.
38 M. S. Srivastava and Y. Wu, Taguchi's on-line control procedures and some improvements, Technical Report 9121, Department of Statistics, University of Toronto, Toronto, Ontario, Canada, 1991.

Quality through Engineering Design - W. Kuo (Editor)

Quality and Reliability of Products through Proper Design

Nam P. Suh

Department of Mechanical Engineering, Massachusetts Institute of Technology, Cambridge, MA 02139, U.S.A.

Abstract

Statistical quality-control techniques are widely used to improve the yield of manufacturing operations, the quality of the final product, and the reliability of the product in service. However, the best means of assuring good quality and highly reliable products is through the proper design of products, manufacturing processes, and systems. Axiomatic design enables us to approach the design process systematically and make correct design decisions. In this paper, the fundamentals of the axiomatic approach to design are reviewed, followed by a discussion of the implications of the design axioms on quality control. When the design axioms are violated, it is difficult to produce good products through statistical means.

INTRODUCTION

One of the remarkable facts of the last five decades, notwithstanding some of the most tragic events of human history, is the development of modern technologies, especially manufacturing technologies. We all remember the days when three essential elements of human necessity—i.e., food, shelter and clothing— were out of reach to most people in many parts of the world. We have achieved agricultural abundance, efficient manufacture of goods, and application of science and technology to a wide range of societal issues to improve the quality and standard of life of many people in many nations. One of the elements in this continuing improvement of living standards and its diffusion to all parts of the world is an ability to produce goods efficiently and economically. In this sense, manufacturing will continue to be one of the most important fields of human endeavor for many decades to come.

As the competition for market share increases, attention to the quality of products (in terms of functionality, reliability, dependability, and cost) has increased in all industrialized nations. In a sense, it is amazing that we can manufacture such complex products as automobiles, airplanes, and nuclear power plants with a degree of reliability and efficiency that we could not dream of only five decades ago. For instance, the failure rate for many mass-produced parts has been reduced to a few parts per million. Some of these improvements are due to the use of statistical quality-control techniques [1] and pro-active management techniques. Others are due to improvements in design practice.

Proper design is one of the key ingredients for assuring high-quality products because quality cannot be derived solely from the shop floor—rational design of products, manufacturing processes, and manufacturing systems have a direct bearing on the delivery of high-quality products. When attempts are made to improve the quality of products and the yield of manufacturing operations, poor design of products, processes, and manufacturing systems are often as much responsible for defects and low yields as the inability to operate the manufacturing system efficiently. Therefore, the questions we must answer include: How does an engineer, designer, or manager develop good designs? How can correct and rational design decisions be made so as to avoid lengthy trial-and-error approaches? How does one identify faulty design decisions when they are made? We have been studying these issues at the Massachusetts Institute of Technology over the past 15 years by using an axiomatic approach to design [2]. In this presentation, a brief introduction to axiomatic design is made, followed by a discussion of its implication on quality control of products.

ELEMENTS OF AXIOMATIC DESIGN

The design process consists of several steps:

1. establishing design goals to satisfy a given set of perceived needs
2. conceptualizing design solutions
3. analyzing the proposed solution
4. selecting the best design from among those proposed
5. implementing the design

These activities occur between and in different design domains.

The design domains are illustrated in Fig. 1. The consumer domain is where customer needs reside. These customer needs must be mapped into the functional domain where they are translated into a set of functional requirements (FRs), which constitute a characteristic vector. These FRs are then mapped into the physical domain, where the design parameters (DPs) are chosen to control or satisfy the FRs. The DPs in the physical domain are mapped into the process domain in terms of the process variables (PVs). In the case of product design for manufacturing, the PVs are the processes used to produce the physical product. In all cases, the relationship between the adjacent domains is that the domain on the left is *What we want* and the domain on the right is *How we will satisfy what we want*. Going from one domain to another is called *mapping,* which is the synthesis phase of the design process.

Another important concept in axiomatic design is the fact that a hierarchy exists in each domain. For example, in the functional domain, the FRs can be decomposed and formed into a hierarchical tree. However, the decomposition in each domain cannot be made independent of the hierarchies in other domains. In fact, in order to decompose a given FR at a given level of the FR hierarchy, we must first move over to the physical domain and conceive a design solution

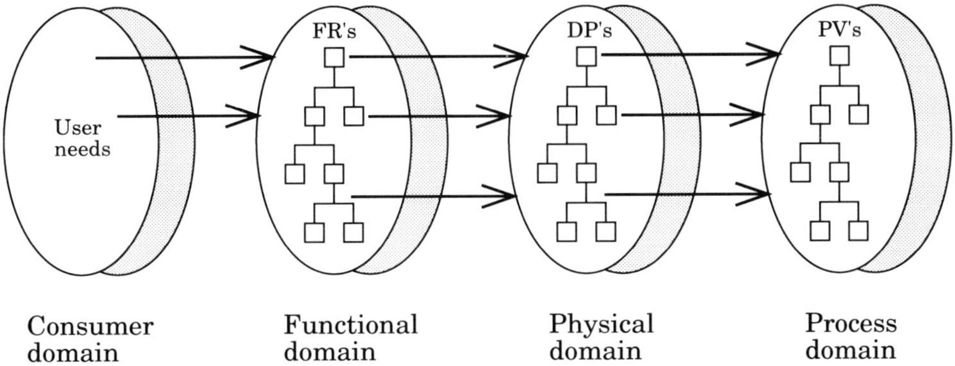

Fig. 1. Concept of domains and mapping.

that is characterized by a set of DPs, and then move back to the functional domain and decompose the FR. This is illustrated in Fig. 2. Therefore, the concepts of decomposition and hierarchy are closely related to the concept of zig-zagging between domains to proceed with the decomposition from one level to the next lower level.

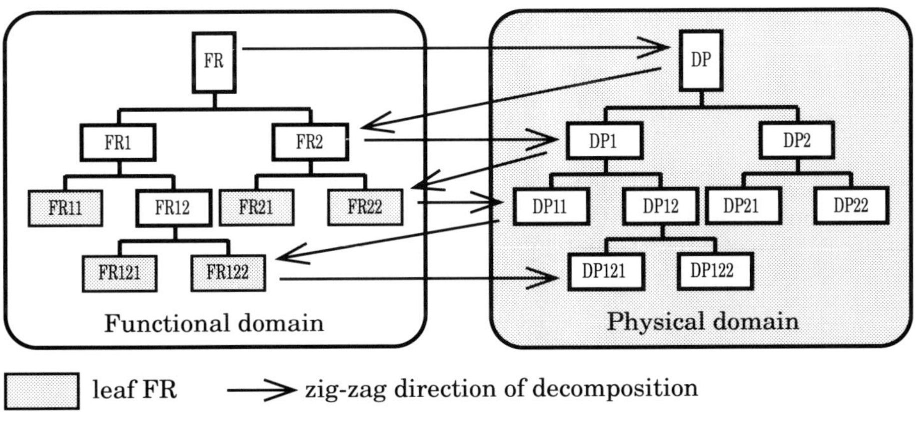

Fig. 2. Hierarchical tree structures of functional requirements and design parameters.

Finally, the most important concept in axiomatic design is the existence of the design axioms that must be satisfied during the mapping process to come up with acceptable design solutions. The nature of the mapping between adjacent domains depends on the designer's individual creative process and knowledge. Consequently, there can be an infinite number of plausible solutions and mapping techniques. Figure 1, for instance, suggests a one-to-one mapping between domains; however, a one-to-many mapping may also be plausible. In the absence of design principles or axioms, the designer must depend on empirical or ad hoc techniques for identifying a good design among all plausible solutions. The first design axiom is known as the independence axiom and the second axiom is known as the information axiom. They are stated as follows:

Axiom 1 *The Independence Axiom:* Maintain the independence of functional requirements

Axiom 2 *The Information Axiom:* Minimize the information content

The first axiom provides the criterion for acceptable design during the mapping process. For each level of the hierarchical tree, the product design, which is the mapping process between the functional domain and the physical domain, can be represented by a design equation as

$$\{FR\} = [A]\{DP\} \tag{1}$$

where {FR} is a vector that describes the functional requirements of the product in terms of its independent components FR_i, {DP} is a vector that describes the parameters that define the product in terms of its effect on {FR}, and [A] is a product design matrix. The elements of the product design matrix, A_{ij}, are given by

$$A_{ij} = \frac{\partial FR_i}{\partial DP_j} \tag{2}$$

which is a constant in linear design. In order to satisfy the independence axiom, [A] must be a diagonal or triangular matrix. The design that has a diagonal matrix is called an uncoupled design. When it is triangular, it is a decoupled design, which also satisfies the independence axiom provided that DPs are changed in a specific sequence. All other designs are coupled designs.

A similar equation can be written for mapping between the physical and the process domains as:

$$\{DP\} = [B]\{PV\} \tag{3}$$

where [B] is a process design matrix and {PV} is the vector that describes the process variables in the process domain. The elements of the [B] matrix are partial derivatives of DP with respect to PV. Again, [B] must be either diagonal or triangular to satisfy the independence axiom. Equation (3) can be substituted in Eq. (2) to relate {FR} to {PV} directly.

In order to determine the degree of independence (or conversely, the degree of coupling) associated with a particular FR to DP mapping, two quantitative measures, reangularity (R) and semiangularity (S), have been defined [3] as

$$R = \prod_{\substack{i=1,n-1 \\ j=1+i,n}} \left\{ 1 - \frac{\left(\sum_{k=1,n} A_{ki} A_{kj} \right)^2}{\left(\sum_{k=1,n} A_{ki}^2 \right) \left(\sum_{k=1,n} A_{kj}^2 \right)} \right\}^{\frac{1}{2}} \tag{4}$$

$$S = \prod_{j=1,n} \left\{ \frac{| A_{ii} |}{\left(\sum_{k=1,n} A_{kj}^2 \right)^{\frac{1}{2}}} \right\} \tag{5}$$

R measures the orthogonality among DPs, and S measures the correspondence among pairs of DP and FR variables. The denominators of Eqs. (4) and (5) are normalizing factors that are equal to unity if the columns of the design matrix are normalized. Normalization is necessary in order to make R and S insensitive to changes in the scale of DPs (e.g., changing the metric of one of the DPs from millimeters to inches should not influence R and S). A design is uncoupled when R = S = 1, and the values of R and S decrease as the degree of coupling increases. In the limit when R = 0, two or more of the DPs influence the same set of FRs and the design is completely coupled. The design is decoupled when R = S.

The information content is defined in terms of the probability of successfully achieving FRs or DPs, depending on whether the evolving design is in the physical or process domain. Information content is defined as

$$I_i = \log_2 \left(\frac{1}{p} \right) \tag{6}$$

where p is the probability of achieving the functional requirement FRi. When there are n FRs to satisfy, the total information content of a design is defined by

$$I = \sum_{i=1}^{n} I_i \tag{7}$$

The best design is the one with the minimum total information content.

In any design situation, the probability of success is given by what the designer wishes to achieve in terms of tolerance (i.e., design range) and what the system is capable of delivering (i.e., system range). As shown in Fig. 3, the overlap between the designer-specified design range and the system capability range, system range, is the region where the acceptable solution exists. Therefore, in the case of a uniform probability distribution function, Eq. (6) may be rewritten as:

$$I = \log_2\left(\frac{\text{System Range}}{\text{Common Range}}\right) \tag{8}$$

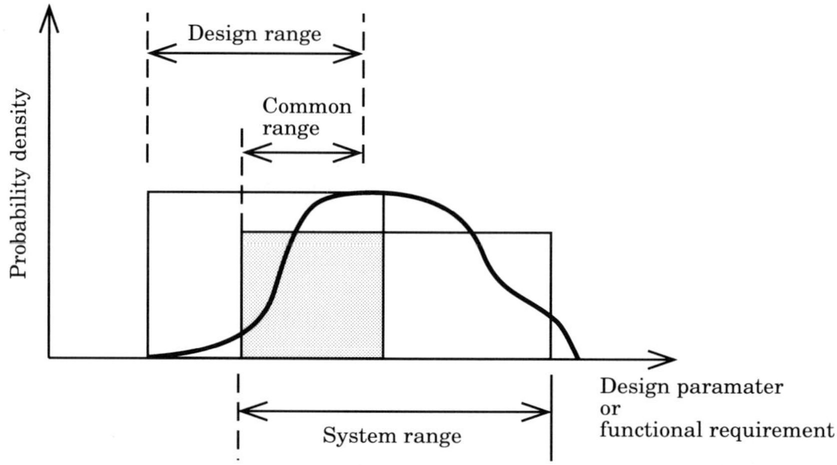

Fig. 3. Probability distribution of a system parameter. The curved line is for the case of nonuniform p-distribution. In the case of a product design, the system parameter is the FR to be achieved, whereas in the case of a process design, it is the DP which must be controlled through the proper choice of PVs.

In design, there are always constraints, which are similar to functional requirements at a given level of the design hierarchy, except that they do not have to be independent of FRs or other constraints. Sometimes, constraints propagate from the process domain to the product domain to the functional domain. In the development of microcellular plastics, for example, the FRs for the product [2] were defined as

FR_1 = Reduce materials consumption by 20%
FR_2 = Maintain toughness of the plastic part to equal or exceed that of impact-grade polystyrene.

The selected DPs for the product [2] may be written as

DP_1 = Volume fraction of microvoids.
DP_2 = Size of microvoids.

For this solution, material consumption is reduced by introducing microvoids into impact-grade polystyrene. Furthermore, the microvoids can enhance the toughness of the original material by acting as crack arresters, provided that the size of the microvoids is smaller than the critical flaw size in the polystyrene. After defining FRs and DPs, a manufacturing process was designed. External pressure, temperature, and time are examples of the PVs selected for controlling the forming process [2]. In this application, the overall cost of the manufacturing process was treated as a constraint because cost is affected by many FRs, DPs, and PVs. Consequently, its independence cannot be assured throughout the design process.

Corollaries and theorems that can be derived from the basic axioms may be used more directly in specific design situations [2]. A large number of theorems and corollaries are given in the Appendix. Corollaries are direct consequences of the axioms and tend to have the flavor of design rules. Good designers have used some of these corollaries intuitively. However, once intuition is replaced with axioms or laws, even inexperienced designers should be able to quickly master the basic tools of the trade. Theorems, on the other hand, are propositions that follow from axioms or other propositions that have been proven. There can be a large number of theorems, depending on the specific design problem under consideration. The proof for the theorems presented in the Appendix can be found in Ref. [2].

IMPLICATIONS OF AXIOMATIC DESIGN TO QUALITY CONTROL

The important implications of the independence axiom and the information axiom on quality control and product reliability are the following:

1. As we search for a design solution to satisfy a given set of FRs, the independence axiom requires the design matrix to be diagonal or triangular and the number of DPs must be equal to the number of FRs in an ideal design. Similarly, when there are many DPs to be satisfied by PVs, the independence of the DPs must be maintained through the design of uncoupled or decoupled manufacturing processes.

   Suppose we have to satisfy three DPs— DP_1, DP_2, and DP_3. Then the desired design equation is of the form:

$$\begin{Bmatrix} DP_1 \\ DP_2 \\ DP_3 \end{Bmatrix} = \begin{bmatrix} x & o & o \\ o & x & o \\ o & o & x \end{bmatrix} \begin{Bmatrix} PV_1 \\ PV_2 \\ PV_3 \end{Bmatrix} \tag{9}$$

   Therefore, we should look for PV_1 that only affects DP_1, etc. However, in many cases, we may be unable to do so. In this case, we have to develop a decoupled design as described in item 2.

2. A design that is less desirable than an uncoupled design but still acceptable is a decoupled design. In the aforementioned example involving DP_1, DP_2, and DP_3, we should first look for a PV_1 that satisfies DP_1. Then, as we seek PV_2 to satisfy DP_2, we must satisfy the condition that PV_2 does

not influence DP_1. Finally, when PV_3 is chosen as the primary input variable for DP_3, it must not affect DP_1 and DP_2 . This strategy may then yield the following design equation:

$$\begin{Bmatrix} DP_1 \\ DP_2 \\ DP_3 \end{Bmatrix} = \begin{bmatrix} x & o & o \\ ? & x & o \\ ? & ? & x \end{bmatrix} \begin{Bmatrix} PV_1 \\ PV_2 \\ PV_3 \end{Bmatrix} \tag{10}$$

The elements of the design matrix designated by "?" can be either zero or nonzero since the independence axiom is still satisfied.

3. When the design is a decoupled design, the governing design equation provides the basis for process control. For example, assume that the design matrix of Eq. (10) has the following form:

$$\begin{Bmatrix} DP_1 \\ DP_2 \\ DP_3 \end{Bmatrix} = \begin{bmatrix} x & o & o \\ x & x & o \\ x & o & x \end{bmatrix} \begin{Bmatrix} PV_1 \\ PV_2 \\ PV_3 \end{Bmatrix} \tag{11}$$

Equation (11) indicates that the three DPs are influenced by PV_1. If we vary PV_1 first, then the value of DP_1 can be satisfied. We can then independently vary the values of PV_2 and PV_3 so as to satisfy DP_2 and DP_3, respectively.

4. In some manufacturing situations it may appear that we have more potential PVs than DPs (especially in analyzing an existing manufacturing system). In this case, Axiom 1 tells us that we have to search for a set of PVs that has the largest influence on a given set of DPs and yields a diagonal or triangular matrix. Taguchi [1] developed a statistical technique for this search process. However, unless the design is uncoupled or decoupled, a statistical approach cannot overcome the fundamental flaws of the design. This point will be discussed later.

5. Many existing manufacturing processes are redundant designs, i.e, there are more PVs than DPs. In this case, the principal PVs must be fixed so as to reduce variability and eliminate bias. For example, suppose there are two DPs and four PVs as follows:

$$DP_1 = A_{11}PV_1 + A_{12}PV_2 + A_{13}PV_3 + A_{14}PV_4$$
$$DP_2 = A_{21}PV_1 + A_{22}PV_2 + A_{23}PV_3 + A_{24}PV_4$$

In addition, suppose that a statistical analysis has identified the following conditions: (1) $A_{12} = A_{21} = 0$, and (2) all other A_{ij}'s are nonzero. In this case, if we let PV_3 and PV_4 vary, the probability distribution functions for DP_1 and DP_2 will have a large variance as shown in Fig. 4 (i.e., system range). However, if PV_3 and PV_4 are fixed, DP_1 will be a function only of PV_1, and DP_2 will be a function only of PV_2. Consequently, the variance of the system range for DP_1 will be much smaller as shown in Fig. 5, and the bias between the system range and the target value can be corrected by adjusting PV_1.

6. In practice, the relationship between the FRs and DPs is nonlinear. That is, the values of the FRs increase or decrease nonlinearly as the values of the DPs are changed. As a result, even an uncoupled design may become coupled when the values of its DPs or PVs are outside a given range, and the independence axiom may only be satisfied at specific design points. For a nonlinear design, we must search for the *design window* where the independence axiom can be satisfied, although there is no prior assurance that such a window exists. There are two methods for performing this search process: the functional analysis and a statistical approach.

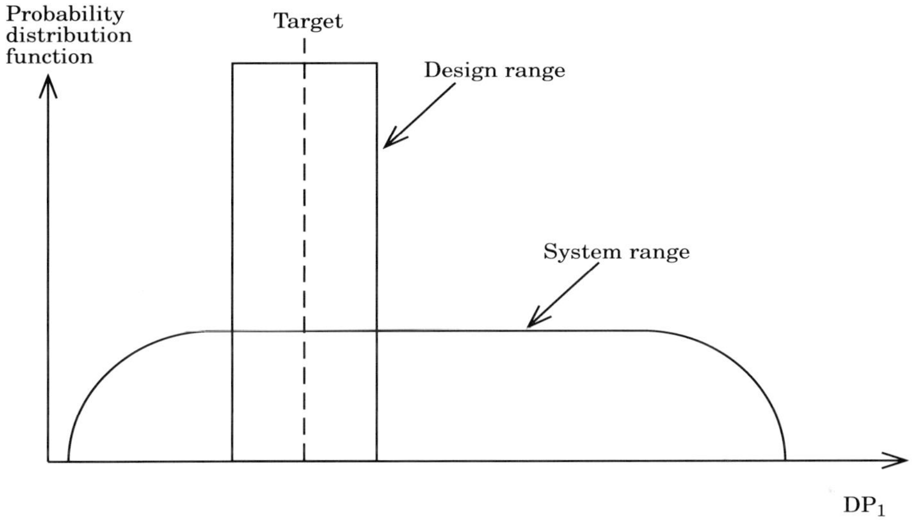

Fig. 4. Variance associated with the probability distribution of a system parameter.

The functional analysis may be expressed as follows: Given the design equation relating functional requirements to design parameters

$$\{FR\} = [A]\,\{DP\}$$

find the specific values of DPs subject to the following conditions:

$$\text{i)} \quad \frac{\delta^2 R}{\delta DP_j^2} < 0 \qquad \text{for } j = 1, \ldots n$$

$$\text{ii)} \quad \frac{\delta^2 S}{\delta DP_j^2} < 0 \qquad \text{for } j = 1, \ldots n$$

$$\text{iii)} \quad \frac{\delta R}{\delta DP_j} = 0 \qquad \text{for } j = 1, \ldots n$$

$$\text{iv)} \quad \frac{\delta S}{\delta DP_j} = 0 \qquad \text{for } j = 1, \ldots n$$

v) Design constraints are met (12)

If the search process yields R = 1 and S = 1, it is an uncoupled design. If R = S, then it is a decoupled design.

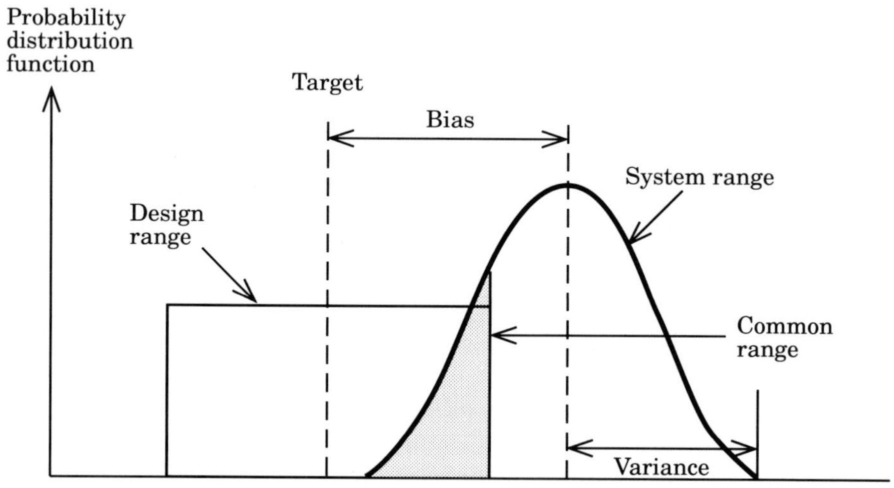

Fig. 5. The relationship among the design range, system range, common range, variance, and bias for a design.

The statistical approach, which is discussed in the next section, relies on statistical experimentation to identify the range in which the signal-to-noise ratio (SNR) is the maximum.

STATISTICAL APPROACH

Taguchi [1] developed a statistical approach to quality control when there are many possible DPs in an existing design, similar to the case studied above in item 4.

In order to identify the principal DPs that control the FRs in a product design, the variance σ^2 of a given FR_i can be determined by varying a DP. If we let DP_i assume a set of n different values around a nominal $(DP_i)_0$, the resulting FR_i will also consist of a set of n FR_i values. The variance of these n FR values is given by

$$\sigma^2 = \frac{1}{n-1} \sum_{p=1}^{n} \left(FR_i^p - \overline{FR_i} \right)^2 \tag{13}$$

where FR_i^p is the pth individual value of FR_i corresponding to the pth value of DP_i. The mean value of the n FR_i^p is denoted by $\overline{FR_i}$ which may be expressed as

$$\overline{FR_i} = m = \frac{1}{n} \sum_{p=1}^{n} FR_i^p \tag{14}$$

The sensitivity of a given FR_i to variations in DP_j must be determined in order to identify the most important (i.e., principal) DPs. Taguchi [1] defined the signal-to-noise (SN) ratio to measure outcome sensitivity. It is defined in terms of the variance and the mean of the n FR_i^p as

$$SN = 10 \log_{10} \left(\frac{m^2}{\sigma^2} \right) \tag{15}$$

where SN is given in decibels (dB). The DP that gives the largest value of SN has the largest effect on the FR and hence is the principal DP for that FR. Through the evaluation of the SN values corresponding to all potential DPs, the principal DPs corresponding to the given set of FRs can be determined. For an ideal design, the number of DPs must be equal to the number of FRs. Redundant designs (i.e., the number of DPs is greater than the number of FRs) may be desirable in some cases, provided that care is taken not to create a coupled design.

In general, the mean value m of the n FR_i^p will be different from the desired value of $(FR_i)_0$. The difference between the two is called the bias b; it may be expressed as

$$b = (FR_i)_0 - m \tag{16}$$

To improve the quality of a product, we must reduce b to zero by choosing a correct value for $(DP_i)_0$. It is easy to see that if the design is completely uncoupled, the value of DP_i can be changed to eliminate the bias.

In a strict sense, this statistical technique is only valid when the design satisfies Axiom 1, i.e., an uncoupled or decoupled design. It is not valid when the design is coupled because two or more FRs share the same principal DP. Furthermore, the accuracy of this statistical technique is dependent upon the sampled output response. A sample of responses must be large enough to fully represent the broad range of possible outcomes. However, it should not be so large that the experimentation becomes overly time consuming and impractical. Orthogonal arrays have been used to design experiments so that a broad range of outcomes are possible without running all possible combinations of variables.

CONCURRENT ENGINEERING

It is now very popular to talk about concurrent engineering as a strategy for reducing both the time taken for development of a new product and the associated cost. In the context of axiomatic design, concurrent engineering focuses on the sequence of mappings from the functional to the physical to the process domains as shown in Fig. 1. The manufacturability of a product is assured when both the design matrix for the product and the design matrix for the manufacturing process satisfy the independence axiom [4]. For instance, if the design equation for a product is

$$\{FR\} = [A]\{DP\} \tag{17}$$

and that for the associated manufacturing process is

$$\{DP\} = [B]\{PV\} \tag{18}$$

the relationship between FRs and PVs is given by

$$\{FR\} = [A][B]\{PV\} \tag{19}$$

In order to satisfy the independence requirement for FRs, the product [A][B] must be diagonal or triangular at all levels of the FR, DP, and PV hierarchies. This condition implies that [A] and [B] must also be either diagonal or triangular matrices.

CONCLUSIONS

Quality must be built into the product through correct design of products, processes, and manufacturing systems. The design axioms provide the basic principles that can be used to design a better product of higher quality as well as to design a more effective manufacturing system. Furthermore, the design axioms and methodologies that satisfy the design axioms can be used to control product design and manufacturing processes more effectively. These basic principles minimize the need to rely on iterative, trial-and-error approaches to product development. Moreover, the design axioms and the basic concepts of the axiomatic approach to design are valid in many different disciplines, e.g., design of products, processes, software, materials, organizations, constructed systems, and manufacturing systems. In addition, the theorems and corollaries presented in the Appendix are useful "design tools" that can help designers to think correctly and make the right decisions in a variety of situations.

REFERENCES

1 G. Taguchi, System of Experimental Design Engineering Methods to Optimize Quality and Minimize Cost, American Supplier Institute, Dearborn, Mich., 1987.
2 N. P. Suh, Principles of Design, Oxford University Press, New York, 1990.
3 J. R. Rinderle and N. P. Suh, J. Eng. Ind., Transactions ASME, 104, no. 4 (1982) 383–388.
4 N. P. Suh, CIRP Annals, 37, no. 2 (1988) 559–567.

APPENDIX: COROLLARIES AND THEOREMS

Corollary 1 *(Decoupling of coupled designs): Decouple or separate parts or aspects of a solution if FRs are coupled or become interdependent in the designs proposed.*

Corollary 2 *(Minimization of FRs): Minimize the number of FRs and constraints.*

Corollary 3 *(Integration of physical parts): Integrate design features in a single physical part if FRs can be independently satisfied in the proposed solution.*

Corollary 4 *(Use of standardization): Use standardized or interchangeable parts if the use of these parts is consistent with FRs and constraints.*

Corollary 5 *(Use of symmetry): Use symmetric shapes and/or components if they are consistent with the FRs and constraints.*

Corollary 6 *(Largest tolerance): Specify the largest allowable tolerance in stating FRs.*

Corollary 7 *(Uncoupled design with less information): Seek an uncoupled design that requires less information than coupled designs in satisfying a set of FRs.*

Corollary 8 *(Effective reangularity of a scalar): The effective reangularity R for a scalar coupling "matrix" or element is unity.*

Theorem 1 *(Coupling due to insufficient number of DPs): When the number of design parameters is less than the number of FRs, either a coupled design results or the FRs cannot be satisfied.*

Theorem 2 *(Decoupling of coupled design): When a design is coupled due to the greater number of FRs than DPs (i.e., m > n), it may be decoupled by addition of new DPs so as to make the number of FRs and DPs equal to each other if a subset of the design matrix containing n x n elements constitutes a triangular matrix.*

Theorem 3 *(Redundant design): When there are more DPs than FRs, the design is either a redundant design or a coupled design.*

Theorem 4 *(Ideal design): In an ideal design, the number of DPs is equal to the number of FRs.*

Theorem 5 *(Need for new design): When a given set of FRs is changed by the addition of a new FR, or substitution of one of the FRs with a new one, or by selection of a completely different set of FRs, the design solution given by the original DPs cannot satisfy the new set of FRs. Consequently, a new design solution must be sought.*

Theorem 6 *(Path independence of uncoupled design): The information content of an uncoupled design is independent of the sequence by which the design parameters are changed to satisfy the given set of FRs.*

Theorem 7 *(Path dependency of coupled and decoupled design): The information contents of coupled and decoupled design parameters depend on the sequence by which the design parameters are changed and on the specific paths of change of these DPs.*

Theorem 8 *(Independence and tolerance): A design is an uncoupled design when the designer-specified tolerance is greater than*

$$\left(\sum_{\substack{j=i \\ j=1}}^{n} \left(\delta FR_i / \delta DP_j \right) \Delta DP_j \right)$$

in which case the nondiagonal elements of the design matrix are negligible.

Theorem 9 *(Design for manufacturability): For a product to be manufacturable, the design matrix for the product [A] (that relates the FR vector for the product to the DP vector of the product) times the design matrix for the manufacturing process [B] (that relates the DP vector to the PV vector of the manufacturing process) must yield either a diagonal or triangular matrix. Consequently, when any one of these design matrices, i.e., either [A] or*

[B], represents a coupled design, the product cannot be manufactured.

Theorem 10 *(Modularity of independence measures): Suppose that a design matrix [DM] can be partitioned into square submatrices which are nonzero only along the main diagonal. Then the reangularity and semiangularity for [DM] are equal to the products of their corresponding measures for each of the nonzero submatrices.*

Theorem 11 *(Invariance): Reangularity and semiangularity for a design matrix [DM] are invariant under alternative orderings of the FR and DP variables, as long as the orderings preserve the association of each FR with its corresponding DP.*

Theorem 12 *(Sum of information): The sum of information for a set of events is also information, provided that proper conditional probabilities are used when the events are not statistically independent.*

Theorem 13 *(Information content of the total system): If each DP is probabilistically independent of other DPs, the information content of the total system is the sum of information of all individual events associated with the set of FRs which must be satisfied.*

Theorem 14 *(Information content of coupled versus uncoupled design): When the state of FRs is changed from one state to another in the functional domain, the information required for the change is greater for a coupled process than for an uncoupled design.*

Theorem 15 *(Design/manufacturing interface): When the manufacturing system compromises the independence of the FRs of the product, either the design of the product must be modified or a new manufacturing process must be designed, used, or both to maintain the independence of the FRs of the products.*

Theorem 16 *(Equality of information content): All information contents that are relevant to the design task are equally important regardless of their physical origin, and no weighting factor should be applied to them.*

Quality through Engineering Design - W. Kuo (Editor)

Design and Analysis of Circuits and Networks—The Role of Mathematical Concepts

T. S. Arthanari

Quality Management and Operations Research Division, Indian Statistical Institute, Madras, India

Abstract

The goal of circuit theory is to predict the electrical behavior of physical circuits. The purpose of these predictions is to improve their design and in particular to decrease their cost and to improve their performance under all operating conditions, for example, temperature effects, aging effects, and possible fault conditions. In a fundamental way, graph algorithms find their use in circuit design optimization. This paper highlights the use of mathematical methods and concepts deriving from mathematical programming and discrete optimization to build quality into the design of circuits and networks.

INTRODUCTION

The world is witnessing great changes in the concepts in design and manufacturing from the quality standpoint. Quality function deployment (QFD) and Taguchi's robust design through parameter and tolerance designs using orthogonal arrays have introduced newer and more relevant objectives for design parameter selection. At the same time, the heuristics for optimizing such objectives proposed by QFD or the Taguchi methods have implicitly or otherwise given the impression that mathematical programming or the combinatorial approach for design quality are unnecessarily sophisticated or inappropriate and that simple analysis and experimentation with the system design and idea generation in research and development (R&D) are sufficient.

The goal of circuit theory is to predict the electrical behavior of physical circuits. The purpose of these predictions is to improve their design and in particular to decrease their cost and to improve their performance under all operating conditions, for example, temperature effects, aging effects, and possible fault conditions [1]. In a fundamental way graph algorithms find their use in circuit design optimization.

This paper highlights the use of mathematical methods and concepts deriving from mathematical programming and discrete optimization to build quality into the design of circuits and networks.

LINEAR COMPLEMENTARITY IN CIRCUIT DESIGN

The geometry of the solution set of a piecewise-linear (PWL) resistive circuit is studied extensively in nonlinear circuits literature, (see Refs. [2–4]; Ref. [5] lists many other useful references). Piecewise linear approximation is useful in analyzing circuits made up of nonlinear resistors and helps in understanding the nonlinear behavior of circuits qualitatively.

It is well known that a piecewise linear resistor can be synthesized with linear elements, constant sources, and ideal diodes. An ideal diode is a nonlinear resistor whose v-i characteristic consists of two straight-line segments on the v-i plane (see Fig. 1) satisfying

$$\{(v,i)\ :\ vi=0,\quad \begin{array}{l} i=0 \text{ for } v<0 \\ v=0 \text{ for } i>0\,\} \end{array} \tag{1}$$

This equation expresses the complementarity between v and i, namely that at most one of them can be nonzero.

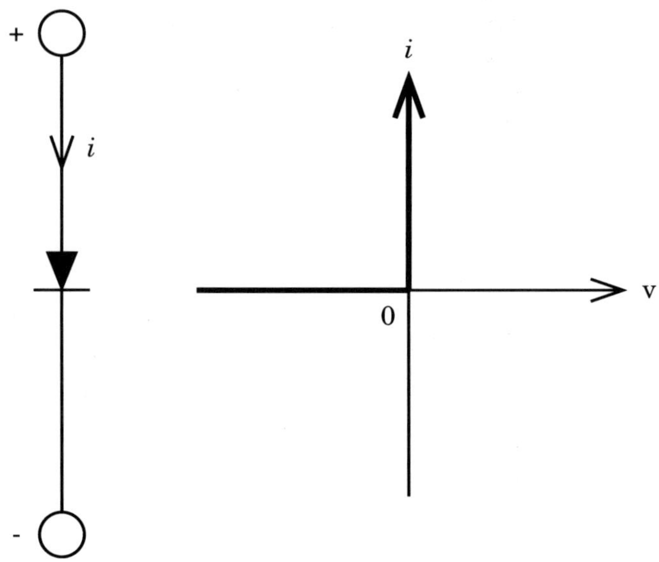

Fig. 1. Symbol for an ideal diode and its characteristic.

The linear complementarity programming (LCP) problem can be stated as follows: Find w, z $\in R^n$ such that

$$w - Mz = q;\ \ w, z \geq 0;\ \ w'z = 0 \tag{2}$$

where M $\in R^{n\times n}$ and q $\in R^n$. LCP has wide-ranging applications both as a methodology and as a unifying theory [6]. Stervens and Lin [3] and Eyndhoven [4] analyze PWL resistive networks by using the complementarity pivot algorithm of Lemke. Every voltage-controlled (n + 1) segment PWL characteristic can be represented by

$$i = a_o + a_1 v + \sum_{j=1}^{n} b_j \left| v - E_j \right| \tag{3}$$

where $E_1 < E_2 \ldots < E_n$ are the break-point voltages and

$$a_1 = \tfrac{1}{2}(m_o + m_n);\ b_j = \tfrac{1}{2}(m_j + m_{j-1});\ \text{and } a_o = i(0) - \sum_{j=1}^{n} b_j \left| E_j \right| \tag{4}$$

where m_o is the slope of the first linear segment from the left and m_j is the slope of the (j+1) linear segment. Current-controlled resistors can be represented similarly. Applying the equivalence between LCP and the set of PWL equations [7], we see that another LCP formulation of the PWL circuits is possible.

A new method for finding all solutions to a generalized version of LCP is applied in Ref. [5] in order to find the set of all solutions of a PWL resistive circuit. Consider a one-dimensional PWL curve in R^2 characterized by a set of n + 1 break points $x_o, x_1, \ldots, x_n$ and two extreme directions $x_{-\infty}$ and $x_{+\infty}$ (see Fig. 2). Let λ^+, λ^- be the positive part and negative part of λ.

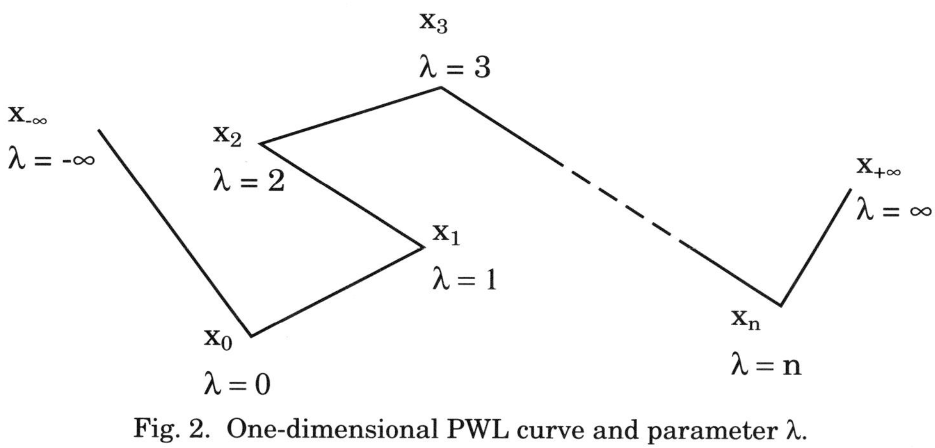

Fig. 2. One-dimensional PWL curve and parameter λ.

Let $\lambda_j = \lambda - j$ with λ_j^+ and λ_j^- the positive and negative parts of λ_j, ie., $\lambda = \lambda^+ - \lambda^-$ and $\lambda^+, \lambda^- \geq 0$ with $\lambda^+ . \lambda^- = 0$.

$$\lambda_j^+, \lambda_j^- \geq 0;\ \lambda_j^+ . \lambda_j^- = 0 \text{ for all j} \tag{5}$$

Then x can be written in terms of λ_j as

$$x = x_o + x_{-\infty}\lambda^- + (x_1 - x_o)\lambda^+ + \sum_{k=2}^{n}\left(x_k - 2x_{k-1} + x_{k-2}\right)\lambda_{k-1}^+ + \left(x_{+\infty} - x_n + x_{n-1}\right)\lambda_n^+ \tag{6}$$

with the defining constraints on $\lambda^{\pm}$ and $\lambda_j^{\pm}$, which include complementarity restrictions.

We use this motivation for a circuit with N nonlinear resistors. We extract all nonlinear elements out of the circuit, and the resulting N-port contains only linear resistive elements and independent sources (Fig. 3).

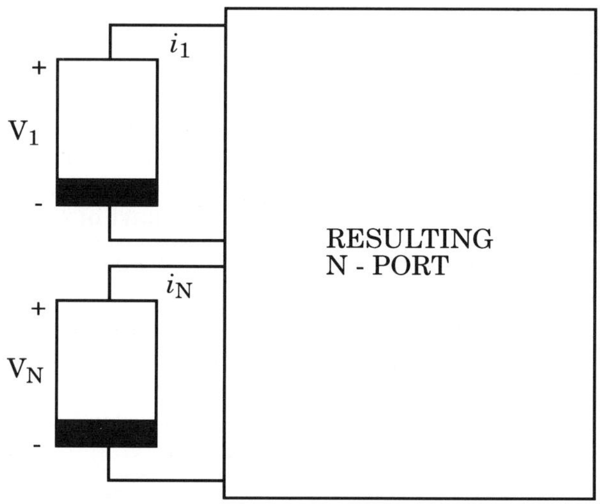

Fig. 3. Nonlinear resistors are extracted and the resulting N-port is linear.

Constraints of the resulting N-port can be written as

$$Cy = b \tag{7}$$

where $y = [y_1, \ldots, y_n]'$ with $y_j = [i_j, v_j]'$ and C is a $1 \times 2N$ matrix with 1 depending on the rest of the circuit.

The PWL parameterization of each nonlinear resistor is

$$Y_k = \begin{bmatrix} i_k \\ v_k \end{bmatrix} = \begin{bmatrix} i_{k0} \\ v_{k0} \end{bmatrix} + X_k^- \Lambda_k^- - X_k^+ \Lambda_k^+ \tag{8}$$

$$E_k\ \Lambda_k^+ - E_k\ \Lambda_k^- = e_k \tag{9}$$

$$\Lambda_k^-\ ,\Lambda_k^+ \geq 0\ ;\ \left(\Lambda_k^-\right)'\Lambda_k^+ = 0 \tag{10}$$

where $\Delta_k^+ = (\lambda^+, \lambda_1^+, \ldots, \lambda_n^+)_k$ and Δ_k^- is defined similarly.

Here E_k and $X_k^{\pm}$ are appropriately defined matrices in accordance with Eqs. (5–6), and $e_k = (1, \ldots, n_k)'$ where the PWL parameterization for the kth resistor has $n_k + 1$ break points. Substituting for Y in Eq. (7), we get a formulation of the PWL circuit with complementarity variables Λ^+ and Λ^-. Find w and z such that

$$Nw\ -\ Mz\ =\ q,\ w,\ z\ \geq\ 0,\ w'z\ =\ 0 \tag{11}$$

This formulation is called the generalized LCP problem because it is identical with the LCP problem, except the coefficient matrix of w is not I and possibly N and M can be rectangular. See Ref. [5] for examples of this formulation and an algorithm to find all the solutions for Eq. (11).

PLACEMENT OF ELEMENTS IN A CIRCUIT

Since the arrival of very large-scale integrated (VLSI) technology, the metal-oxide-semiconductor (MOS) transistor has emerged as a dominant device in microelectronics. In analog-integrated circuits, the individual devices are placed on the silicon area and then interconnected. The parameters of some devices have a strong influence on circuit performance, and thus the devices need to be placed optimally so as to minimize the effect of parameter deviations of the individual components on the circuit performance. Even though computer-aided network tuning is possible after manufacture of the chips [8], it is much less expensive to place the devices optimally during the design stage. This ensures off-line quality of the circuit.

Let n be the number of transistors and m be the number of parameter types (such as threshold voltage). A sensitivity vector S describes the linear influence of all transistor parameters on a global circuit parameter ρ, which measures circuit quality. Let

$$\Delta_p = (\Delta p_1, \ldots, \Delta p_m)' \tag{12}$$

be the vector of deviations of the transistor parameter from their nominal values for each parameter type, then

$$\Delta p_i = (\Delta p_{i1}, \ldots, \Delta p_{in})' \tag{13}$$

gives these deviations for the ith parameter type for each transistor. Circuit quality is reflected by

$$\rho = S\,\Delta_p = \sum_i \sum_j S_{ij} \Delta p_{ij} \tag{14}$$

where $S = (S_1, \ldots, S_m)$ is the sensitivity vector with

$$S_i = (S_{i1}, \ldots, S_{in}) \tag{15}$$

Let b be the $k \times 1$ vector containing all distinct absolute values of the sensitivities S_{ij}.

Therefore an S_{ij} can be written with $k \times 1$ vector C_{ij} with elements C_{rij} as given below:

$$C_{rij} = \begin{cases} 1 & \text{if } b_r = S_{ij} > 0 \\ -1 & \text{if } -b_r = S_{ij} < 0 \\ 0 & \text{otherwise} \end{cases} \tag{16}$$

so that $S_{ij} = \sum_r b_r C_{rij}$.

Now we obtain a $k \times mn$ matrix $\hat{C}$, whose columns are the C_{ij} vectors. Let $\hat{C} = (C_1, \ldots, C_m)$ with $C_i = (C_{i1}, \ldots, C_{in})$. Each transistor j is located at the point (x_j, y_j) in the chip area, and its parameters depend on this position in a linear fashion.

$$\Delta p = GW + g \tag{17}$$

where $W = \begin{bmatrix} X \\ Y \end{bmatrix}$ with $x = (x_1, \ldots, x_n)'$ and $y = (y_1, \ldots, y_n)'$

$$G = \begin{bmatrix} G_{11} & G_{12} \\ \vdots & \vdots \\ G_{m1} & G_{m2} \end{bmatrix} \text{ and } g = \begin{bmatrix} g_1 \\ \vdots \\ g_m \end{bmatrix} \tag{18}$$

where $G_{ij} = \Gamma_{ij} I$ is of size $n \times n$ and $g_i = \begin{bmatrix} y_i \\ \vdots \\ y_i \end{bmatrix}$ is of size $n \times 1$. Thus we have the optimal placement condition $\rho = 0$ for all G, g, and b, given by

$$b\hat{C}\,(GW + g) = 0 \tag{19}$$

Since b is any arbitrary vector, we shall have for all G and g

$$\hat{C}(GW + g) = 0 \tag{20}$$

which implies

$$\textit{for all } G : \hat{C}GW = 0 \tag{21}$$

$$for\ all\ g\ :\ \hat{C}g = 0 \tag{22}$$

Equation 22 shows that $\hat{C}g = 0$ is a condition to be satisfied by $\hat{C}$ in order that the optimal placement problem has a solution. This may be the case in practice.

Since C_i is a $k \times n$ and G_{ij} is an $n \times n$ matrix equal to $\Gamma_{ij}I$,

$$\hat{C}\ GW = (C_1, \ldots, C_m) \begin{bmatrix} G_{11} & G_{12} \\ \vdots & \vdots \\ G_{m1} & G_{m2} \end{bmatrix} \begin{bmatrix} x \\ \vdots \\ y \end{bmatrix} = 0 \tag{23}$$

each C_ix and C_iy must vanish separately. Similarly $\hat{C}g = 0$ implies that $C_ie = 0$ for each i separately.

Let $\underset{n \times 2}{X} = (x \mid y)$ and C be the matrix obtained from $\hat{C}$ after dropping linearly dependent rows and transposing at the block level.

Thus we have

$$CX = 0 \tag{24}$$

$$Ce = 0;\ \text{where}\ e = (1, \ldots, 1)' \tag{25}$$

Example: Suppose $b = (b_1, b_2, b_3)$ and the sensitivity vector $S = (b_1, -b_1, -b_1, b_1; b_2, b_3, -b_2, -b_3)$. Then

$$\hat{C} = \begin{bmatrix} 1 & -1 & -1 & 1 & 0 & 0 & 0 & 0 \\ 0 & 0 & 0 & 0 & 1 & 0 & -1 & 0 \\ 0 & 0 & 0 & 0 & 0 & 1 & 0 & -1 \end{bmatrix} \tag{26a}$$

and

$$C = \begin{bmatrix} 1 & 0 & -1 & 0 \\ 0 & 1 & 0 & -1 \end{bmatrix} \tag{26b}$$

obtained from Eq. (26a) after transposing and getting $\overline{C}$, a 6×4 matrix, and then dropping the linearly dependent rows. Solving Eq. (24) yields the placements that achieved $\rho = 0$. $Ce = 0$ implies that each row of C is a contrast, in the language of experiment design. Since in practice the transistors are placed at distinct places on a rectangular grid, the further restriction that x_j, y_j are positive integers, bounded above, is also imposed.

By removing symmetry, such solutions of interest are produced by Sackinger and Fornera [9], who ingeniously use the results from symmetric groups. Since a large number of solutions that render the circuit quality measure ρ to be zero are obtainable in this fashion, other criteria to make the design robust can be searched for among these solutions. Sackinger and Fornera [9] have applied such an approach to an integrated differential difference amplifier circuit obtaining 48 insensitive placements from among the possible arrangements of eight components in $4 \times 2 = 8$ sites.

ROBUST DESIGN AND OPTIMIZATION

Quality engineering and assurance aims at minimizing deviation of a product's functional characteristics from the target values. To withstand changes in the external and internal sources of variation, Taguchi advocates using parameter design and tolerance design to build robustness into products [10–11].

Traditionally, yield maximization or cost minimization were chosen as objectives in their own right. The concept of robustness is definitely an advancement in the selection of right objectives at the design stage. However, the use of orthogonal array heuristics for optimum seeking, used by Taguchi, definitely requires discussion [12]. Choosing the right objective is like increasing effectiveness by doing the right thing. However, a black box type of experimental approach for optimizing the right objectives might lead to inefficient solutions or an incomplete understanding of the underlying structure [12].

The problem of minimizing the variance of a nonlinear function of several random variables is central to such problems as the parameter design problem (PDP). Let $X = (X_1, \ldots, X_n)'$ be a vector of independent normal random variables with

$$E(X_i) = \mu_i \text{ and } Var(X_i) = \sigma_i^2 \tag{27}$$

$$\mu = (\mu_1, \ldots, \mu_n)' \tag{28}$$

Let $G : R^n \rightarrow R$, be a given real-valued function of $x_i, \ldots, x_n$, called the transfer function. Let $Y = G(X)$ where Y is also a random variable, and let μ_y and σ_y^2 denote the mean and variance of y.

Suppose $\sigma_i = g_i(\mu_i)$, where g_i is a given real-valued function from R to R, for $i = 1, \ldots, n$. Then it is possible to estimate μ_y and σ_y^2 by using simulation.

In parameter design the dependence of σ_y^2 and μ_y on μ_y on $\mu_1, \ldots, \mu_n$ is recognized and an attempt is made to choose μ in such a manner that μ_y or $G(\mu)$ is close to target value and simultaneously $\sigma_y^2(\mu)$ is as small as possible. While y measures the signal at μ, σ_y^2 measures the noise at the same setting. Hence an appropriately defined, combined measure of signal-to-noise ratio Z, as given in Eq. (29), is considered by Taguchi to maximize

$$Z = 10 \log y^2 / s^2 \tag{29}$$

where y and s^2 are the mean and variance of the characteristic under study at a particular setting in the inner orthogonal array, which corresponds to a certain level $\mu_1, \ldots, \mu_n$ of the factors under consideration.

Since the number of trials required under the orthogonal array heuristics of

Taguchi generally can be large, this method is suitable only when the transfer function is available. Assuming G is available, continuous, and twice differentiable, the response y can be approximated by a quadratic function q(x) given by

$$y \simeq q(x) = \tfrac{1}{2}\, x'\, Hx + C'x + \alpha \tag{30}$$

where H = Hessian of G at a given point ξ, C = gradient of G at ξ—(Hessian of G at ξ)ξ, and $\alpha = G(\xi) - \nabla\, G(\xi)'\xi + \xi' H \xi$. Then the PDP problem can be approximated by the following: Minimize z = $V(\mu_1, \ldots, \mu_n)$ subject to G $(\mu_1, \ldots, \mu_n) = \tau$

$$\mu_1, \ldots, \mu_n \in \mathrm{M} \tag{31}$$

where $V(\mu_1, \ldots, \mu_n)$ has an expression in terms of H and C as given by Eq. (32).

$$\begin{aligned} V(\mu_1,\ldots,\mu_n) \;=\; & \tfrac{1}{2}\sum_i \sum_j h_{ij}^2\, \sigma_i^2\, \sigma_j^2 + \sum_i \sum_j \sum_k h_{ik} h_{kj} \mu_i \mu_j \sigma_k^2 \\ & + \sum_j C_j^2 \sigma_j^2 + 2 \sum_i \sum_j h_{ij} c_i \mu_i \sigma_i^2 \\ & (\text{All } \textstyle\sum \text{ are over 1 to n}). \end{aligned} \tag{32}$$

If $\sigma_i = g_i(\mu_i)$ = constant, then $V(\mu_1, \ldots, \mu_n)$ turns out to be a convex function. In this case, methods available in the mathematical programming literature with convex objectives can be used to solve this problem [13].

The optimal solution to Eq. (31) need not be optimal for the PDP problem; hence the quadratic approximation of G at the optimum can be taken up as the problem to be solved by replacing ξ by the optimum found for Eq. (31). This iterative process can be continued and stopped according to some specified termination criterion. Such a procedure is suggested in Ref. [13] for attacking the parameter design problem. Modern optimization procedures recognize the need to safeguard iterative methods by techniques such as line searches or trust region methods. Even this procedure might lead only to a local optimum of the PDP; thus the search needs to be continued by using some global optimization approach such as simulated annealing [12].

GLOBAL OPTIMIZATION AND LARGE CIRCUIT PARTITIONING

In VLSI design where a large circuit with N nodes is partitioned into layers of smaller circuits, it is advisable to minimize the number of arcs between layers [14]. This problem is similar to the problem of grouping parts into families and then assigning families to different flexible manufacturing cells in flexible manufacturing systems (FMS). Transportation between facilities is considered a waste because it does not add value to the product. Therefore minimizing the total number of movements of jobs between machines is supposed to produce better system performance in FMS.

A general problem of this nature can be formulated as follows: Given a graph with arc weights, partition the graph into n nonempty components where each component has a capacity of c nodes. The idea is to minimize the total arc weight between components. Let i, j denote the index of the nodes on the graph G. Let m denote the index of the machines, where each machine has a capacity of c nodes and the machines are identical.

$$\text{Let } x_{im} = \begin{cases} 1 \text{ if node } i \text{ is assigned to machine } m \\ 0 \text{ otherwise.} \end{cases} \tag{33}$$

Therefore, we require

$$\sum_m x_{im} = 1 \text{ for } i = 1, \ldots, N \tag{34}$$

$$\sum_i x_{im} \leq c \text{ for } m = 1, \ldots, n. \tag{35}$$

The operational sequence of each product p can be represented by $G_p = (N_p, A_p)$ where each node $i \in N_p$ represents an operation required by product p and each arc $(i, j) \in A_p$ denotes a pair of consecutive operations. Let $G = (N, A)$ where $N = \bigcup_{p \in P} N_p$ and each $(i, j) \in A$, n_{ij} represents the number of products requiring operation, i and j to be consecutive. Let $S = (s_1, \ldots, S_m)$ be a partition of N, and S_m denote the set of operations to be performed on machine m. For any feasible partition S, a cut set $A(S) \subset A$ is a set of arcs, such that the arc $(i, j) \in A(s)$ if nodes i and j belong to distinct sets. The total number of movements between machines incurred by all the arcs in the cut set A (S) is

$$C(S) = \sum_{\substack{(i,j) \in A(s) \\ i < j}} n_{ij} \tag{36}$$

The objective of the operation partitioning problem (OPP) is to minimize C(S). To check whether the arc (i, j) is in the cut set, we have

$$y_{ij} \geq x_{im} - x_{jm} \text{ for each } m; \text{ for } i, j, \; i < j \text{ and } (i, j) \in A \tag{37}$$

$$Y_{ij} \geq 0 \text{ for each } i, j \tag{38}$$

The OPP can then be formulated as

$$\text{minimize} \sum_{i}\sum_{\substack{j\\ i<j}} Y_{ij}X_{ij} \tag{39}$$

subject to Eqs. (33–35, 37–38). Since the OPP is a combinatorial problem for which it is easy to find a local optimal solution but computationally hard to find the global optimum, a simulated annealing heuristic is applied to find nearly optimal solutions to OPP [15].

In addition, simulated annealing has been found to be useful in the global optimization arising in parameter design problems related to robust designing [12].

FAULT TOLERANT COMMUNICATION IN NETWORKS

The design of multiprocessor networks aims at linking processors through communication lines in order to accommodate a large number of processors with power connecting lines, while at the same time keeping the length of the communication path between any two processors small and above all preventing the network from degrading rapidly if a few processors are down or faulty. This problem is a good example of off-line activities that ensure better operational quality of communication networks, for which corrective actions after construction of the network are either impossible or prohibitively costly, especially when fixed paths are used for communication between the nodes.

Combinatorial objects like graphs and experimental design arrangements provide substantial insight into problems of this nature. Let G = (V, E) be a graph with the correspondence of nodes V to the set of processors and edges E to the links for communication. Let d(x, y) be the length of a shortest path between x and y belonging to V. Let diameter, d(G) be maximum of d(x, y) over all pairs of nodes. Let $F \subset V$ be a set of faulty nodes in G. The surviving graph G/F has V–F as the node set and edges. $E \cap (V - F) \times (V - F)$.

Fault tolerant behavior of the network can be measured by different criteria, such as the number of disjoint communication paths between any two nodes of the network and the length of the communication path not increasing substantially in the presence of faulty nodes, i.e., d(G/F) – d(G) should be small. When a fixed path for communication between the nodes is used, the network is supposed to have a fixed routing. Failure of a node can destroy the routing between some other nodes X, Y in such networks. Communication between such nodes is possible by X sending the message at the end of each route. An important criterion in fault-tolerant networks with fixed routing is minimizing the maximum number of times the message for communication between pairs of nodes is broadcast in the presence of faults [16–17].

Use of balanced incomplete block designs (BIBD) is made in Ref. [18] in the design of fault-tolerant communication networks. BIBD is a combinatorial arrangement used in the statistical design of experiments and is denoted by (b, n, r, k, λ), where a set S of n symbols or treatments is given and b blocks containing exactly k elements from S are formed such that each element of S appears in exactly r blocks and every pair of distinct elements of S is contained in exactly λ of these blocks.

A design of a network model (G, b, n, r, k) consists of b copies of $G_1 \ldots G_b$ of graph G and nodes $v_1 \ldots v_n$. Node v_j is connected to a node of graph G_j if block B_i of (b, n, r, k, 1) design contains v_j.

Example : A BIBD design (12, 9, 4, 3, 1) is given by the blocks:

126	234	369	459	567
135	258	378	468	
147	279			
189				

Suppose G = (V, E) = K_3 = the complete graph on three nodes. Now consider (K_3, 12, 9, 4, 3) as described above. (K_3, 12, 9, 4, 3) consists of 12 copies $G_1 \ldots G_{12}$ of K_3 and nodes (1, . . . , 9). Node 2 is connected to G_5, for example, if B_5 = 234 which contains 2. This network (K_3, 12, 9, 4, 3) has 45 nodes.

Reference [18] shows that such a network (K_k, b, n, r, k) has diameter 4 or 5 on n(r + 1) nodes. For any set F of (k – 1) or fewer, faulty nodes, the diameter of the surviving graph is at most 7. Also when a fixed routing is used with routes being the shortest path between x and y in the network (K_k, b, n, r, k), the surviving route graph of the above network has a diameter less than or equal to 2. Thus a highly fault-tolerant network design has been achieved through combinatorial construction such as BIBD. (See Ref. [18] for other fault-tolerant networks.)

CONCLUSIONS

Circuits and network design optimization problems have been tackled using mathematical programming and concepts from combinatorial mathematics. While the concept of robustness in design to ensure trouble-free product performance is to be given due recognition, the drawback of ad hoc or simplistic approaches for design parameter selection cannot be overlooked. This paper also illustrates how structured knowledge available in the file of mathematical programming and combinatorics is inevitable for solving certain design and analysis of circuits and networks. Methods available in the optimization literature (see Ref. [20–22]) can be used, depending on the nature of the problems encountered in design and analysis of circuits.

Currently multicriterion optimization approaches are being used in circuit design [19]. New approaches to finding global optimal solutions to nonlinear programming problems promise to be implemented in real time via an associated NLP circuit [23].

With these developments, it is expected that circuits and network design problems will be solved more efficiently in the future.

REFERENCES

1 L. O. Chua, C. A. Desoer and E. S. Kuh, Linear and Non-Linear Circuits, McGraw-Hill, New York, 1987.
2 L. O. Chua, T. Matsumoto and S. Ichiraku, IEEE Trans. Circuits Syst., CAS-27 (1980) 577–603.
3 S. N. Stervens and P. M. Lin, IEEE Trans. Circuits Syst., CAS-28 (1981) 429–441.
4 J. T. J. Van Eyndhoven, SIAM J. Control Opt., 24 (1986) 1050–1062.
5 L. Vanderberghe, B. L. DeMoor and J. Vanderwalle (1989) 1382–1391.
6 K. G. Murty, Linear Complementarity, Linear and Nonlinear Programming, Heldermann-Verlag, Berlin, 1988.
7 B. C. Eaves and C. E. Lemke, Math. Oper. Res., 6 (1981) 475–484.
8 J. F. Pinel, IEEE Trans. Circuit Theory, CT18 (1971) 192–194.
9 E. Sackinger and L. Fornera, IEEE Trans Circuits Syst., 37 (1990) 1052–1057.
10 G. Taguchi, Introduction to Quality Engineering, Designing Quality into Product and Processes, Kravs Int. Pub., White Plains, N.Y., 1986.
11 G. Taguchi and Y. Wu, Introduction to Off-line Quality Control, Central Japan Quality Control Association, Nagoya, Japan, 1980.
12 T. A. Arthanari, First Int. Conference on Optimal Design and Analysis of Experiments, Neuchatal, Switzerland (1988).
13 Y. Fathi, Oper. Res. Letters, 10 (1991) 461–466.
14 R. Raltson, Encyclopedia of Computer Science and Engineering, Van Nostrand and Reinhold, New York, 1983.
15 R. H. Ahmadi and C. S. Tang, Oper. Res., 39 (1991) 824–835.
16 A. Esfahanian and S. L. Hakimi, IEEE Trans. Computers, 34 (1985) 777–788.
17 A. Broder, M. Fischer, D. Dolov and B. Simons, in: Proc. 1615 Annual ACM Symp. on Theory of Computing (1984) 536–541.
18 J. Opatrny, N. Srinivasan and V. S. Alagar, IEEE, Trans. Circuits Syst., 36 (1989) 23–29.
19 C. Charalambous, IEEE Trans. Circuits Syst., 36 (1989) 773–784.
20 R. Fletcher, Practical Methods of Optimization, 2nd ed., John Wiley and Sons, Ltd., Chichester, U.K., (1987).
21 P. E. Gill, W. Murray and M. H. Wright, Practical Optimization, Academic Press, London, 1981.
22 J. E. Dennis and R. B. Schnabel, Numerical Methods for Unconstrained Optimization and Nonlinear Equations, Prentice-Hall, Englewood Cliffs, N.J., 1983.
23 L. O. Chua, IEEE Trans. Circuits and Syst., 37 (1990) 966–969.

Chapter 2
Management through Strategic Considerations of Design

As illustrated in the "Quality in software development—an example" section in the Introduction (p. 4), product and process should be designed strategically, including not only the tools and methodologies employed but also the design philosophy carried out by the designer and the associated life-cycle cost along with its optimization. Considering total quality control and management, we present in this chapter many case studies ranging from systematic fool proofing techniques to off-highway dump truck and spacecraft mechanics practiced in a number of industries in Japan, Canada, and India.

Quality through Engineering Design - W. Kuo (Editor)

Steps for Developing New Products Viewed from the Standpoint of Total Quality Control

Yasutoshi Washio

Department of Administration Engineering, Keio University, 3-14-1, Hiyoshi, Kohoku-ku, Yokohama 223, Japan

Abstract

This paper describes the steps for developing new products from the total quality control viewpoint. By faithfully implementing the steps, the following effects can be expected: (1) shortening the time period for new product development, (2) reducing problems during the initial production period of new products, (3) reducing the cost for new product development, and (4) enabling the new products to become market-oriented.

INTRODUCTION

Manufacturers must continue to develop and produce new products in order to profit and grow. New products are needed for two reasons: one is the change in market and customer needs and the other is technological innovation. Therefore, new-product development is one of the most important activities of manufacturing companies. This paper outlines the necessary steps for developing new products from the total quality control viewpoint with an emphasis on the design stage.

STEPS FOR DEVELOPING NEW PRODUCTS

Gather market information

In the past, companies were product oriented, and they primarily developed new products on the basis of the company's circumstances. Now, however, they are more market oriented and base their product development on market needs. The question now is, "How does a company grasp market needs?" One way to accomplish this task is by conducting customer satisfaction surveys. Potential needs cannot be understood unless they are approached from the company's standpoint. Another way is to compare the quality of the company's product with competing products. Since complaints are different from needs, it is dangerous to develop new products by looking only at complaints.

In addition to assessing market needs, developing new demand or markets on the basis of the company's current technology is also important. Companies should prepare a map of their current technology, and where it is insufficient, they should develop it. This activity is independent of the development steps for individual new products. The main quality control techniques used at this stage are quality function deployment (QFD) and reliability techniques such as Weibull analysis.

The basic tools for QFD are shown in the following three tables. Table 1 shows the relationship between the customers' requirements, which are quality items required by the market expressed in the users' terms, and quality characteristics, which are those characteristics that will encompass the customers' requirements. Showing the information in matrix form clarifies the relation between the two sets of information and thereby aids the conversion of the users' language into technical language.

Table 1. Quality Table: an example of handy cleaner.

Customer's Requirements \ Quality Characteristics		1st level	Motor		Load		Body			
1st level	2nd level	3rd level \ 2nd level	Power	Location	Torque	Angle of the Fans	Material	Strength	Thickness	Density
cleaning is easy	feel comfortable while using	less noisy	⊙		○		△		△	
		less variation	△				○		⊙	
		exhaust air does not blow to users								
		exhaust air is weak	△							
		exhaust air does not contain dust								
	stable and strong absorption	absorb all dust	⊙		○	○				
		keep strong absorption even when dust bag is full	△							

⊙ strong relationship ○ medium relationship △ weak relationship

Decide the product concept

The next step in developing new products is to decide the product concept, which involves considering the market needs and the company's technical capability.

Quality may be classified into characteristics that a product "must have" and those that make it "attractive." A new product must always have an attractive quality, but it is necessary to clarify just what that is. At this stage, companies must target the appropriate market and determine the market size, and target the amount of sales desired, the reliability level, and the cost. Then a schedule must be developed.

Nishihara and Yokokura [1] point out that all recent hit products in Japan are the result of successful product concepts. For example, film with lens/flash from Fuji Photo Film Co., Ltd., shows that it is possible to develop new types of products by applying ordinary technology. Likewise Hitachi's low-noise electric washer came about when the company studied and compared washer types worldwide and then chose noise-reduction as their sales point.

Development and product planning

The main quality characteristics (technical characteristics) are decided at the development and product planning stage. Here the customers' needs and product concepts are converted into technical terms. For example, a lively accelerator pedal is a product concept. In order to convert this concept into technical terms we have to specify the relation between the stepping power and the stroke in designing the accelerator pedal.

It is important to anticipate critical problems or bottle-necks in the technology that will appear in the future and ask the research and development (R&D) division to solve them. QFD should be used at this stage.

Product design

At this step, more detailed and concrete plans for deciding the product concept and developing and planning the product are made, and drawings are prepared. Here it is necessary to grasp the relation between the design specification ($s_1, s_2, \ldots, s_p$) and the quality characteristic y.

$$y = f(s_1, s_2, \ldots, s_p)$$

Although technology intrinsic to an industry, such as mechanical engineering for the mechanical industry, plays an important role for this purpose, statistical techniques such as the design of experiments and regression analysis are also used.

Next we must consider what constitutes a good product. A good product is that which achieves a certain function regardless of how the customer uses it. Therefore, the product must be designed from the beginning by taking into account the usage and environmental conditions it will be subjected to during use. For this purpose Taguchi's method is effective and is considered as a tool for technology development.

Take an example of applying parameter design, which is a tool in Taguchi's method, to the design of a weighing scale [2]. Consider the design parameters, such as the material for the springs, reinforcement of the base, and position of the base supporting legs, as the control factors and lay them out in the inner orthogonal array. Also consider the usage and environmental conditions such as floor surface, slant angle, and deviating position as the error factors and the weight of the body to be measured as the signal factor and lay them out in the outer orthogonal array. The signal-to-noise (S/N) ratio β^2/σ^2 is taken as the characteristic value.

Next design reviews (DRs) are carried out at the stages of product planning and product design. DRs are usually performed at three or four different levels. Table 2 shows an example of a DR system. Design reviews are usually carried out in three steps (DR1, DR2, and DR3) and the items checked by DRs include whether

- the planned quality is fully reflected
- the design is such that it leads to product defects
- there are any design mistakes
- manufacturing equipment is taken into consideration in the design
- the product is designed for easy manufacturing
- the parts used are standardized and used in common where possible (If the designer is left to his own discretion, he will pursue an optimum design that will dramatically increase the number of new parts.)
- the target cost has been achieved

Table 2. DR system.

DR	Period	Object/Purpose	Division Participated
DR1	After product planning	Checking the concept of merchandize	Sales, Product planning, R&D, Product Design, Production Engineering
DR2	After outline design	↑	Product Planning, R&D, Product Design, Product Engineering, Manufacturing, QA, Purchasing
DR3	After final design	Checking the concept of merchandize, standardization of design, productivity and design errors	Product Design, Product Engineering, Manufacturing, QA, Purchasing

To implement DRs requires participation of related departments and preparation of checklists for the individual design reviews DRl, DR2, and DR3. These checklists should be revised through the implementation of the design reviews (implementation of the PDCA [Plan-Do-Check-Action] cycle).

The next problem to consider is how to reduce variation in implementing intrinsic technology from engineer to engineer and from division to division. One method is to set standards or develop design manuals.

A feature of total quality control (QC) in Japan is group-wide QC as opposed to company-wide QC. One way to implement this approach in the product design stage is to ask major suppliers to participate in new product development from the product design stage.

Other ways include carrying out failure mode and effect analysis (FMEA) and fault tree analysis (FTA) to prevent mistakes in design from occurring. Table 3 shows an example of an FMEA sheet for a design part.

Prototype manufacturing

Table 3. FMEA Sheet for a Design Part.

Part Name (Part Number)	Part Function	Failure Mode	Effect(s) of Failure	Cause(s) of Failure	Existing Conditions			Recommended Action(s) and Status	Resulting			
					Occurrence	Severity	Risk Priority Number		Action(s) Taken	Occurrence	Severity	Risk Priority Number
					8	6	48			2	6	12
					4	2	8					

Once the design has been determined, manufacturing a prototype enables the project team to see how the design works. At this stage, design drawings can be evaluated in terms of the product (DR4). In addition, quality evaluation of durability and reliability can be determined through accelerated testing.

Production preparation

Products are made by the 4Ms (machine, material, method, and man). Therefore, the purpose of this step is to establish these 4Ms.

First, machines and equipment must be procured or prepared with sufficient process capability to do the job. In order to determine that level, process capability studies, Cp/Cpk analysis, should be carried out. Next the process and equipment must be designed and evaluated by conducting process and equipment FMEAs. Finally, jigs, tools, and measuring instruments must be acquired and maintained.

Second, raw materials and parts must be acquired. Raw materials will, of course, need to be purchased, but acquiring parts involves deciding whether to make or buy them and determining how to control the quality of purchased items. Purchasers must understand the process capability of outside suppliers and fit the method of ordering to that capability. When necessary, this will involve guiding and instructing suppliers to upgrade their process capabilities.

Determining operation methods involves establishing technical and operating standards for production, deciding the process control method, determining inspection items and standards, and devising automatic inspection equipment. At this time the QC process chart is also prepared.

Finally workers must be trained to carry out the planned production and maintain quality. The primary techniques used at this stage are Cp/Cpk analysis, the QC process chart, experimental design, regression analysis, and sampling inspection.

Initial run

With the initial run, the goal is to carry out production using the 4Ms, pick out any portions in the design or 4Ms that go wrong, and correct them. This step should be carried out at the regular plant and not at a pilot plant. After the initial run, the criteria needed to switch from the initial run to mass production can be clarified.

Production

During the production phase, "quality consciousness" of the workers needs to be stressed. Such techniques as using a QC circle and a suggestion box and establishing a fool proof system can be used.

Implementing process control, that is, maintenance of the process, is also important. This step involves applying technical and operational standards and QC process charts, taking corrective measures to prevent a recurrence of any failures in the manufacturing process, and revising or setting new standards as needed. A process irregularity report should be issued to track problems. Implementing problem solving in the workshops to improve the process is also helpful.

The main quality control techniques used at this stage are the QC process chart, control chart, design of experiments, QC 7 tools (histogram, stratification, Pareto diagram, cause-and-effect diagram, check sheet, graph, and scatter diagram), testing hypothesis, and regression analysis.

After-sales service

After-sales service is an important step in developing new products. Here information is collected from the user's product evaluation, and then the results are fed back to the technical divisions. Establishing systems of after-sales service and complaints processing can also provide the means to feed back the data on troubles or complaints to the technical divisions. The technical divisions must take two kinds of actions: (1) emergency measures and (2) measures to prevent recurrence. One way of implementing item (2) is to change the checklists of DRs.

Another technique is surveying customer satisfaction regularly and informing all the divisions about the results. Using field data to understand the quality situation while the product is in use is also important.

The main quality control techniques used at this stage are QFD, mean time between failure (MTBF)/mean time to failure (MTTF), Weibull analysis, hazard analysis, sample survey, and multivariate analysis.

REFERENCES

1 R. Nishihara and C. Yokokura, Total Qual. Control, 41 (1990) 1166–1168 (in Japanese).

2 Parameter Design in New Product Development, Japanese Standard Association, Tokyo, 1984, pp. 177–194 (in Japanese).

Quality through Engineering Design - W. Kuo (Editor)

Management of Variation Reduction Investigations

Bovas Abraham and James B. Whitney

Institute for Improvement in Quality and Productivity
University of Waterloo, Waterloo, Ontario N2L 3G1, Canada

Abstract

Issues arising in managing and implementing variation reduction investigations are considered. The discussion is organized around a suggested sequence of steps in a structural problem-solving approach. Two levels of management responsibilities are discussed to increase the probability of success of the variation reduction investigation.

INTRODUCTION

Meeting and exceeding customer needs and expectations are the driving forces behind continuous process improvement and the production of a quality product in the eyes of the customer. In many instances, one of the basic expectations of the customer is conformance to design [1]. That is, the customer expects output characteristics to be on target with minimum variation. Thus in many process improvement efforts [2], variation reduction is a key element.

There are many sources of variation in a process output. Variation reduction investigations focus first on isolating important sources and then proceed to try and eliminate them or compensate for them. Variation reduction investigations are examples of continuous process improvement efforts through data-based problem solving.

Decisions about process changes must be based on facts supported by data [3]. In variation reduction investigations the collection of the appropriate data and its proper analysis are critical. Hence these investigations must be embedded in a structured approach to problem solving. Such an approach requires careful management and usually a team effort.

This paper discusses some issues that arise in the management and implementation of a structured problem-solving approach. The discussion is organized around a suggested sequence of steps for a structured approach. Within each step both management and implementation issues are considered.

A STRUCTURED APPROACH TO PROBLEM SOLVING

Although there are many versions of a step-by-step approach to problem solving, these versions are basically different packaging of the following steps.

STEPS IN THE STRUCTURED APPROACH

PROBLEM IDENTIFICATION

- define broad objective
- narrow the focus
- describe the process
- measure process outputs
- develop base line
- organize time
- prepare schedule

↓

POTENTIAL CAUSES

- consider all possible causes
- prioritize

↓

PLAN FOR DATA COLLECTION

- check measurement systems
- decide what, who, when, where, how

↓

IDENTIFY CAUSES

- collect data
- analyze data

↓

SOLUTION GENERATION

- identify possible solutions
- test and select solution

↓

IMPLEMENTATION

- standardize new process
- monitor improvement

The use of a system to solve problems is important because it enforces discipline, defines roles and responsibilities, and in the long run, is more effective.

The difficulty in completing each step will vary with each investigation. Some steps such as “potential causes” and “planning for data collection” must draw on a wide knowledge base for the process under investigation. This makes a team approach essential. At various steps in the problem-solving process, the team composition may change as the required knowledge base changes. However, to preserve the constancy of purpose of the investigation, there must be a core group who remain on the team for the duration of the investigation.

For effective implementation of the step-by-step approach to problem solving, it is important that the team members span a wide enough knowledge base and work well together. In addition, the team leader must be able to foster enthusiasm and manage the team activities toward the completion of each step.

There are two levels of management responsibilities in implementing a structured approach:

1. the management of the problem-solving process (by an individual or group) to ensure that the team follows the process and that road blocks to the team's progress are removed or at least reduced
2. the management of the team by the team leader to ensure that each problem-solving step is completed.

It is the role of the person or group of individuals managing the problem-solving process to take the necessary steps to ensure that the team members have had the proper training so that they will function effectively as a team. In this paper we will not discuss how this is done. Much literature is available on the subject, and a good starting point would be Scholtes [4].

PROBLEM IDENTIFICATION

For this step, management issues are important. It is essential that all interested parties of the investigation buy into the problem to be solved and agree on the goals, the types of solutions to be considered, and the resources needed to achieve the stated goals.

It is the responsibility of the group or person managing the problem-solving process to ensure that problems important to the organization and its customers are considered. Thus their duty is to lay out the broad objectives of the variance reduction investigation and assist in forming a team for the investigation. Having done this, the managing group and the team together must lay out a general mandate for the investigation that states

- the types of solutions that are to be considered first
- the type of documentation that should accompany suggested solutions
- the broad objectives that are reasonable to achieve and how progress towards them can be measured

Following this process, the team leader's job is to assist the team in completing the following tasks:

1. understanding the process at which the investigation is directed
2. focusing the investigation on a specific problem (The next two steps in the problem-solving process, "potential causes" and "planning for data collection," are process- and problem-specific.)
3. determining the goals of the investigation of the specific problem and how progress towards these goals can be measured

4. justifying the usefulness of solving the specific problem in meeting the broad objectives of the general mandate
5. estimating the amount of time required of the team members to achieve the objectives of the specific problem.

Numerous tools are available that are useful in assisting in the above tasks: flow charts of the process, pareto charts, histograms, run charts, etc. Historical data may be of use or data may have to be collected. As part of describing the specific problem, three important things must be established:

1. the outputs to be measured for the investigation of the specific problem
2. a measurement process for the outputs that can be trusted
3. a baseline for the specific problem and if possible a baseline for the more general problem of the mandate.

When tasks 1 to 5 are completed and documented, the team or team leader presents the information to the group managing the problem-solving process for discussion. This step is necessary because the group responsible for managing the problem-solving process needs to have some assurance that the problem is worthwhile and some rough idea of the resources that need to be allocated to complete the investigation before they give final approval for investigating the specific problem.

POTENTIAL CAUSES

The team has the responsibility to list and prioritize all the sources of variation in the output under consideration. The team leader is responsible for making sure that this happens. Carefully managed, brain-storming sessions are required for soliciting the list of sources from the team members.

To assist the team in reaching a consensus about prioritizing the sources to investigate first, one can systematically organize the list of sources by cause-and-effect diagrams, and classify the sources as controllable or not controllable, easy or hard to change, costly or cheap to change, etc. In addition to soliciting information from the team members, it is also the team leader's responsibility to try and ensure that resource people outside the team are contacted for their input.

At the end of this step, there will be a systematically compiled list of sources of variation and a short list for immediate investigation. At this point, the specific problem under investigation and the time requirements for the investigation should be reviewed and updated if necessary. Any changes must be approved by the group managing the problem-solving process.

PLANNING FOR DATA COLLECTION

Completing the previous step provides a short list of sources of variation. The sources in this short list are to be studied to determine the importance of their effect on the outputs being measured. The main issue in this step is the careful development of a data collection strategy to achieve this goal. An important aspect of the strategy development is the assessment of whether or not the resources and expertise are available to carry it out.

In general, two approaches can be taken to collect data: an observational study or an experimental study. An observational study is a carefully planned collection of data from the process to assess the effect of sources of variation without making deliberate process changes, while an experimental study is one in which data are collected to assess the effect of sources of variation through carefully planned deliberate process changes.

Our objective in this discussion is not to discuss the technical details of these types of studies but to touch briefly on some of the managerial and implementation issues that the investigative team must consider. No matter which type of study is used the same detailed planning involved in the completion of the first two problem-solving steps is required.

Observational studies

These studies are useful for studying the effects of such sources as shifts, days, operators, batches, or spindles on the outputs being measured or assessing the association between sources of variation such as temperature or pressure and the measured outputs. Two of the most critical considerations for implementation of these types of studies are

- that the input value of the source can be measured reliably
- that the measured input value of the source can be traced to a measured output value

Designing the appropriate sampling procedure to investigate a particular source of variation is fundamental to observational studies. This aspect will not be discussed here. For a discussion on this topic see Hamada, MacKay, and Whitney [5].

Some advantages of observational studies that the investigative team should consider include: (1) no risk of producing poor quality products through deliberate process changes, (2) low implementation costs, (3) less personnel training required, and (4) relatively straightforward management of the sampling plan to collect the data. However, with these studies process changes cannot be assessed. Also, the natural range of variation of the source may not be large enough to detect an important effect. Moreover the duration of the study may take longer than an experimental study and the confounding of effects may present interpretation problems.

Experimental studies

The idea behind the experimental investigation is to study the effect of the sources of variation in the short list on the measured outputs by simultaneously changing the input values of some or all of them. Before embarking on an experimental study, the team should make sure that it is the right tool.

A discussion concerning the development of a specific data collection scheme for these studies is beyond the scope of this paper. Our intention is to discuss some of the more general issues related to these studies. These studies do require more training and expertise than observational studies. Before the investigative team decides to embark upon such a study, the team leader and the group managing the problem-solving process must be sure that adequate training will be given to those who require it or that the required expertise can be obtained. For more information on developing the data collection schemes, see Box, Hunter, and Hunter [6] and Abraham and MacKay [7].

Some of the first implementation considerations that arise are the following:

- Can the input values of a particular source be changed when desired?
- When the input value is changed, can it be held long enough to get measurements on the outputs?
- What input values are to be used for each source?
- Is there a measuring system in place to ensure that the inputs are at their selected values?

It is the team leader's responsibility to assist the team in fully addressing these issues; otherwise the implementation of the study is in jeopardy (see the discussion of a foundry experiment by Young, Abraham, and Whitney [8]).

An experimental study involving a large number of sources may require the cooperation of personnel and the coordination of activities from a number of departments. This requires very careful organization and management and will very likely require the assistance of the group managing the problem-solving process. In practically all of these experimental studies and especially for those involving a large number of sources of variation, trial experimental runs are a must. These trial runs (1) check out input values that might be questionable, (2) indicate where bottlenecks could arise, (3) give an idea of the time needed to collect the data from the study, (4) refine the communication channels between groups, and (5) generally familiarize those involved with the experimental procedure. For successful implementation, clear instructions must be given for each experimental run of the study.

Associated with each run is an individual run sheet which answers such questions as

- What is the run number?
- What is measured?
- When and how is it measured?
- What are the source input values?

The sheet will also have spaces for day, time, measured responses, and unexpected happenings. Trial runs give feedback on how well these instructions have been laid out and how easy the data collection part is to use.

When the trial runs are finished and this stage of the planning is complete, a written document describing the experimental plan is presented by the team to the managing group and all other interested parties. This document states the objectives of the study and contains specific details about the design such as

- the exact experimental runs to be done (the exact combination of input values to be used in each run)
- the way the runs are blocked and what randomization has been done
- the number of observations to be taken during each run and how they are to be taken
- the time period over which the experimental study is to be done
- the individual run sheets for each run

This presentation is required to ensure that all parties involved in the implementation of the experimental study understand and agree upon the plan. This presentation could result in a modification of the plan before it is finally implemented.

DATA COLLECTION

The processes for data collection are well defined by the sampling plan for observational studies and the detailed experimental plan for experimental studies. It is the responsibility of the team leader or a designated member of the team to ensure that the agreed-upon plan is carried out. They should ensure that (1) the data collection process established before is strictly followed, (2) any deviation is documented, and (3) no factors other than the planned ones are changed (no tampering).

SOLUTION GENERATION

The team should make sure that the solution is based on the analysis of the data collected in the previous step and is explainable on the basis of process knowledge. A first pass at data analysis can be simple (graphical) if the previous steps are executed properly. In some cases a more detailed and in-depth analysis can be useful in obtaining as much information as possible from the data. However, if the planned steps have not been executed properly, no sophisticated analysis can repair the mistake. The team leader should ensure that (1) the tentative solution and the recommendations are followed by confirmation runs or follow-up studies to arrive at a final solution, (2) considerations of implementation (time, cost, etc.) of the solution and recommendations are addressed, (3) the recommendations are shared by the team members, and (4) a final report containing the solution, results of confirmation runs, recommendations, and a cost-benefit analysis is produced.

IMPLEMENTATION

The group managing the problem-solving process plays a critical role here. They should ensure that

1. the contribution of the team is recognized and the results publicized
2. if the solutions are within the guidelines laid down in the general mandate, the recommendations are promptly implemented and adequate monitoring is put in place to hold the gains
3. if solutions are not implemented, direct feedback for not doing so is given to the team

CONCLUSIONS

In this paper we discussed some management issues involved in a structured approach to variation-reduction investigations. Two levels of management responsibilities were considered:

1. managing the process of variation reduction
2. managing each step in the structured approach.

If either of these levels is ignored, the probability of success of the variation reduction investigation is substantially decreased.

REFERENCES

1 D. A. Garvin, Managing Quality, Free Press (a division of Macmillan, Inc.), New York, 1988.

2 H. J. Harrington, Business Process Improvements, McGraw-Hill, Inc., New York, 1991.

3 W. E. Deming, Out of the Crisis, Massachusetts Institute of Technology, Cambridge, 1986.

4 P. R. Scholtes, Team Handbook, Joiner Associates, Madison, Wis., 1988.

5 M. Hamada, R. J. MacKay and J. B. Whitney, Process Improvement through Observational Studies, Research Report (in press), I.I.Q.P., University of Waterloo, 1991.

6 G. E. P. Box, W. G. Hunter and J. S. Hunter, Statistics for Experimenters, John Wiley and Sons, Inc., New York, 1978.

7 B. Abraham and R. J. MacKay, Designed Experiments For Reduction Of Variation, Research Report, RR-91-09, I.I.Q.P., University of Waterloo, 1991.

8 J. C. Young, B. Abraham and J. B. Whitney, Qual. Engr. 3, no. 2 (1990–91) 167–180.

Quality through Engineering Design - W. Kuo (Editor)

Implications of Fool Proofing in the Manufacturing Process

Yoshikazu Tsuda

Faculty of General Education, Rikkyo University, Nishi-Ikebukuro, Toshima-ku, Tokyo, Japan

Abstract

To realize perfect product quality, we have to provide systematic fool proofing techniques for all company activities. This paper describes various fool proofing techniques that were developed for product design and manufacturing processes. Then this paper tries to show that the evolution of fool proofing is the way to achieve completely automated manufacturing. In this context, we conclude that developing a systematic approach to fool proofing has opened a field of engineering we call quality engineering.

PREFACE

To error is human. No one can avoid mistakes. When a person works at a simple job or even more so a simple, monotonous job, mistakes are common. In companies with many employees, simple mistakes often cause trouble, requiring work to be redone and resulting in a loss of income. Finding mistakes calls people's attention to their work and raises their awareness of quality. If we rely only on employees' attention and awareness, however, it is impossible to eliminate mistakes. To prevent mistakes, we have to install fool proofing systems into all company activities.

Fool proofing, a system to detect and prevent mistakes, consists of four components: mistake prevention environment, mistake detection, mistake prevention., and prevention of the influence of mistakes.

Many ideas for fool proofing have been developed, tested, and improved during the last decade as part of quality improvement activities on the production floor. When we follow the development of fool proofing from a systematic point of view, we conclude that a new field of engineering, which we call quality engineering, has come into being.

For many years, very keen attention has been paid to fool proofing in Japan, especially in automobile-related industries. Designing products and production processes to minimize errors has been widespread in order to ensure product quality and minimize the risk of product failure.

Problems with industrial products come primarily from two sources; one is the user's misuse and abuse of products, and the other is the manufacturer's mistakes in conceiving, designing, or manufacturing a product. Both of these sources are important and can benefit from fool proofing. In light of the theme for this conference, however, I will discuss fool proofing only as a countermeasure for mistakes that may occur in the manufacturing process. In the course of this discussion, I will try to show the implications of fool proofing in engineering.

TYPES OF FOOL PROOFING

As I have mentioned, to err is human. However, not only humans but also automatic systems are prone to error. While errors are common, however, consumers are requiring higher and more stringent product quality. Often mistakes in product design and manufacturing become the object of prosecutors' investigations when defective products cause serious accidents. The resulting compensation for these failures is also skyrocketing.

Demand has increased not only for quality consumer products but also for industrial products. In the electronics and car industries today, the competitive level of quality (i.e., the number of defects of incoming products) is at most 10 PPM (parts per million), the same fraction of defects they ask of their own production line.

Lowering the fraction of defects to less than 10 PPM in a production process and building perfect quality into products are impossible if quality control and statistical process control are the only techniques used. To achieve such a high standard in quality, we need to install double and triple fool proofing systems into the production process and install a system to limit the influence of mistakes on production.

Fool proofing to minimize errors has been carried out in all functions of Japanese companies as a part of the total quality management (TQM) system. When we see the depth and breadth of fool proofing activities in technological and production areas, I believe there is enough content to constitute a field of engineering that we can call quality engineering.

Just as with quality control, quality engineering is a mixture of engineering technology and managerial technology. To make the nature of quality engineering clear, we may classify fool proofing in the following four categories:

1. mistake prevention in the work environment
2. mistake detection
3. mistake prevention
4. prevention of the influence of mistakes

Mistake prevention in the work environment

Good housekeeping and a tidy environment on the work floor can help workers prevent mistakes and make any mistakes that do happen easy to find. An emphasis on safety on the work floor improves workers' concentration and discipline. Also emphasizing the importance of quality in the work through education and training in quality and safety reduces the chances of mistakes.

Changing the layout of an operation manual from a book format to a single sheet so that only one product is featured on a page instead of two (as shown in Fig. 1) eliminates the possibility of reading a wrong dimension from the description of the second product. Such a manual can be hung overhead in front of the working position for ease of use.

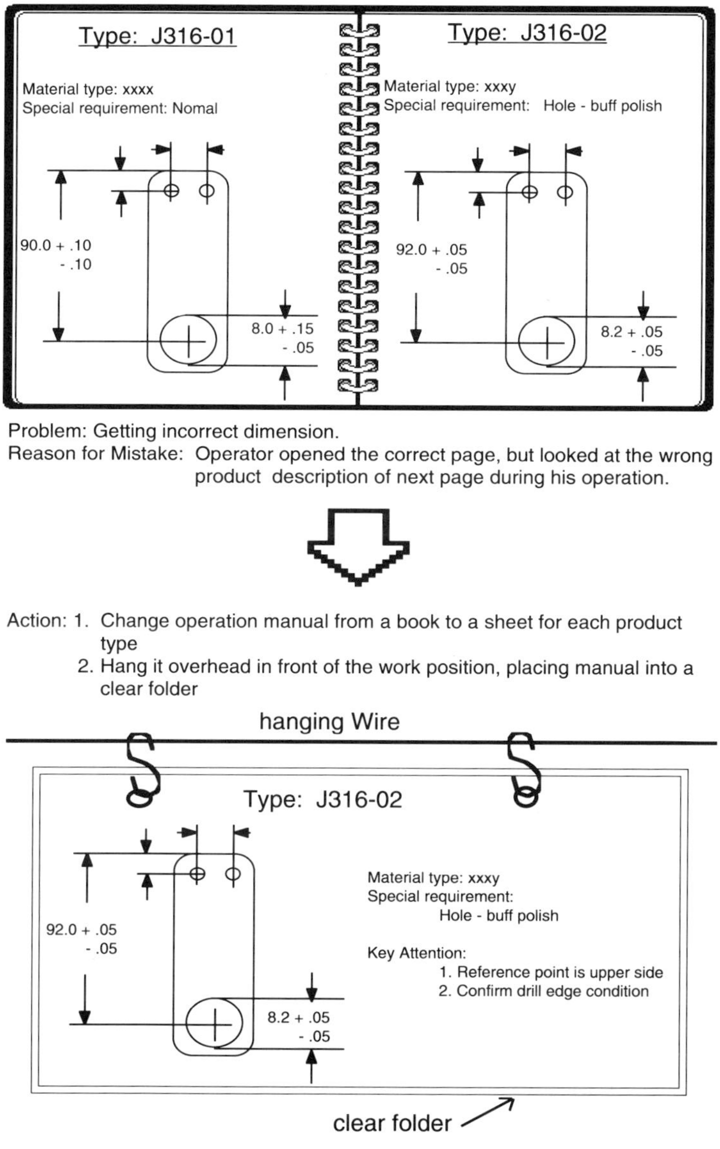

Fig. 1. Improvement of instruction manual.

Figure 2 shows an example of a work procedure that has been simplified to improve accuracy. The instruction manual was changed to incorporate a pocket to hold a tape necessary for the procedure for each product type.

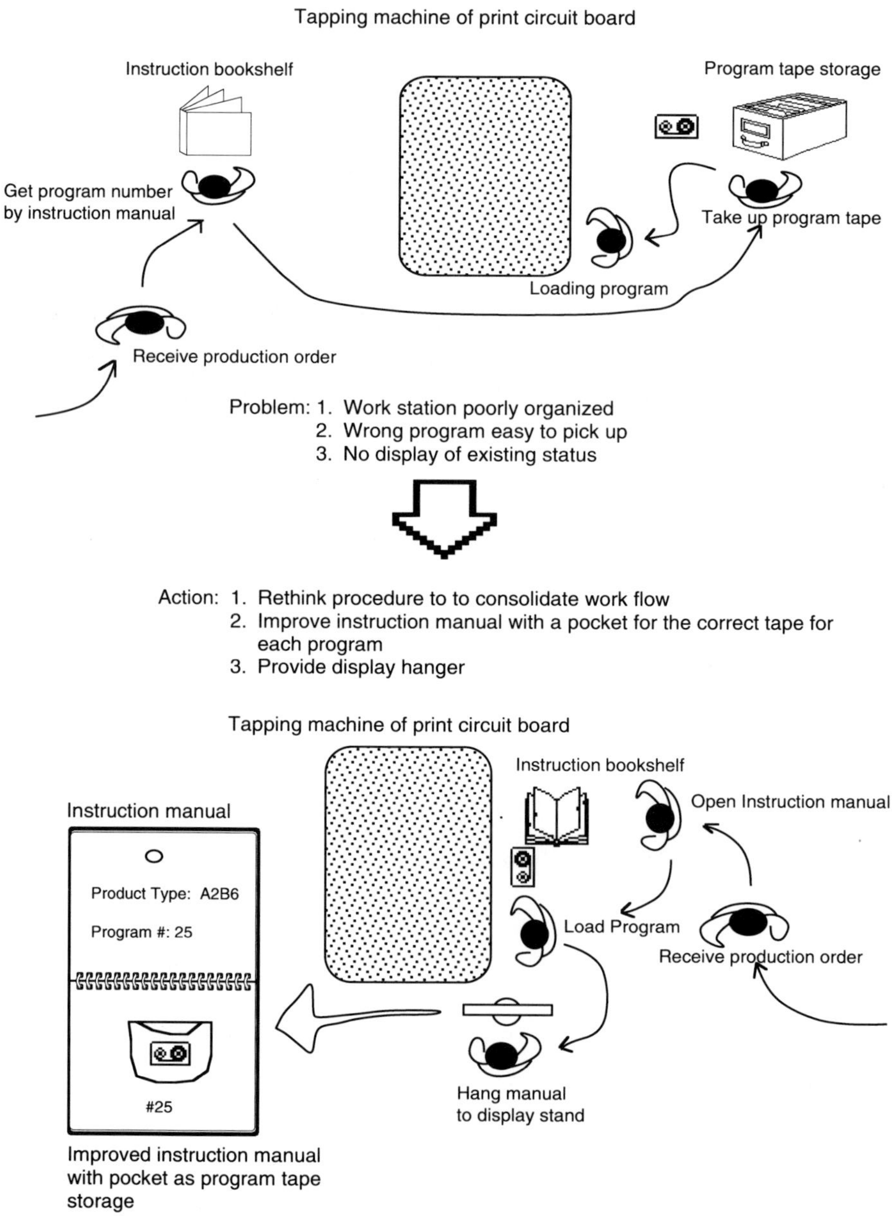

Fig. 2. Simplifying work procedure.

Figure 3 shows that simplifying the quality control for an electric circuit board by including an instrument cover that indicates the tolerance range of each parameter reduces the chance of a worker using the incorrect product specification.

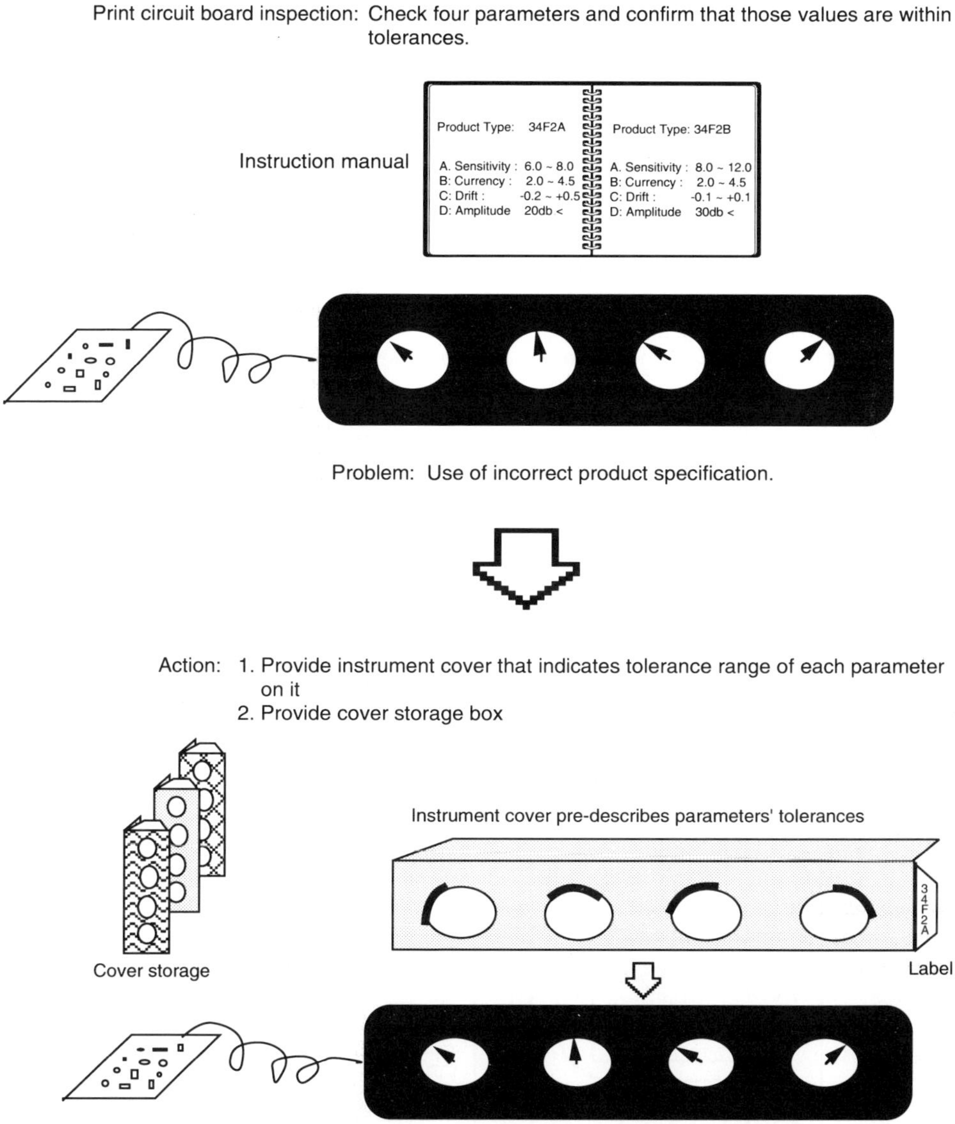

Fig. 3. Simplifying quality control for an electric circuit board.

Rethinking how to simplify the function regulation of an electric current board has shortened the time and increased the accuracy of function regulation as shown in Fig. 4.

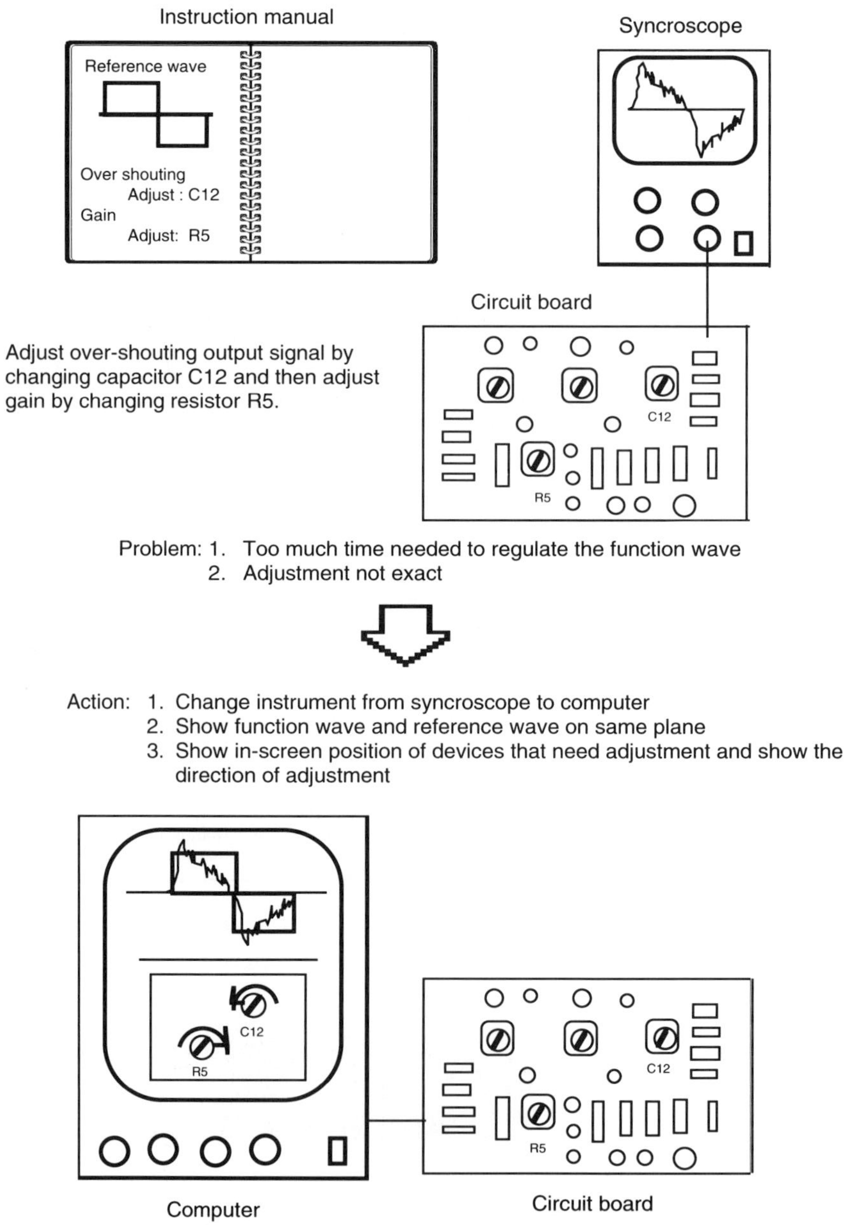

Fig. 4. Simplifying function regulation of an electric circuit board.

Mistake detection

Mistake detection is a system to make it easy to find errors or abnormalities in the work. Depending on where the mistake is detected, the mistake detection system is classified into three types: (1) immediate detection, (2) detection in the next production step, and (3) detection in quality control of the product.

Placing marks on the parts involved enables workers to find incorrect assembly of two parts easily (see Fig. 5). Various sensors installed in the production line detect incorrectly assembled products and automatically reject them.

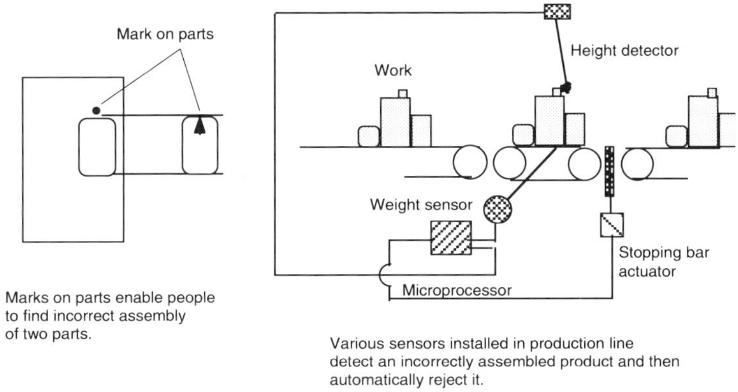

Fig. 5. Mistake detection system for incorrectly assembled parts.

Figure 6 shows an example of detecting air-flow problems in a power wrench. An air-flow detector establishes whether the wrench has been used up to the preset torque level, and a sequence counter determines how many times the wrench has been used by presetting the torque in a standard time. If the frequency detected is not the same as the number of bolts to be fixed, a warning light and sound are given. A missing red mark on the bolt head indicates the forgotten position.

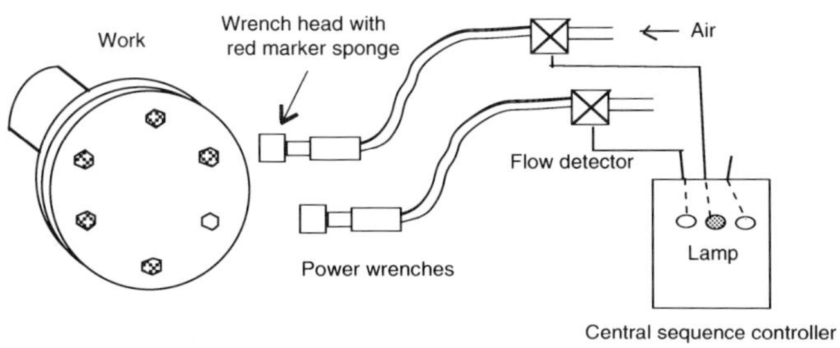

Fig. 6. Air-flow sequence detector for a power wrench.

Application of a vacuum force is often seen in automatic processing, for example, separating sheets of paper in a copy machine. The example in Fig. 7 shows us that applying vacuum force in automatic operations offers many possibilities in detecting mistakes.

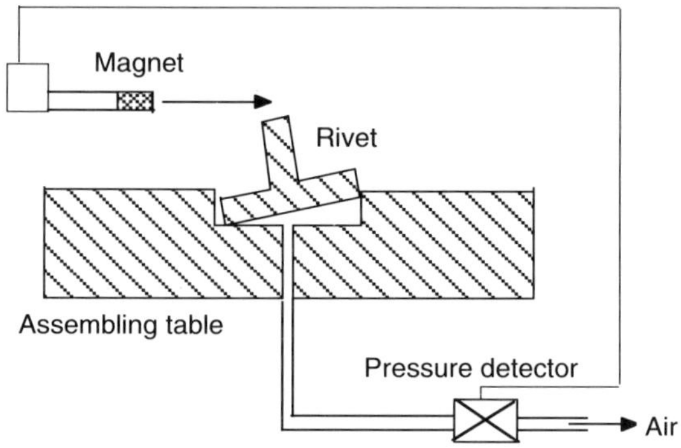

Fig. 7. Rivet position detector in an automatic assembly machine.

Figure 8 shows an example of set operation in packing tape recorders. Before the packing operation started, there were 20 sets of parts, including a certification card, two dry batteries, and a user's manual, to be packed with 20 tape recorders. What sometimes happens in this type of situation is that at the end of the packing operation, materials are either left over, which means that one box is short of materials, or there is a shortage of inserts, indicating a box has extra copies. Set operation, which shows a mistake in assembly after the finish of one lot production, is a widely used technique both for packing and for small quantity batch production (for example, producing an electric circuit board by inserting parts manually).

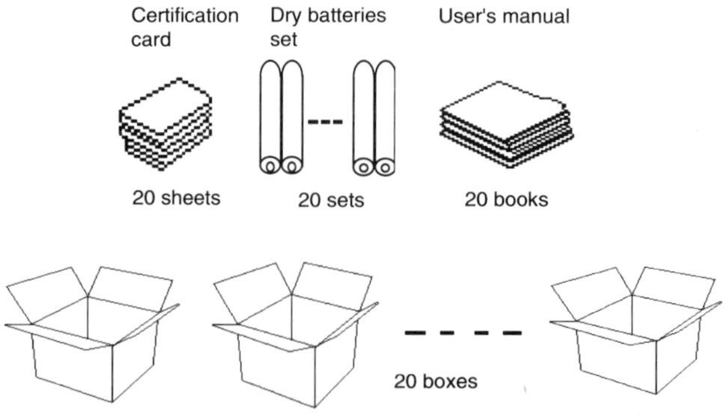

Fig. 8. Set operation for packing.

In the example in Fig. 9, the door of each box is equipped with a contact switch that changes a lamp from red to green. A sequencer detects whether one sequence of doors has been opened, and then it resets all the door lamps from green to red. This method indicates a mistake in the order of assembly or a missing part in an assembly operation and evolved from the set operation of the previous example.

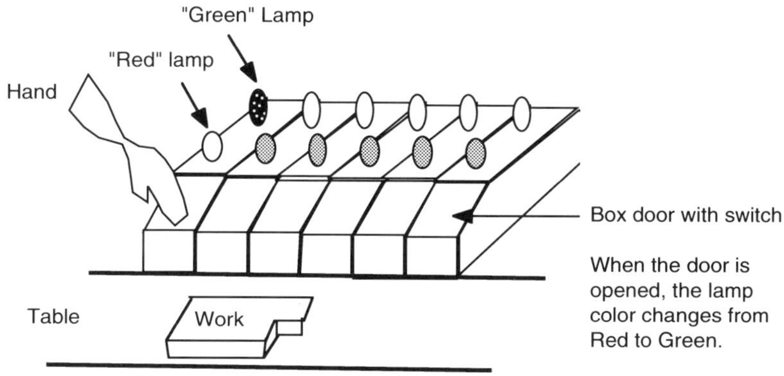

Fig. 9. Parts box door with switch and lamp.

Mistake prevention

Mistake prevention is the system that eliminates any possibility of workers' mistakes. Although mistake prevention is the ideal goal of fool proofing, it is virtually impossible to prevent all mistakes that will happen just by one action of a mistake prevention system. When we focus on only one type of mistake, however, it is not so difficult to develop a mistake prevention strategy for that particular type of mistake. To be successful in mistake prevention, we need exact analysis for mistakes that occurred in the past in order to define the types of critical mistakes that hinder product quality.

The guide pin shown in Fig. 10 is designed to prevent putting number 6 in the position of number 9 on the clock panel. The principle of mistake prevention used here is unsymmetry in design. It is impossible to put number 6 in number 9's position because their positions on the panel are very different and because of the interference of number 10. Unsymmetrical design for the purpose of preventing mistakes is often observed in electric cable connectors.

Another example of applying unsymmetry is shown in Fig. 11. Here, guide pins of the plastic mold push button and the cut slots on levers are each located in a different position to prevent placing the incorrect button on the lever.

Figure 12 shows that depth of the tapped hole is determined in such a way as to prevent a missing washer. If the spring washer is not used, the part will not fit well.

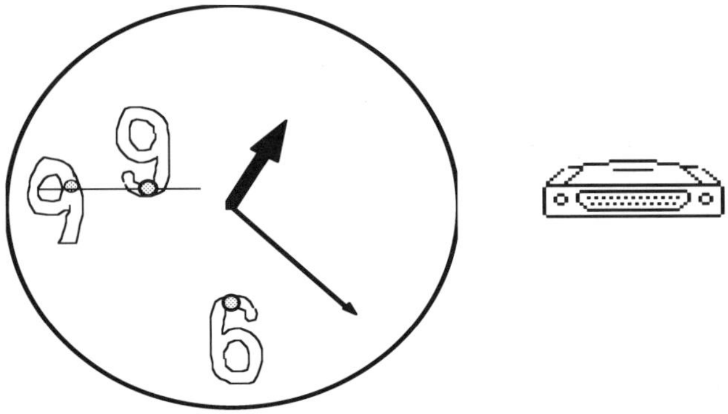

Fig. 10. Mistake prevention using unsymmetry of design.

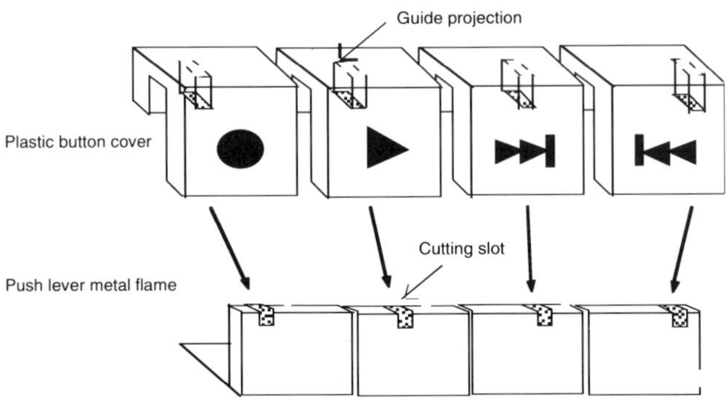

Fig. 11. Push-button guide pin design of video recorder.

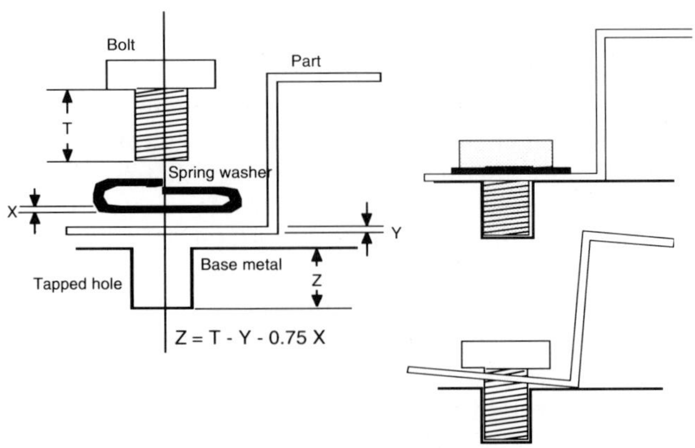

Fig. 12. Using the depth of a tapped hole to prevent a missing washer.

In the system shown in Fig. 13, a height sensor measures the height of the cylinder block, a microprocessor calculates the appropriate thickness of the gasket, and then a signal is sent to the door opener. This is a vast improvement over the early stages of development of this system, where the process was handled manually. The operator measured the height, looked up the correct gasket thickness from a table, and then picked up the gasket for assembly. Without exact measurements, workers erred in obtaining the correct thickness and also in picking up the correct gasket. Later a height sensor was introduced and a microprocessor calculated the gasket thickness and displayed it on screen, but the operator still picked up the gasket. Although this process reduced errors, the operator still made mistakes in picking the correct gasket.

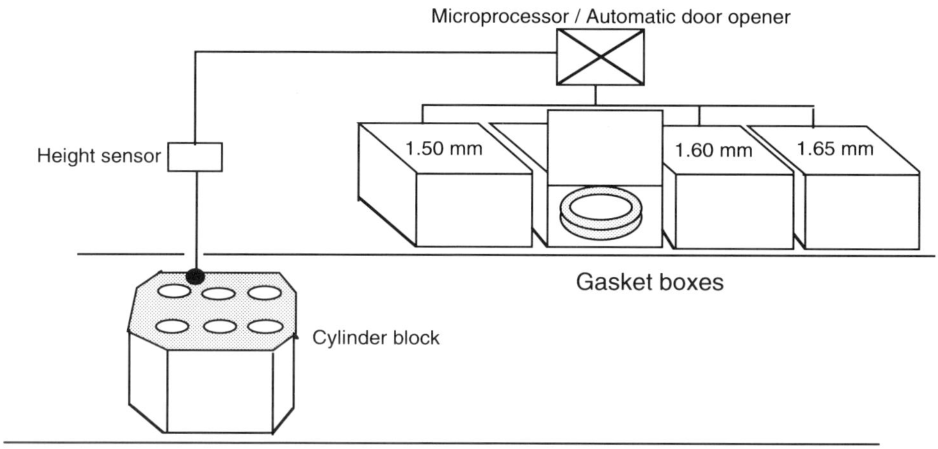

Fig. 13. Parts box door-opening system with height sensor.

Preventing the influence of mistakes

Preventing the influence of mistakes is the system that aims to limit and reduce the influence of a mistake even though it happens. A fail-safe system, which is usually installed in products such as two independent hydraulic control systems of an airplane, is an example of fool proofing of this type. This system is much more costly than the other types of fool proofing. Figures 14–16 show other examples.

A confirmation dialogue box for computer software (Fig. 14) is a redundancy system. It can prevent the influence of mistakes by confirming an order before executing it.

Another example is the anti-skid braking system (Fig. 15) used to prevent complete locking of the wheels. This system reduces the influence of a driver's unskilled operation of the automobile.

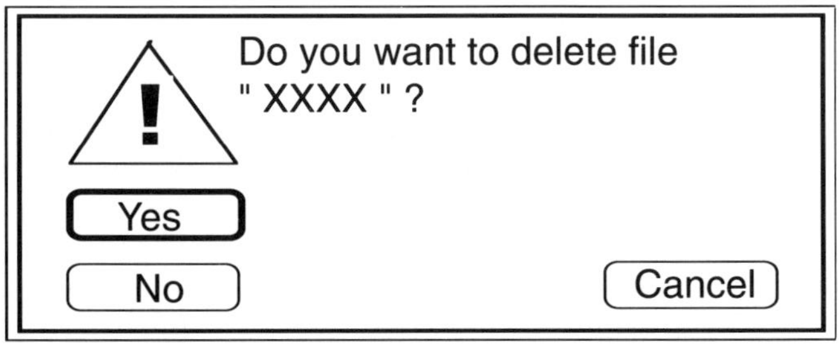

Fig. 14. Confirmation dialogue box for preventing the influence of mistakes.

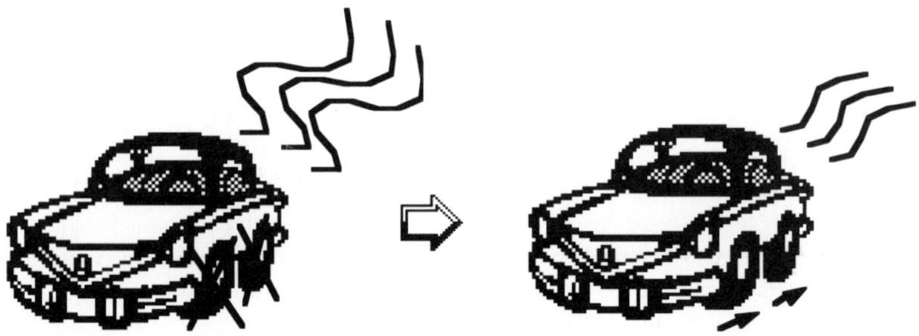

Fig. 15. Anti-skid braking system for automobiles.

In the example in Fig. 16, a heavy concrete-lead vessel is built with the goal of reducing the seriousness of an accident at a nuclear reactor.

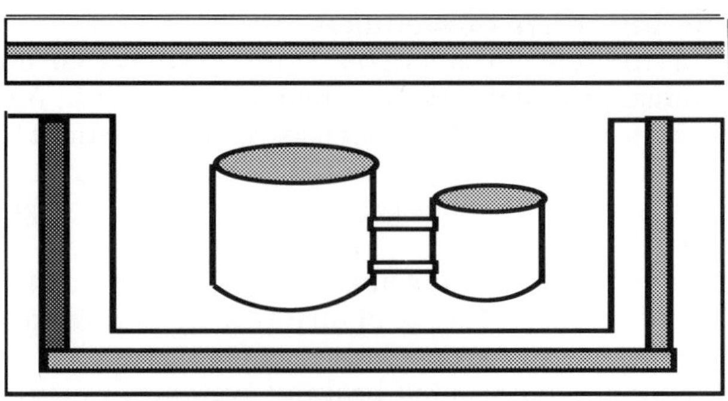

Fig. 16. Heavy concrete and lead vessel and emergency water powering system for a nuclear reactor.

SOURCE OF IDEAS

Where do we find the ideas for fool proofing? What is the general principle used for getting a solution for fool proofing? When we study many examples of fool proofing that have been developed, we can extract several ways of thinking about how to find an idea for fool proofing.

Simplification and visualization

Simplifying procedures and operations, making operations easy, and showing the status of the operation by using a display with pictures and images are all good sources of ideas for fool proofing. (See, for example, Figs. 1–4, 6, 8, 9, and 14.)

This principle is especially effective for preventing mistakes in the work environment and also for improving the quality, productivity, and safety in work. For example, the time needed to change over a heavy press model of a car body has been reduced to 6 minutes from the 6 hours needed 10 years ago. This reduction was realized by an elaborate effort to simplify the work and eliminate unnecessary work in the change-over operation.

Introduction of sensors and application of microprocessors

Introducing various sensors and information (signal) processing by microcomputers are powerful ways of detecting mistakes. (See, for example, Figs. 5, 6, 13, and 15.) Various sensors are applicable for checking the status of operations. Sensor types used in fool proofing include mechanical, electrical, chemical, optical, and biological. Samples of characteristics to be checked are weight, pressure, height, color, temperature, smell, sound, and vibration. Bar codes are also powerful tools for preventing and detecting mistakes.

Unsymmetrical design

Another simple and powerful principle for preventing mistakes is the use of unsymmetrical design in position and in shape. (See Figs. 10–11.) Before numerical control (NC) machine tools and machining centers were introduced on the production line, using unsymmetrical design in metal parts was costly in terms of production and ease of manufacture. Today, however, such disadvantages have been reduced. In plastic injection molding, for example, using an unsymmetrical design in products compares favorably in cost to those using symmetric design. But the expected profit in preventing mistakes by using unsymmetrical design far exceeds the increased cost of using unsymmetry.

EVOLUTION OF FOOL PROOFING

From manual detection to automatic detection

Many companies in a variety of industries have independently developed a large number of ideas for fool proofing. When we classify those ideas according to the type of mistake involved, however, a certain evolutionary direction in fool proofing emerges from the chaos of so many ideas.

When we put two examples of mistake detection in the following order: (1) set operation of packing (Fig. 8) and (2) door switch and lamp for parts boxes (Fig. 9), we recognize that the example shown is Fig. 9 evolved from the one in Fig. 8. We can find missing parts in a package only after we finish one round of packing by the set operation system, but if we apply the method of Fig. 9 to the packing operation, we can find missing parts just after one packing.

In Fig. 9, the detection of mistakes depends on an operator's attention after each packing operation to see if all the lamps of the parts boxes are green. Thus if an operator is not paying attention, mistakes can occur.

Next, adopting the idea of using an air pressure sequence detector (Fig. 6) to find out whether a sequence of taking up parts is completed, we can build up an automatic detection system for missing materials by sending detected information to a stopper that is attached to the side of a conveyer belt to stop the incomplete packing of the box. (See Fig. 17.)

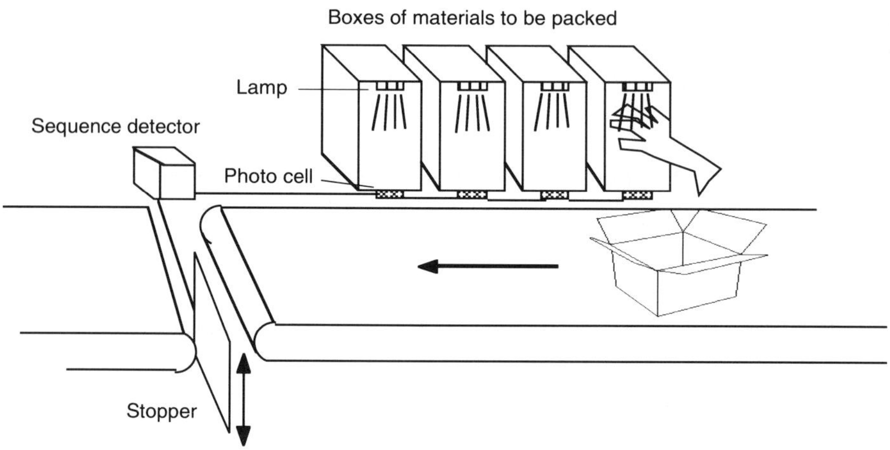

Fig. 17. Automatic detection system for missing materials.

Here, photo cells detect whether material is taken up, a sequencer detects whether a sequence of picking up materials has been done within a standard time, and the stopper prevents an incompletely packaged box from going to the warehouse.

From detection to prevention

The parts box door opener, shown in Fig. 13, evolved from the fool proofing shown in Fig. 9 (parts box door switch and lamp). This is a typical example of evolution in fool proofing from detection to prevention, which is enabled by introducing sensors and microprocessors.

From action for production to action in design

There are many sad stories of trouble shooting and the resulting compensation having to be paid because of missing parts, such as a gaskets, oil seals, or spring washers. The evolution of fool proofing for the missing parts is as follows:

1. Set operation (Fig. 8)
2. Door switch and lamp for parts boxes (Fig. 9)
3. Consideration in the design stage to enable detection of misassembly in the next step of production, for example, by marking the parts (Fig. 5) or the difference of the relative position in the assembled parts.
4. Design that makes an operator unable to perform further assembly (Fig. 12).

The application of unsymmetrical design is also part of the product design stage. Fool proofing installed in product design or production process design is much more cost effective than the fool proofing applied in production. Well-provided engineering standards that are used in product and process design will help design engineers pay better attention to fool proofing in their design. Providing a design review where people from various departments are invited also has an important role in fool proofing because those attending can point out needs concerning mistake prevention in design from their various standpoints.

From manual operation to automation

Taking four examples of "preventing mistakes in the work environment" section as shown in the figures cited, we extract the meaning of each example in production engineering (see Table 1).

Table 1. Evolutionary steps from manual production to automation by implementing fool proofing.

Example	Implication of example in production engineering
Fig. 1	Define operation
Fig. 2	Rationalize operation and material handling (preparation for automatic manufacturing)
Fig. 3	Simplify check of operation results (preparation for automatic in-process quality control)
Fig. 4	Simplify adjustment/correction work (preparation to enable automatic adjustment and correction)

This table shows a typical set of steps to build a production system that can work 48 hours without production people on the floor.

QUALITY ENGINEERING—TECHNOLOGY OF ACHIEVING QUALITY IN PRODUCTS

As we have seen in the discussion on fool proofing, we can extract knowledge from a huge collection of facts that have been accumulated for many years. When we define quality engineering as that knowledge required to realize the goal of quality products, then fool proofing is an element of quality engineering. Quality engineering consists of two technologies; the technology of manufacturability and the technology of fool proofing.

Just as in fool proofing, manufacturability has two action areas: product design and production process design. The detailed concept of manufacturability is shown in Table 2.

Table 2. Quality Engineering:

Technology of Manufacturability:

- Technology to translate a quality goal easily into products in manufacturing
  - Technology to choose principle, mechanism, construction of components, and material of parts to realize aiming function easily
  - Technology to choose manufacturing methods effective to realize aiming function in expected level

- Technology to reduce the influence of disturbance that is originated in manufacturing
  - Technology to choose product parameters and to define those values in order to minimize the influence of disturbances that come from manufacturing
  - Technology to choose process parameters and those values in order to minimize the influence of disturbances that comes from manufacturing

---

Technology of Fool Proofing:

- Technology to detect mistakes in manufacturing
- Technology to prevent mistakes in manufacturing
- Technology to reduce the influence of mistakes in manufacturing products
- Technology to provide an environment to prevent mistakes and to reduce disturbances in manufacturing

---

NOTE ABOUT EXAMPLES

Examples that are referred to in this paper have come from various sources. Many were picked up from my experience working with people in various industries. Fool proofing has been reported on various occasions such as in-company quality control (QC) circles and suggestions discussed in private conferences, QC circle public conferences, and magazines and books. Among the large number of ideas, many are similar although they were reported independently. Examples used in this paper also are similar to ideas reported by other conference proceedings and publications.

Although there are many conference proceedings, magazines, and books on fool proofing, I would mention only one book as reference: *Pokayoke Dai Zukan (Grand Picture Book of Fool Proofing)* [1] The way ideas are expressed by comprehensive graphics for the examples in Figs. 3, 7, and 9 depend on this book.

REFERENCES

1 Koujyou Kanri, (Plant Management) Editorial Department, Pokayoke Dai Zukan (Grand Picture Book of Fool Proofing), Nitkkan Kougyou Shinbun (The Industrial Daily News), Tokyo, 1987.

Quality through Engineering Design - W. Kuo (Editor)

Design Planning of an Off-Highway Dump Truck — A QFD Approach

Omnamasivaya Maduri

Research & Development, Bharat Earth Movers Ltd., Mysore, India

Abstract

Designing a new dump truck or the fundamental modification of an existing one while keeping the end user in mind requires strategic planning and systematic implementation procedures. Planning the design of an off-highway dump truck is so detailed that often no individual can comprehend the whole picture. Moreover, arriving at the design specifications is a difficult task involving large interdependent parameters. This task requires thorough planning, frequent communication, and substantial work effort. Quality function deployment (QFD) is a technique that translates the customer's requirements directly into the design parameters. Besides, the parameters can be deployed systematically down the line through the process planning, manufacturing, assembly, and service departments.

This paper explains the detailed design planning of an off-highway rear dump truck in the 35-ton range, developed at M/s Bharat Earth Movers Limited and making use of the QFD approach.

INTRODUCTION

Many times, efficiently generated products that are sound from a technical point of view, fail to achieve acceptance by the customers [1]. In such cases, the defect likely resides in the organization's perception of its relationship with the market and individual customers. Moreover, the customer often takes for granted the features already existent either in one's product or that of the competitor, and thus their level of expectations increases. In such regard, designers have a very important role to play in obtaining the customer's confidence and satisfaction. At the planning stage, the designer should take into account design, technological, operational, economical, psychological, and social requirements.

These observations imply that at the time of developing a new product, the relationship among customers' quality requirements, design, and manufacturing must be understood. At each step of product planning, trial design, manufacture, assembly, and inspection, the feedback of quality information must be maintained. Hence, a total quality management (TQM) system is required to ensure the desired quality in the product at every stage [2]. The quality function deployment (QFD) technique aids in maintaining the TQM environment and can be used efficiently at the initial planning stages.

DESIGN PLANNING

Once the project is initiated and approved by the top management, the research and development (R&D) department must plan and execute it accordingly. In general, it is a difficult task, particularly in the case of off-highway equipment like dump trucks. A dump truck is a heavy-duty machine used off-the-road for hauling and dumping the overburden or mineral ore from the mine to pit or dump site. Dump trucks are machines, consisting of huge structures like the chassis and body; power elements like the engine, transmission, and drive axle; and operational control through hydraulic, pneumatic, electrical, and electronic circuitry. The design planning is extremely complicated in this case because of the lack of sound technological background and scarcity of design data in this field. Moreover, there is a threat of international competition. Regardless of how the task is represented, it is exceedingly complex and consists of numerous trade-offs. Shared responsibilities and differences in interpretation often result in conflicting priorities. The only solution to this problem is to approach it from the customer's viewpoint; that is, systematically assessing the customer's requirements and providing for them. Failure to provide these requirements can result in total dissatisfaction.

Next is to search out the things attractive to customers that can be used to differentiate one's product from the competitor's. Marketing can play an important role in estimating what the competitor is doing or planning to do in this regard. If required, an additional feature can be given as an option at an added cost. For specific customers a new feature can be developed and promoted. Accordingly, customer awareness must be developed to appreciate the features through advertising, seminars, or marketing techniques. Finally, the customer's requirements must be evaluated and the valid ones converted into physical design parameters that are systematically deployed into functional components or production processes.

THE QFD

According to the Japanese Industrial Standard (JIS) Z8101-1981 on Quality and IS:14004 (ISO 9004) on quality systems [3–5], quality is defined as a system of means to economically produce goods or services which satisfy customer requirements under the proposed production, installation, and operational conditions.

QFD is a practical way of understanding the word quality because it is a technique that translates the "voice of the customer" into the physical design parameters, which are systematically deployed down the line in making the product. QFD makes this happen through its key documents, namely, the product planning matrix, the product and component deployment matrix, and the operating instruction sheet.

In this paper, the design planning of a dump truck is explained through QFD and its key documents [6].

Product planning matrix

The purpose of the product-planning matrix is to translate valid customer requirements into important design features. The product-planning matrix (Fig. 1) incorporates such customer requirements as stability, riding comfort, and productivity, on one side and the alternate design features closely related to the customer needs such as chassis, engine, and transmission on the other. This step is the most crucial part of the entire process because it requires obtaining and expressing what the customer really wants and not what we assume that he expects. Since there are varying degrees of correlation between customer needs and design features, a significant relationship should be established by developing a weighted scale (1–5). The individual customer needs are ranked for importance and the cumulative effect on each of the design features is obtained. The percent rating shows the priority of the feature. Also, the current product status and the future improvement plan can be marked for reference. Moreover, the mutual impact of the design features can be seen in the correlation matrix.

RELATIONSHIP/WEIGHTAGE

5 ---> STRONG
3 ---> AVERAGE
1 ---> WEAK
0 ---> NIL

CUSTOMER NEEDS	RANKING	DESIGN FEATURES: CHASSIS	ENGINE	TRANSMISSION	DRIVE AXLE	TYRES	STEERING	HYDRAULICS	BRAKES	SUSPENSIONS	DIMENSIONS	CABIN	SYSTEM REFILLS	OPERATING WEIGHTS	COMPUTER DESIGNED	FATIGUE TESTED	PAYLOAD/GVW	COST	PERFORMANCE	SAFETY GADGETS
STABILITY	5	5	1	1	3	3	5	3	5	5	5	0	0	3	0	3	3	3	5	5
RIDING COMFORT	3	1	5	5	5	5	5	5	0	5	1	5	0	5	3	0	5	3	0	0
AUTOMATION	3	0	1	1	1	0	1	1	3	1	0	3	0	1	0	0	1	3	0	0
PRODUCTIVITY	5	1	5	5	5	3	1	0	0	0	5	1	3	5	1	5	5	5	3	0
LONG LIFE	5	3	1	1	1	1	1	1	1	5	0	1	1	5	3	5	5	5	5	0
SAFETY	5	3	3	3	3	3	5	5	3	3	5	1	0	5	0	1	1	3	5	5
COMPACTNESS	3	5	5	5	5	0	5	5	0	1	1	5	0	5	0	1	0	5	5	1
OWNING COST	5	1	1	3	3	1	1	3	3	3	3	1	1	1	1	1	1	5	0	0
SERVICEABILITY	3	3	3	3	3	3	3	3	3	3	3	3	3	0	3	0	0	5	0	0
MANOUVERABILITY	3	5	0	0	0	0	5	0	0	3	3	0	0	0	0	0	0	3	0	0
OPERATING COST	5	1	1	3	3	1	1	0	0	1	1	0	1	1	0	0	0	5	0	0
RELIABILITY	5	3	3	3	3	1	3	3	5	1	1	1	0	0	0	0	0	5	5	0
RATING		137	127	127	147	89	148	117	103	129	124	73	39	133	43	78	93	212	130	53
PERCENT RATING		6.51	6.04	6.04	6.99	4.23	7.04	5.56	4.90	6.13	5.89	3.47	1.85	6.32	2.04	3.71	4.42	10.1	6.18	2.52
PRODUCT STATUS		5	3	5	3	5	5	3	5	3	5	5	3	5	1	5	5	5	3	5
IMPROVEMENT PLAN		5	5	5	5	5	5	5	5	5	5	5	5	5	5	5	5	5	5	5

COMPARATIVE ANALYSIS

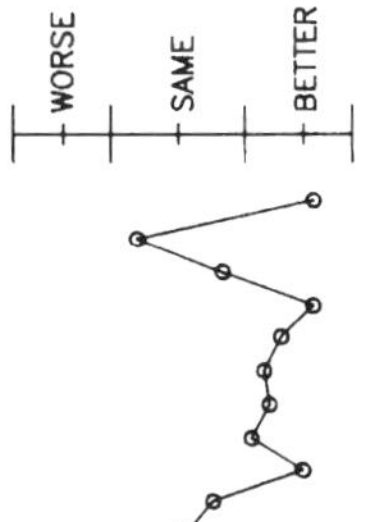

Fig. 1. Planning matrix (dump truck).

The functional dimensions of the equipment are obtained by using the planning matrix (see Fig. 2). The figure shows that wheel base has the highest priority and that the turning radius has an impact on the wheel base. Thus, both wheel base and turning radius have to be considered together on a priority basis. If it is difficult to decide the value of an individual parameter, an index can be arrived at by using the related equations. Some of the indexes thus generated are listed in Table 1. These indexes can also be used in future developments of related projects.

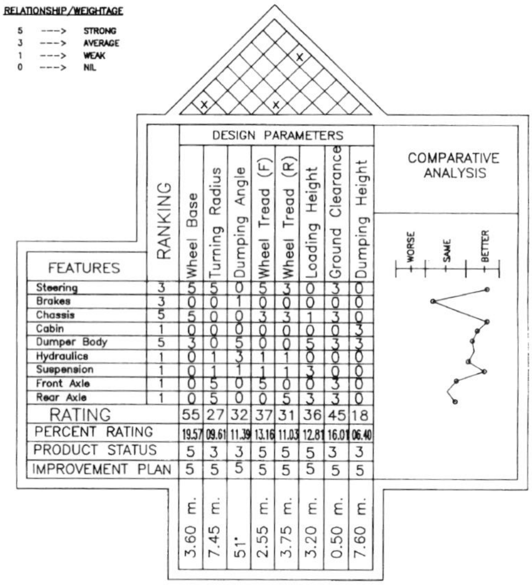

Fig. 2. Planning matrix (dimensions).

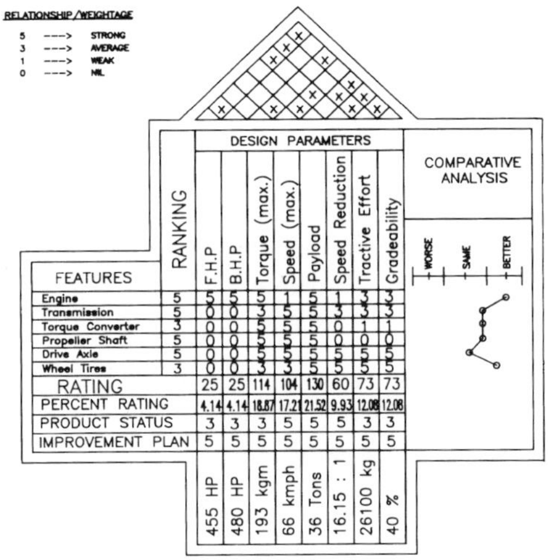

Fig. 3. Planning matrix (power train).

The planning matrix shown in Fig. 3 helps us to arrive at a suitable power train configuration. The "payload" parameter has a dominance rating, which means that it is more important than others. If required, indexes similar to Table 1 can be arrived at for the parameters shown here.

INDEX			KOMATSU Japan	DRESSER USA	TEREX USA	PAYHAULER USA	CATERPILLAR USA	BEML India
TR/WB			7.5	7.4	6.2	7.8	7.2	7.5
HP/GVW			0.55	0.53	0.58	0.57	0.50	0.54
PL/GVW			2.1	2.4	2.2	2.8	2.2	2.0
AXLE LOADS %	FRONT	Empty	48	49	52	70	47	47
		Loaded	35	33	33	50	33	33
	REAR	Empty	52	51	48	30	53	53
		Loaded	65	67	67	50	67	67
Heap Ratio %			36	40	27	34	35	34
Body Weight/ Struck (kg/cu.m)			340	437	434	393	400	432

Table 1. Specification indexes.

Product-deployment matrix

Once the planning matrix is completed, then a deployment matrix is made for each of the product features down to the subsystem and component level. This shows to what extent, if any, the relationships between component and product characteristics are critical and affordable. Most importantly, this matrix can be used not only for technical parameters but also for certain managerial strategies. For example, to assess the application and extent of the computer's usage in product design (see Fig. 4).

COMPUTER DESIGNED \ APPLICATION	LAYOUTS	DIMENSIONAL CHECKS	DRAFTING	STRENGTH EVALUATION	KINEMATIC ANALYSIS	SYSTEM DYNAMICS	METHODS PLANNING	MANUFACTURING	AESTHETICS	STABILITY
POWER TRAIN	X	X	X			X	X			X
CHASSIS	X	X	X	X				X	X	X
STEERING SYSTEM	X	X	X		X			X		X
SYSTEM PIPINGS	X	X	X					X	X	
DUMPING EFFICIENCY	X	X	X			X		X		X
SUB–SYSTEMS	X	X	X	X			X			X
COMPONENTS	X	X	X	X					X	X

Fig. 4. Deployment matrix (computer designed).

Component-deployment matrix

If a component is very critical, then it is further deployed and monitored in the design, production planning, and control. The component deployment matrix expands the list of the components or the exact parameters required to design a complete component. These parameters should relate directly or indirectly to the customer's requirements. The matrix shown in Fig. 5 shows the details regarding the engine BHP and chassis, while the matrix in Fig. 6 details the hydraulic tank specifications.

Operating-instruction sheet

The operating instruction sheet is the final key document that basically defines the operator requirements as determined previously by the actual process requirements, the process check points, and the quality control points. This is more or less identical to the traditional process sheets being used.

On the whole, QFD has helped to effectively evolve the specification [7] dimensions (Fig. 7) and the aggregate planning of the dump truck. Presently, the dump truck (prototype) is working satisfactorily at TISCO mines (Fig. 8), West Bokaro, India, and has clocked around 1,000 hours. QFD can be applied not only for new product development but also to identify the needs of internal customers, that is, individual departments [8]. In general, QFD is useful wherever a customer-vendor relationship exists.

COMPONENT	PARAMETER	VALUE	REASONS REMARKS
ENGINE (B.H.P)	F.H.P	455	Design
	FAN	18.5	"
	PUMPS	9.5	"
	ALTERNATOR	3.0	"
	COMPRESSOR	3.0	"
CHASSIS	SECTION	140 X 400	Design
	MODULUS	1480 Cu.m.	"
	KICKUP	10°	"
	FACTOR OF SAFETY	3	"
	MATERIAL	35 KSc	"

Fig. 5. Deployment matrix (engine & chassis).

PARAMETER	VALUE	REASONS REMARKS
OPERATING PRESSURE	0.6 KSC	SYSTEM LIMITS.
FACTOR OF SAFETY	10	DESIGN CRITERIA
MATERIAL STRENGTH	410 KSC	DESIGN & COST
PLATE THICKNESS	6 mm	DESIGN & COST
TANK VOLUME	326 LITRES	DESIGN
GAUGE VOLUME	225 LITRES	MAINTENANCE
LENGTH/WIDTH	1.3 : 1	AESTHETICS
LENGTH/HEIGHT	1.1 : 1	AESTHETICS
LENGTH	775 mm	DESIGN
WIDTH	596 mm	DESIGN
HEIGHT	705 mm	DESIGN
GAUGE HEIGHT	487 mm	MAINTENANCE

Fig. 6. Deployment matrix (hydraulic tank).

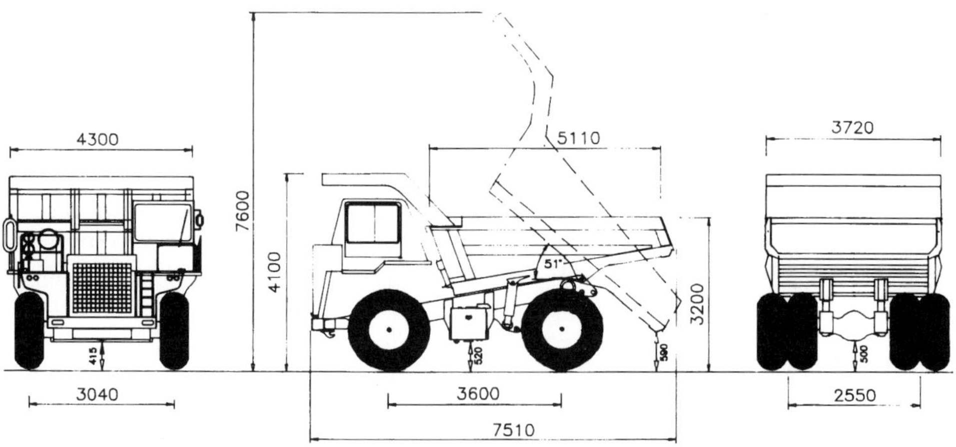

Fig. 7. Dump truck specifications.

CONCLUSIONS

In this paper, the guidelines for using the QFD concept in establishing the specification parameters and a case study for the development of new dump truck are explained. The QFD's planning matrix points out clearly the cross impact and conflicting design parameters at the specification stage itself and generates some of the specification indexes that can be used for future developments. In addition, QFD provides a clear operational definition and promotes collective thinking, setting common priorities, and focusing attention from time to time. Also, QFD promotes a very good team approach by fostering a better understanding of requirements and what is needed to meet them.

Fig. 8. Dump truck working at TISCO.

In the future, it is planned to extend the technique to other applications in order to reduce prototype development cycles, improve overall quality, and lower the costs. Finally, the applications of QFD are limited only by the degree of an individual's imagination, and implementing QFD will be a challenge for any company. The best way to gain a better understanding of this technique is to recognize the overall concept and then undertake a case-study assignment.

ACKNOWLEDGMENTS

The author thanks the management for permission to present the paper. The opinions expressed or conclusions drawn in the paper are entirely those of the author and not of the management. Any correspondence in this regard should be directed only to the author.

REFERENCES

1. M. Oakley, Managing Product Design, Weidenfeld and Nicolson Publishers, London, 1984.
2. D. A. Ginder, Automot. Eng., 98, no. 10 (1990) 18–19.
3. L. P. Sullivan, Qual. Prog. (June 1986) 39–50.
4. N. E. Morell, Quality Function Deployment, American Supplier Institute, Dearborn, Mich., 1990.
5. Indian Standards, Quality Systems-Guidelines on Elements of Quality Management Systems, BIS, Delhi, IS:14004 (1989).
6. O. Maduri, J. Qual. Part., 15, no. 1, (1992) 64–69.
7. Society of Automotive Engineers, Technical Specifications of Off-Highway Equipment, SAE J1233, SAE Inc., Warrendale, Pa., Jan. 1985.
8. G. D. Stecker, Qual. Prog. (Jan. 1991) 120.

Quality through Engineering Design - W. Kuo (Editor)

Quality through Design Practices in Spacecraft Mechanisms

N. C. Bhat and A. V. Patki

ISRO Satellite Centre, Airport Road, Bangalore-17, India

Abstract

Large appendages on spacecraft need deployment in orbit. Very high reliability of performance is expected from these deployment mechanisms in view of their critical mission. This paper describes the methodology adopted for the design of a typical mechanism to ensure the quality and acceptability of deployment mechanisms.

INTRODUCTION

Most satellites have large appendages, including communication dish antennas, solar arrays for power generation, and certain special devices such as long booms. In the launch phase, they are stowed in the satellite structure because of volume constraints. Later they are deployed to their final configuration (Fig. 1). The deployment is usually initiated through radio commands, either by time-tagging or ground controllers. The remote operation of the system, therefore, requires a high degree of reliability of performance because deployment systems are critical to the mission. This high reliability can only be achieved through an integrated quality approach right from the design phase of the system.

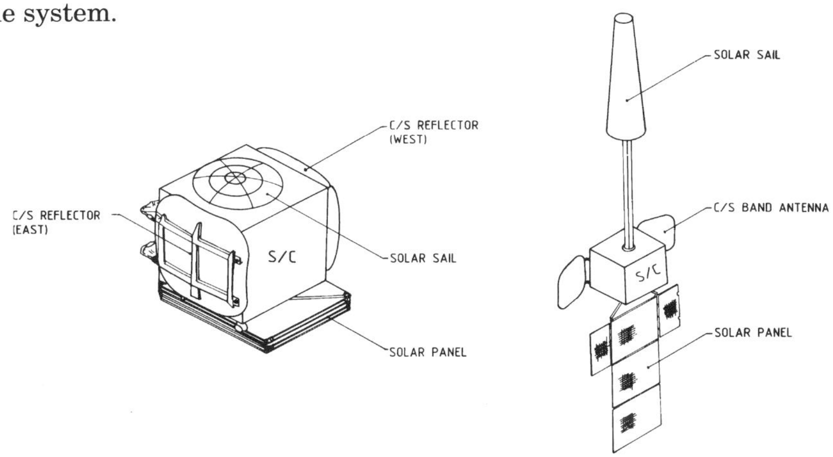

Fig. 1. INSAT-II stowed and deployed view.

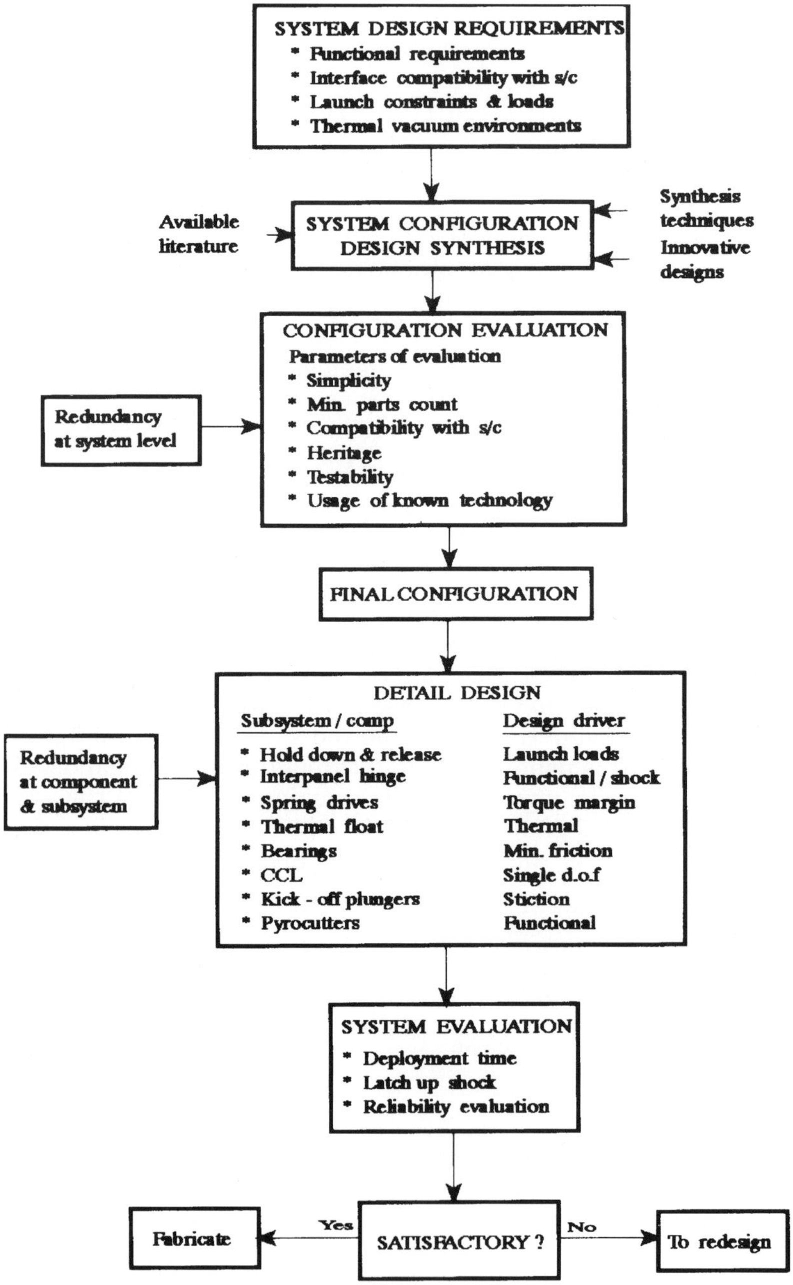

Fig. 2. Design methodology for deployment mechanism.

This paper describes the design philosophy as applicable to space mechanisms. It is an iterative procedure beginning with specification and extending to the final blue print. The discussions are restricted to the most common deployment systems, namely the solar array and the reflectors. The procedure is highlighted in Fig. 2.

SYSTEM REQUIREMENTS AND CONSTRAINTS

The overall demand on the deployment system is mainly governed by the spacecraft's mission requirements and the launcher used to put the spacecraft into orbit. The deployed configuration, area, geometry, and stowed volume availability lead to a design that is specific for a particular mission. Compliance of interface requirements is one of the major goals of the system. Further, satisfying minimum weight and size is a guiding factor in the design of the mechanism. The basic goals are defined in terms of restraining the appendages in the desired compact stowage and achieving positive deployment by a command during orbit. The system has to endure all the anticipated environments without undue degradation. The environments encountered by the mechanism are split into two types: (1) mechanical environments such as ground handling and launch loads and (2) orbit environments such as space, vacuum, and thermal properties.

Ground handling and launch loads

Ground handling loads arise from assembly, integration, handling, and transportation. During these operations, the appendages are usually kept in the stowed position so that loads on the hinge bearings and other sensitive elements are low. Appropriate fixtures are employed for these operations. During deployment trials, however, latch-up shocks are induced in the bearings. In addition, launch loads are induced during lift-off, transonic atmospheric flight, stage separation, and injection into orbit. These loads govern the design of hold-down and release mechanisms when appendages are in the stowed mode.

Vehicle performance studies lead to standard test specifications for the spacecraft on the ground. The test loads are categorized into static and dynamic components. Although static loads are defined in terms of inertia (g) loads, dynamic loads are composed of sinusoidal vibrations, random vibrations, acoustic noise, and shock spectra. The dynamic loads of the appendages, which are important for the design of hold-down and release systems, are derived by dynamic response analysis involving the total spacecraft. Accordingly, the inertia loads (g) form the basis for the design of the hold-down system.

Space environment and its effect

Space is characterized by a high vacuum level, extreme thermal conditions, high-intensity radiations including UV and X-rays, cosmic radiations, and meteorites. The vacuum level varies from 10^{-3} Torr at 200 km altitude to 10^{-13} Torr at geostationary orbital heights (36,000 km). The most important effect of such a high vacuum is outgassing of materials over a long duration. The part of the evolved gases that has a tendency to condense is a source of contamination and risk to the spacecraft operation [1]. Hence, a standard acceptance test for material outgassing (TML) and condensed products (CVCM) as adopted by the U.S. National Aeronautics and Space Administration (NASA) is followed.

Most of the metallic materials used in the spacecraft meet these specifications. Organic materials, however, exhibit greater levels of TML and CVCM, and the selection of lubricants is of special interest. There are many lubricants available for space applications, and data need to be assessed carefully to ensure the desired tribological properties under space environments.

A second important aspect for moving joints in space is cold molecular welding. Under high vacuum, joints tend to fuse together and thereby increase friction; this can make the joint ineffective. Proper selection of joint material and lubrication is important to ensure that cold welding does not occur.

Spacecraft temperatures are extreme, often varying with time and creating gradients. These extremes, combined with the different expansion coefficients of the various materials, can cause distortions of appendages and subsequent loads on the hinges and other moving elements. These factors must be considered in the design of the hold-down system. Another factor of the orbital environment to consider is the absence of gravity loads. While beneficial from the design perspective in terms of reducing loads on the moving elements and thereby leading to light-weight design, such a light-weight design puts stringent restrictions on the zero-g (gravity) simulations during ground deployment tests.

CONFIGURATION SYNTHESIS

The overall system specification forms the basis for the configuration study (see Table 1). The system is expected to perform two functions, stowage during launch and deployment on orbit. Accordingly, the system has two major components, a hold-down and release system and the deployment system. The design techniques used for a typical solar array deployment are described.

Hold-down and release system

The geometry of the spacecraft (S/C) and the vehicle envelope available determine the stowed configuration of the appendages. The number of hold-down points and locations depend on array size, number of hard structural mounts available, launch loads, and the accessibility for assembly. The aim is to keep the resonance frequency of the appendages away from the spacecraft's first-resonance frequencies (typically above 45 Hz) in order to avoid high dynamic loads likely to be experienced during vibration tests. Although a greater number of hold-down points increases the stowed-mode resonance frequency, it also increases the complexity and thereby reduces reliability. A compromise is thus made to meet the structural requirement with a minimum number of hold-down points. The hold-down joints are preloaded to maintain positive contacts during launch. An accurate estimation of preloading is needed to avoid physical separation at these joints.

Table 1. Solar array mechanism design specifications.

Deployed array config.	3 hinged panels of 1.1M x 1.46M with tip 5.26M from s/c
Length of yoke	750.65 mm
Stowed stack height and interpanel gap	219.2 mm, 20 mm
Functional	Positive deployment at worst case environment
	No hardware to shadow the cell side of array
Loads	Withstand launch inertia loads of stowed panels without permanent deformation or release
Resonance frequencies stowed on s/c	≥ 50 Hz
Fully deployed	≥ 0.19Hz
Functional temperatures	– 150° C to + 100°C

Once design loads and locations for the hold-down points are defined, the synthesis of the hold-down configuration is initiated. Two systems, a swing-lever and a bolt restrained by a wedge mechanism have been used for the IRS-1 and INSAT-II solar array. Other hold-down methods are available, including the use of pyrobolts. Hold-down methods using swinging clamps or a wedge mechanism with central cable cutters are preferred because they avoid any single-point failure, unlike a scheme in which a number of pyrobolts are used, one at each hold-down point.

Deployment system

The function of the deployment mechanism is to positively deploy the solar array after the release of the hold-down mechanism and final locking. A positive driving torque is required throughout the deployment phase. To lock at the end of the deployment, we need a latch mechanism. All these functions are combined in the hinge mechanism (see Fig. 3). Hinge-mechanism selection is dependent on the final stiffness of the deployed configuration and the alignment stability of the appendages. All the hinges are interconnected by closed control loops (CCL) to give the system a single degree-of-freedom motion (see Fig. 4). The CCL has an averaging effect on drive torque and frictional resistance at each hinge and hence increases the reliability of deployment.

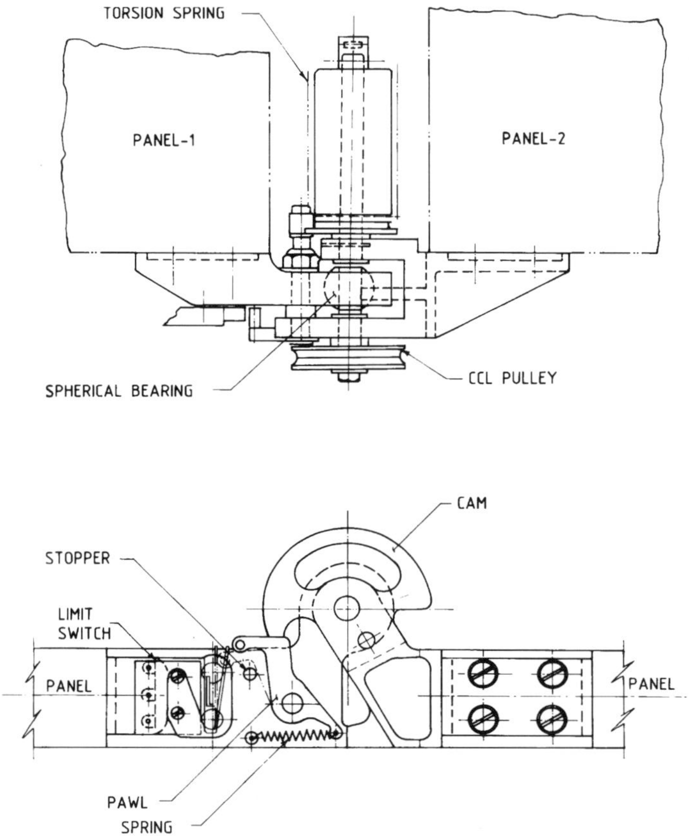

Fig. 3. Typical hinge mechanism.

System evaluation and trade off studies

The final design of a hold-down and release and deployment system is chosen after careful study of different evaluation parameters to achieve maximum reliability. The following parameters are of interest for the evaluation.

- simplicity
- minimum number of parts
- testability on the spacecraft for demonstration
- history of components in space application
- reliability of the components used
- possibility of redundancy for release function

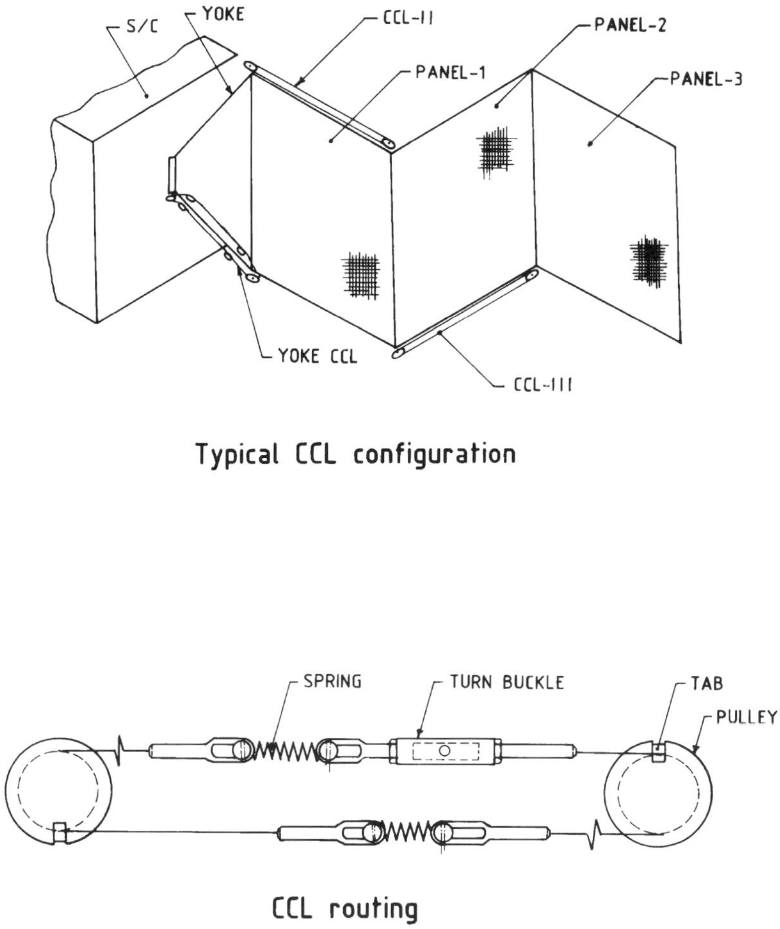

Fig. 4. Closed control loop (CCL) for array deployment.

SUBASSEMBLY AND COMPONENTS DESIGN

Detail design follows the selection of the configuration. The design loads for each system are specified in detail. The governing mechanical load for the hold-down system is the maximum preload to be applied and is derived from the launch loads. The hold-down system also should ensure positive release. The hinge design is dictated by the latch-up shock loads and the deployment torque margins of the drive system.

Design consideration for the launch restraint

In the launch-restraint scheme for a typical solar-panel array in the stowed mode, the major loads to be resisted by the restraints are the out-of-plane vibration loads and the inplane loads of the solar panels. The clamps effectively resist the out-of-plane loads whereas the inplane loads are handled by frictional resistance at the interfaces of panels at the clamp location. To improve the frictional resistance requires the serrated aluminum pads to press against softer Delrin pads. This scheme is efficient because it achieves a good frictional coefficient and avoids the stiction problem. The clamp is preloaded to counter maximum inertia loads without preload loss or slip.

Design for the extreme thermal environment

Mechanisms are designed to deploy under extreme temperatures. These temperatures, combined with materials of different coefficients of expansion, can cause thermo-elastic distortions if restraints do not permit slippage of panels. To overcome this problem, researchers make the panels slip selectively on one side (see Fig. 5). The serration angles are selected such that the frictional resistance of the Delrin-Al serration pads is quite low along the slip direction. Once the panels are allowed to slip in the selected direction, the two hinges should also be allowed to expand in the panel inplane direction. For this reason, axial freedom or float is given for the hinge, commensurate with the allowance for slip of the panel.

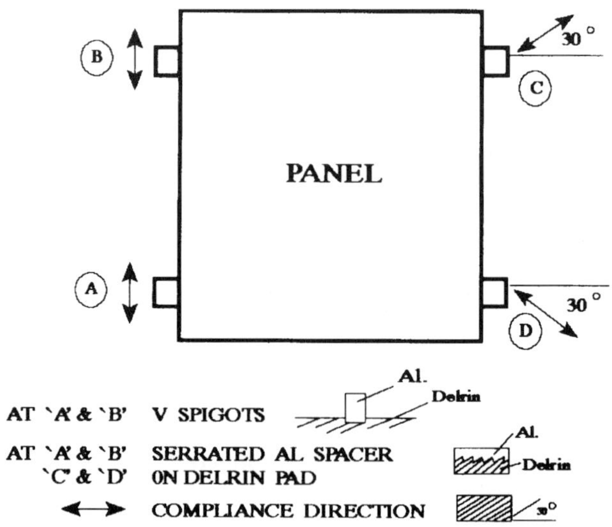

Fig. 5. Expansion compliance for temperature effect.

The hold-down wire loops and the CCL are generally made of steel. The relative thermal expansions can affect the performance of the system by inadvertent release or high loads. Compensator springs, generally helical tension or bellvelle springs with suitable spring characteristics, are put in the loop to contain the variation of tensions within allowable limits.

Design for the positive release

The initiation of the release is done by cutting the tensioned release loop with a pyrocutter. Pyrocutters with redundant charge cartridges are used to improve the reliability. The cable movement is transferred through the pulley systems to the loading levers and to the release clamp. All the moving elements are dry-lubricated by molybdenum disulfide (MoS_2) with inorganic binders or are fitted with low-friction journal bushes (trade name DU bushes).

All the levers, including clamping levers and release stirrups, are provided with return springs to ensure safe retraction. The return springs are designed to provide positive return torque, even for the worst frictional conditions. The chance of cold-welding is eliminated by avoiding contact of bare metals. Because the panels are held with compressive force, stiction problems can arise between the interpanel spacers of two adjacent panels. Aluminum serrated pads and Delrin pads for two mating surfaces are used to reduce this possibility. In addition, a kick-off plunger that pushes the panel after release is incorporated to overcome this problem.

Design for positive deployment

After the release of the clamps, the CCL controls deployment and ensures that the system will behave with a single degree of freedom. In order to ensure deployment, a positive drive torque is ensured throughout thc deployment phase. The interpanel hinges incorporate drive springs (see Fig. 3). Self-aligning spherical bearings with dry lubricated surfaces are used to minimize the effect of misalignment which is likely to occur for large solar panels. The hinge shaft and the inner bore of the monoball are dry lubricated and are made to rotate in addition to the rotational freedom between the monoball and the outer race. This arrangement effectively gives redundancy in the rotational freedom.

Because temperature affects bearing clearances, minimum radial bearing clearances are kept for the worst temperatures. As explained earlier, axial clearance is given at one of the hinges to account for thermal float of the panel. The drive springs are made of helical torsion springs and are mounted in redundant fashion along the hinge line to increase drive reliability. All the rotating elements, such as hinges and pulleys, are dry lubricated to reduce the friction.

As the drive springs aid deployment, frictional torques resist deployment (see Fig. 6). Friction torques are caused by spring mounting loads and CCL spring bias loads. The torque margin should be positive for all the anticipated deployment environments. As this margin increases, the reliability of deployment increases. A greater torque margin, however, manifests as greater kinetic energy at the end of deployment and results in more serious latch-up shocks.

Therefore, a safe torque margin for the design purposes is specified. Generally, the factor between drive torque and friction torque is two at the end of the deployment and three at the beginning. The friction torques are calculated for the worst frictional coefficients (generally 0.3), and the drive springs are designed to achieve the desired torque margins.

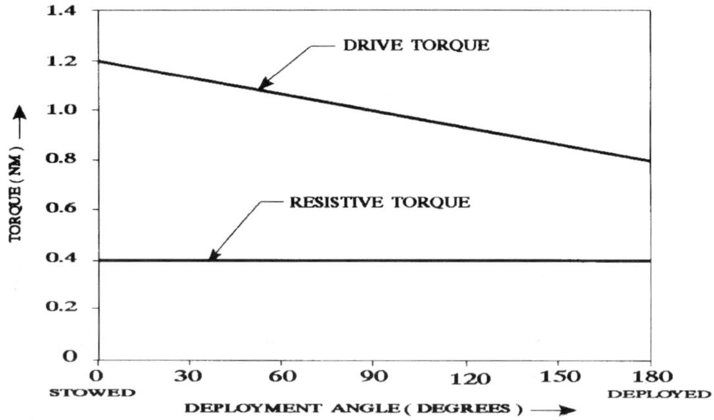

Fig. 6. Typical torque margin diagram.

Material selection, secondary treatment, and lubrication

For most of the structural components, materials with high specific strength and stiffness are recommended. Aluminum alloys like 2024-T4 and 6061-T6 are used for brackets, levers, and hinge arms. Titanium alloys and stainless steels are used for components where severe bearing stresses are induced, as in pins and rollers. For wire-loops, however, stainless steel wire ropes with end swagging are extensively used. The springs are made of beryllium-copper alloy wires because of their high resilience. The spherical bearings used at the hinges are available in high carbon chromium stainless steel (SS-440C).

The journal bushes for pulleys, levers etc., incorporate steel impregnated with PTFE (poly-tetra-flouro-ethylene) to reduce the friction. Aluminum alloy parts are generally anodized or polished, depending upon the thermal control requirements. Wherever toleranced fits are required, anodized layers are removed. The anodized layer also acts as a base for the dry lubrication coating. Hard anodizing treatment is done for components requiring wear resistance, such as pulleys. Lubrication by coating MoS_2 with suitable binders is generally adopted for all the surfaces that need low friction.

Strength design of components

The qualification load forms the basis for the strength design of components. The component loads, in turn, are derived from the governing subsystem loads, and the failure modes and stresses are calculated. A positive margin of safety is required beyond the yield strength of the materials used. A smaller margin is acceptable for the bearing stresses than for such critical components as wire ropes, release levers, or release bolts. Springs are designed with margins of 0.5, while normally prestressed release wires have a factor of safety of 3 to 5. The strength of these components is usually derived from actual test data.

DYNAMIC ANALYSIS OF THE DEPLOYMENT

After the detailed design, a deployment-dynamics study is conducted, incorporating the system parameters that affect deployment: geometry, appendage mass, moment of inertia parameters, drive torques and friction torques. The dynamic study is conducted for two cases: on-orbit deployment and ground deployment with a zero-g simulation fixture. The mathematical model for ground deployment incorporates extra frictional forces because of overhead zero-g trolleys, air drag, and bearing friction. Results of the ground tests are correlated with the corresponding mathematical model, and predictions are then made for the on-orbit condition. The deployment time, latch-up velocity, and shock moments at the end of deployment are predicted from the analysis. The shock moments are used in updating the hinge-strength margins, whereas the deployment time becomes the controlling parameter for all the deployment test programs.

RELIABILITY PREDICTION

Reliability prediction needs a statistical analysis using probability theory. Unlike electronic parts, mechanical parts are designed for a particular application and are not standardized. Hence, failure rates are difficult to obtain for these parts. In broad terms, however, reliability of the system is the product of reliability of release, reliability of deployment, and reliability against structural failure. A reliability block diagram is made accordingly. The failure mode and effects analysis (FMEA) is conducted on the system under the expected environmental conditions to identify all possible failure modes [2]. Critical failure modes are further investigated and included in the reliability block diagram. Because the deployment mechanism is used only once, it is not always effective to perform life tests on parts, but the characteristics of the parts are considered. The probability of deployment depends on the probabilistic distribution of driving torques and friction torques. Likewise, the probability of structural failures depends on the distribution of mechanical loads and strength. It is generally observed that the main contribution to unreliability comes from pyrocutters, failure modes of release, and the torque margin of deployment. These components need utmost quality control to achieve high reliability of the system.

CONCLUSIONS

The quality and reliability requirements for spacecraft deployment mechanisms cannot be overemphasized. Established design methodology to achieve quality by the design process is essential. The constraints imposed by the spacecraft, the environments of space, and the launcher govern the configuration, detail design, materials, and lubrication. The design practices to account for thermal vacuum effects, positive release, and deployment are discussed and presented as guidelines for arriving at an efficient system. A reliability evaluation is needed to assess the sensitivity of various parameters affecting the system. A good design practice eliminates failures likely to occur at a later stage of development and thereby saves overall cost and time.

ACKNOWLEDGMENTS

Authors wish to thank the Director of ISRO Satellite Centre for permission to publish this material. Special mention is due to Mr. M. N. Sathyanarayan, under whose guidance the work has been in progress. Finally, all the members of Mechanisms Division of ISAC have contributed to different aspects of the development report and are duly acknowledged here.

REFERENCES

1 T. J. Patrick, Vacuum, 31 (1981) 351–357.
2 M. Masayoshi and T. Yasaka, Spacecraft, 26, no. 3 (1989) 181–187.

Quality through Engineering Design - W. Kuo (Editor)

Product Design and Development under Constraints of a Developing Industry: A Case Study

M. A. Parameswaran

Department of Mechanical Engineering, Indian Institute of Technology, Madras 600 036, India

Abstract

The design and development of quality engineering products in a developing industry are severely inhibited by problems arising mainly from the customer's lack of confidence in the product design and development capabilities of indigenous industries, the designer's lack of confidence in the manufacturer, the limited production facilities, the limited availability of materials, and the constraints on time and money.

However, concerted efforts in product design and development in which the prospective user, the manufacturer, and the designer collaborate can be a rewarding experience, resulting not only in a quality product but also in general growth in the technological level of the developing industry. A case study is presented.

INTRODUCTION

At the present rapid rate of technological advances, the industries in the developing nations face great difficulties related to product design and engineering. While demand exists for better and more sophisticated products in the local market as well as an attractive potential for growth through export of quality products, the general conditions prevailing in a developing industry and economy are not conducive to the design, development, and production of new and improved products. With the help of world-wide communications and information technology, users, whether engineers or not, are usually well-informed about international trends in product development and quality specifications. They thus lay down specifications and performance guarantees for the product they want to make that match these trends. Although developing industries are also quite aware of the international state of the art, they generally do not embark upon product design and development because of the following constraints:

1. the prospective user's lack of confidence in the product design and development capabilities of the indigenous industries

2. the designer's lack of confidence in the indigenous manufacturing capabilities
3. the limited production facilities available
4. the limited availability of raw materials as well as finished and semifinished items
5. the limited time and money available for product design and development (including prototype fabrication and testing)

In the absence of continuing product design and development activities, manufacturing industries in developing nations sustain themselves only on borrowed technology. The technology improvement process then makes the full cycle from borrowing technology, over manufacturing, use and consumption, increased technological demands, and back to renewed borrowing of technology.

A vital break from this tedious cycle occurs only through a concerted effort of product design and development in which the prospective user, the manufacturer, and the designer collaborate. Such an effort leads to a quality product measuring up to the actual requirements and to an improved technological level of the developing industry.

CONSTRAINTS

The development of an industry from a mere workshop into an independent manufacturer offering quality products of its own design is inhibited by the constraints mentioned in the previous section. They arise out of the practical situation found in a developing economy.

User's lack of confidence

The prospective user's lack of confidence in the product design and development capabilities of the indigenous industry is the natural outcome of his desire for an efficient and reliable product. Estimates of component and product performance and reliability made during the design stage and often during the prototype testing are essentially conjectures, and many innovative designs are known to go through stages of improved prototype and production models before getting perfected. The ultimate proof of reliability is demonstrated through actual use. The market is thus wary of new design claims, especially when made by new enterprises.

Designer's lack of confidence

The designer's lack of confidence in the capabilities of indigenous manufacturers arises from his concern for ensuring quality in manufacturing. Although quality of manufacture is a prerequisite to the success of the product, it is also the first casualty if production facilities are insufficient or if the production planning and execution are not carried out meticulously. Such adverse situations are also more likely to exist in developing industries.

Limited production facilities

The limited production facilities available in developing nations compel the designer to adapt his product design to production equipment and methods that are either insufficient or outdated in comparison to the modern state of the art. Examples include not only machining, forming, casting, welding, heat treatment, and surface treatment and preparation but also inspection and testing. Even if state-of-the-art facilities are available in a few cases, the availability of properly trained personnel to operate and maintain them can often be difficult. The net result is that much effort is required to plan and to follow up the design and manufacture to reach internationally acceptable standards.

Limited availability of materials

In many developing countries, local markets frequently do not have special raw materials and finished and semifinished items, such as special steels and alloys; plastics, bearings, and sealing devices; and electric, electronic, and fluid power elements. Many of these are neither indigenously produced nor allowed to be freely imported. The task of designing and developing equipment or systems is thus manifoldly increased as each individual component has to be accounted for either by the industry's own design and development efforts or by identifying reliable external sources. Not only are the efforts to be put in and the responsibility to be borne by the design team increased inordinately but also the time required to complete the design and development work. The product quality also suffers on two counts:

1. Little choice is available in procuring raw materials and other items from outside, and thus one has to accept what is available.
2. It becomes practically impossible to properly coordinate the total design and development activity that extends over a large variety of materials and components.

With processes, equipment, and systems becoming more and more diversified and sophisticated, the individual qualities of each material, component, and subsystem used in them assume great importance.

Limited time and money available

The limited time and money available for product design and development is a very discouraging factor for the design team, especially when it is inexperienced. Many intuitive and empirical improvisations may be within the abilities of a well-experienced design team, but the design work in developing industries is usually carried out by engineers who do not possess such experience. They would prefer more of a "text-book" approach in their work, dwelling long over possible alternatives and testing each detail. Hence, lack of sufficient time and money can decrease their confidence level. Apart from this psychological effect, detailed computer analyses and prototype testing, which are so essential for the success of many modern products, get curtailed. This in turn, can lead to not achieving the expected level of quality.

CASE STUDY

Because of the above-mentioned constraints and because it is easier to borrow or buy technology than to develop it, most industries in developing nations do not encourage in-house design and development. Tying in with a well-established industry from one of the more technologically advanced nations usually occurs in the form of a license for the use of design drawings and manufacturing "know-how." Less frequently, the developing industries work with a local industry that is sufficiently developed to extend technical assistance. Quite a different situation is the one in which the customer, the manufacturing firm, and a design consultant put in their joint efforts for design and development. The customer contributes by drawing up the specifications and carrying out the acceptance tests, the design consultant does the detailed design and analysis, and the firm carries out manufacture and assembly. This delineation of responsibility is not as rigid as it might appear. More often than not, the development project is the first one of its kind for each of the three parties, and hence it is executed essentially in a cooperative spirit with constant exchange of views and suggestions. The design specifications are usually drawn up in mutual consultations: two or more design reviews are conducted before the final acceptance of the design; and any difficulties arising during manufacturing and testing are often discussed together. This type of technical cooperation has the advantage that teamwork is encouraged, technical know-how and skills of several members are pooled together, and some of the constraints can be overcome to a certain extent. Each member gains more experience and knowledge in the process as an additional benefit.

Two design and development projects relating to two different antenna systems were recently completed under the collaborative efforts of the Indian Space Research Organization as the customer's representative, M/S Prabhakar Products, a small-scale industry with experience in the fabrication and erection of small antenna systems, and The Indian Institute of Technology, Madras as the design consultant for the structural and mechanical portions of the antenna systems.

For the drive and control portions, another organization competent in the area was also involved. One of these projects concerned the design and development of a 14-m-diameter tracking antenna (Fig.1) with both elevation and azimuth movements, the most interesting as well as rewarding part being the high-reduction gear box. The following discussion deals with some aspects of the design and development work of this gear box.

Specifications

Although the initial estimates of the drive-torque requirements showed an appreciable difference between the elevation and azimuth movements, the team agreed that both sets of reduction gear boxes could be identical because there were two numbers per movement. This lightened the task of design and development. The reduction ratio required in the gear box was to be about 3,000/1, the exact value to be given to the drive-control group as early as possible. Such high-reduction gear reducers of compact design are not readily available in the Indian market. Although they can be imported from foreign suppliers, the import formalities probably cost as much time and money as indigenous development. Hence, once the mutual confidence was built up among the team members, the decision was made to go ahead with the design and development. Another reason for taking this decision was the possible easy and ready

Fig. 1. General view of 14m antenna.

coordination of the structural and mechanical design of the antenna system. If the gear boxes had been imported from a distant supplier, on the other hand, the antenna design group would have had the added constraints of possible difficult communication with the gear box supplier. Such a difficulty was encountered by the design group with respect to the selection of another major mechanical item, namely the large diameter slewing ring bearings for the azimuth and elevation rotational movements of the antenna.

In view of the restricted space availability, the team agreed after much discussion to develop a multistage planetary gear box. The input torque rating decided upon during one of the earliest discussions had to be lowered later because of the space constraints. The lowered rating was backed up with more exact drive torque estimates and a consideration of the statistics of wind-load data. The finally agreed-upon design specifications included the following:

- Rated input torque for continuous duty was 22 Nm.
- Total reduction ratio was 3,231.36.
- The gear box should be capable of withstanding 43 Nm input torque in the event one box failed.
- The inertia, backlash, and elastic deformation effects due to the gear box should be kept to a minimum, the last-mentioned being essential to satisfy the stringent pointing accuracy requirement of the antenna.

Interaction of the groups

The manufacturer is a small-scale unit with good experience in quality fabricating and testing antenna dishes and backup structures but with less experience in fabricating mechanical items, like the gear box to be developed. Out of all the groups involved, only the design group could claim some expertise in gear design. Thus it was quite fortunate that mutual confidence and cooperation among the groups reached the necessary level. The important factor, however, was that a good understanding and rapport could be established among the participants during the preliminary discussions. The initial meeting to finalize the general specifications as well as one preliminary meeting and one critical design-review meeting were required in the contract. However, several meetings took place in which various aspects of the total antenna project were discussed. In these meetings, the gear box always elicited much inquiry and discussion not only because of its novelty but also because of the general anxiety that the antenna project could suffer a set back if the gear box performance did not come up to expectations.

Manufacturing facility

Since the manufacturer possessed facilities only for general machining, gear cutting, and assembly, he had to locate reliable firms to do such tasks as heat treatment, gear finishing, and gear inspection. Typical of the problems arising was one based on the assumption that the internal gear teeth would be induction hardened. When the gears were ready, however, the induction-hardening equipment was found not to be functional, and discussions took place about alternative heat-treatment methods. Fortunately, after much improvisation, the manufacturer could recommission an old induction-hardening machine in his works. Another question related to specifying the quality grade for the gears that were to be finish-machined and inspected externally. As a precautionary measure, the quality specified on the drawings was one grade higher than what was required. This extra caution on the part of the designers proved itself when the final inspection was done. The actual accuracy grade attained was satisfactory, but not what was specified on the drawing.

Materials and bought-out items

In order to reduce the size and inertia of the gears, it was imperative to use high quality, case-hardened steels for the external gears and induction-hardened steels for the internal gears. Since such special steels are not readily available in the Indian market, a study of the availability of steels preceded the choice of material. Even then, it became necessary later to substitute a different steel for the actual manufacture of the external gears because the initially selected one became scarce in the meantime. The internal gears were machined out of ring forgings from an outside supplier. A delay on this supplier's part resulted in a longer delay in the manufacturing schedule. Part of the needed roller bearings were readily available in the local market and part had to be imported. Because of the very compact arrangement of the gears, the choice for some of the bearings was limited. Finally, since some of the initially selected bearings could not be procured in time, the design had to be modified to suit available bearings. Another special item that took some effort to locate was a rubbing-contact oil seal, which had to be very reliable because its failure meant that the lubricant would completely drain out of the gear box. Apart from the extreme care taken in the design of the components and in the manufacture by the firm, bought-out items also had to be selected for high reliability. Any loose component, let alone a single metal piece from a possible breakage, could lead to major damage of the gear box.

Time schedule and testing

Fortunately, sufficient time was available for the design and development, and the pressure of time was not felt very much. Prototype testing had to be ruled out because the contract amount did not provide for it. Also, if done, the testing could have delayed the total antenna project inordinately. Hence the design, once made and procured, necessarily had to function satisfactorily. Apart from no-load running of the assembled gear boxes, the only other test that was carried out was an improvised one with two boxes coupled back-to-back. One of the high-speed shafts was driven by an electric motor and the other loaded with a rope brake. This test was carried out successfully, much to the satisfaction of all concerned. When the whole antenna system was shop assembled, extensive tests were successfully carried out for the system acceptance. See Fig. 2.

Fig. 2. A view of the partly assembled internal portion of the gear box in which three of the four planetary stages can be seen.

CONCLUSIONS

Design and development of quality engineering products are severely inhibited in a developing industry by various factors that can be mainly attributed to lack of confidence, knowledge, and facilities, made worse by the rapid rate of technological changes in the well-developed industrial nations. Product design and development involve much more than in-house design knowledge and basic production facilities. The quality of externally procured materials, components, and external services used for design, manufacture, and inspection contribute largely to the quality of the final product. The planning and execution of the product design and development require good coordination and judgment. This is often beyond the scope of a single party in a developing industrial situation, whether it is a manufacturer or a design consultant. However, a well-coordinated, cooperative effort in which the prospective user of a product, the manufacturer, and the designer pool their experience and knowledge about the product can succeed in designing and developing a quality product. Such an effort also contributes to the growth of the technological level and capability of the developing industry. The case study presented here is a good example of such a cooperative effort.

ACKNOWLEDGMENTS

The author takes this opportunity to thank the Space Applications Centre, Indian Space Research Organisation, Ahmedabad and M/S Prabhakar Product, Guindy Industrial Estate, Madras for their valuable contribution to the design and development of the gear box that is the subject of the case study.

Chapter 3
Robust Design and Data Analysis

A design is called robust if the product or process performs consistently on target and hence is insensitive to factors that are difficult to control. According to Taguchi's philosophy on quality engineering, three stages are considered in the development of a product or process: system design, parameter design, and tolerance design. Robust design in this chapter includes a generalization of the blocking concept in the design of experiments through stratified replications, suggestions of statistical methods for the target mean problem and variance components estimation, statistical properties of Hadamard matrices of Order 16, application of experimental design to industrial problems, multivariate generalization of Taguchi's signal-to-noise ratio for industrial experiments, and exploratory data analysis for the characterization of high-performance turning machines. This chapter introduces many new methods and techniques, including block plots to examine the interaction among various factors.

Quality through Engineering Design - W. Kuo (Editor)
1993 Elsevier Science Publishers B.V.

Stratified Replications

Raghu Kacker[1]

Statistical Engineering Division, National Institute of Standards and Technology, Gaithersburg, MD 20899, U.S.A.

Abstract

The object of replication is to estimate, for given conditions, certain parameters of the distribution of the characteristic being measured. The causes of variation in the replicate values are called noise factors. The process of replication provides the values of the characteristic at selected points of the noise space. Independence and representativeness of the replicate values are critical to their interpretation. These properties can be improved by modifying the manner in which noise space is sampled. When substantial knowledge of noise factors is either available or can be acquired, one can use stratified sampling to improve these properties. This idea is a generalization of the concept of blocking in the design of experiments.

INTRODUCTION

Replication is an important scientific method. A single experimental result or measurement of a characteristic is almost never adequate in and of itself. For example, without replication there is no way of saying whether a result is real or a fluke. Even if real, a single result is rarely sufficient. The experimenter is generally concerned with the distribution of all possible values rather than a single realization of the characteristic being measured. Since one can conduct an infinite number of replications in principle, a small finite set of replicate values is a sample from the distribution of all possible values of the characteristic. The sample of replicate values is then used to estimate one or more parameters of the distribution.

The process of replication produces two or more measurements of a characteristic under similar conditions or under conditions that are assumed to be as-identical-as-possible. Since it is not possible to reproduce exactly identical conditions, certain scatter (or variation) in the replicate values is generally observed. When the causes of variation are unseen, the experimenter may assume that the scatter in the values of a characteristic is random. Often the experimenters also assume that the replicate values are independent and

---

[1] This article has been prepared by U.S. Government employees as part of their official duties and is therefore not subject to copyright.

representative of the distribution of all possible values associated with the given set of experimental conditions. Some may even go as far as to assume that the replicate values come from a well-defined statistical distribution.

The assumptions of randomness, independence, representativeness, and distributional form are made so that elementary statistical theory may be used to interpret the replicate data. In reality, however, the assumptions may not be reasonable; and the experimenter may have too few replicate values to check the assumptions. Even randomness cannot be assumed arbitrarily. Random values arise either through a natural process that is known to induce random perturbations or through an externally introduced randomization procedure.

The assumptions of independence and representative sample are most critical. For example, suppose one operator has made six measurements using one apparatus over the shortest practical time interval within one day on one sample of a homogeneous material in one laboratory. Let us consider the representativeness of the six replicate values. Before we can speak of representativeness, we have to decide what the distribution or population of interest is. Suppose the population of interest is the measurements made by any operator, on any day, and using any apparatus on the given sample of material in the given laboratory. (For our discussion we can treat a recalibrated apparatus as a different apparatus.) We cannot say that these values are representative of the population of interest. We cannot even say that these values represent measurements made by an operator on a day using an apparatus unless the operator, the day, and the apparatus were selected by a random process from the population of eligible candidates. When the variability due to operator, day, and apparatus is large, the measurements made by the operator on the day using the specific apparatus may be of no interest in and of themselves. Thus representativeness cannot be obtained unless the population or distribution of interest is fully identified and the replication process is designed to adequately cover the population.

Similarly let us consider independence. Suppose nothing that affected the measurements changed in the short period of time that the measurements were taken. Further suppose that the scatter in the replicate values was solely due to human error and the measurements were affected by a learning curve. Then the replicate values will not be independent, regardless of the population of interest.

My experience is that the assumptions of independence and representativeness are imperfect in most real applications. The experimenters may define the problem away by using definitions that are too restrictive in view of the potential sources of variation and the scope of study. The resulting assumptions may thus be an illusion.

To check the validity of the assumptions, one may augment the initial replicate values with additional data to obtain a large sample. The additional data may include measurements made at different times by different people using different experimental units or made using recalibrated or different instruments under different ambient conditions. The experimenter may then see patterns in the large replicate sample that cast a doubt on the assumptions of independence and representativeness of the initial replicate sample.

Nevertheless, independence and representativeness are very desirable properties of replicate sample values. Most of the common statistical methods are based on these assumptions. This paper describes an approach to design replication processes to increase the independence and representativeness of the sample values. The approach is based on the well-known techniques of sample-survey design and design of experiments such as stratification, orthogonal arrays, and fractional-factorial plans.

REPLICATION AS SAMPLING NOISE SPACE

The most common uses of replication include the following: to estimate the margin of error in designed experiments, to define the inductive basis of replications done to evaluate the robustness of a product's performance characteristics, to safeguard against wrong and unlikely results, to estimate one or more components of variation and the combined variation, and to detect presence of systematic causes of variation.

In designed experiments, experimental runs are replicated to estimate the margin of error in the measured values of a characteristic. The standard deviation of the replicate values is often used as a measure of the margin of error. The treatment effects are then compared with an estimate of their standard error, computed from replication standard deviation, to determine whether there is a sufficient basis to declare certain treatment effects to be distinguishable from noise.

The usual assumptions underlying designed experiments presume that the replicate values of a characteristic can be viewed as random, independent, and representative. Randomness and independence mean that the replicate values are free of identifiable patterns that can be attributed to systematic errors. Conceptually this may happen when many individual causes of variation are independently operative but none dominates in such a way that the combined effect appears as a random scatter. The scatter is then called random stationary noise.

In real life a random stationary noise in the replicate sample values is almost never the case in the early stages of a study. It is the desired end result of identifying and removing the special causes of variation. Whether these causes of variation are special or chronic, resulting in a random stationary noise, they usually fall into five categories: inter- and intra-experiment unit variations, errors in experimental variables, variations intrinsic to the measurement method and instruments, external factors such as the ambient conditions and human variations, and deterioration. I will discuss these categories of variables in more detail in the next section.

To evaluate the robustness of a product's performance characteristics, engineers test the product under various test conditions that simulate the intended-use conditions. For example, automobile companies test the performance of prototypes and final products under a wide range of road, weather, and accident-like conditions. The test conditions are usually stretched to demonstrate that the product performs well under a wide range of conditions. Here the performance data under different test conditions are the replicate values. These replicate values are then summarized into a criterion that quantifies robustness.

It is said that W. S. Gosset (who published under the name Student and introduced the t-distribution) conducted experiments in different parts of Ireland to find varieties of barley used by the Guinness Brewery that were insensitive to local environments. The range of the conditions in which these experiments were done defined the inductive basis of the replicate values. A similar replication is done in pharmaceutical companies that test new drugs on a number of animals to determine by induction whether they can be safely tested on humans. The range of the animals tested defines the inductive basis of such replications. In these examples, the replicate values are obtained by deliberately changing the experimental conditions and changing the experimental units to simulate the distribution of interest. This is in contrast to replications done under conditions that are as identical as possible.

Replication is used by scientists to safeguard against wrong and unlikely results. A minimum of three independent results under controlled conditions is usually required before a result can be taken seriously. This recognizes that all measurements are subject to the influence of operative variables that are not effectively controlled.

Replications are often designed to estimate one or more components of variation and the combined variation. A common example is inter-laboratory study of a test method. Here two major components are between- and within-laboratory variations. Usually the laboratories receive randomly selected samples from a homogenous material and follow a common test protocol. Yet significant between-laboratory variations are often observed, which indicates the existence of unidentified variables affecting the test method results that are distributed across laboratories.

Replication is often used in industrial processes to detect the presence of special causes of variation. A common example is control charting where systematic patterns and trends in the replicate values signal the presence of special causes of variation. The essence of process control is to diagnose and remedy these special causes of variation continually.

The examples of replication discussed above demonstrate that for a given set of test conditions, the scatter in replicate values is the effect of certain underlying causes of variation. Whether these causes are identified or not, they are operative nevertheless. We will call all such causes of variation noise factors. The values of the characteristic are thus a function of the levels of noise factors operative at the time of measurement. The nature of the function will in general depend on the given set of test conditions.

For the process of replication to be defined unambiguously the following categories of variables must be fully identified: (1) the response characteristic, (2) the experimental factors, (3) the conditioned factors, and (4) the noise factors.

The response characteristic is the attribute being measured. It is important to specify both the characteristic and the method of measurement because the value of the property being measured often depends on the method of measurement. The experimental factors are those variables whose effects on the measured characteristic are the object of study. Common examples are treatments and product- and process-design variables, each tested at two or more levels in the course of the study. From the view point of replication, the levels of experimental factors are assumed to be fixed at their nominal values.

The conditioned factors are those variables whose levels define the background conditions that limit the scope of the replications. The levels of conditioned factors are kept fixed as far as possible throughout the replication process. The effects of the experimental factors may depend on the levels of the conditioned factors.

Noise factors are all other factors that affect the measured characteristics. An important class of noise factors is the unavoidable and uncontrolled variations in the levels of experimental and conditioned factors. Indeed a large percentage of the noise factors arise because the experimenter is unable to control the experimental and the conditioned factors at their nominal values. In principle the number of noise factors is infinite. But most of the noise factors fall into the following categories:

1. causes of variations inherent in the measurement method and the measuring instruments
2. intra- and inter-experimental unit variations (Examples include variations between and within different units of a class of materials, machines, animals, agricultural plots, etc.)
3. external factors such as environmental variables, factors that change with time, and human variations (Examples include variations in temperature, humidity, pressure, and electrical power. Similarly, measurements on a characteristic made under like conditions at different times or made by different people often show a scatter.)
4. deterioration due to wear and chemical and physical changes due to the natural causes
5. unavoidable or uncontrolled variations in the levels of the experimental and conditioned factors

Although these categories are neither exclusive nor exhaustive, they are useful in identifying important noise factors. Further, despite best efforts, the causes of some part of the scatter in replicate data may not be discernible within the scope of the study.

The set of all possible values of the noise factors will be called a noise space. By definition the noise space is conditioned on the fixed nominal levels of the experimental and the conditioned factors. The errors in the actual levels of the experimental and the conditioned factors about their nominal values are noise factors. The points of the noise space map into the individual values of the measured characteristic. Thus the whole noise space maps into the distribution of the characteristic. For naturally occurring noise factors, this mapping is usually a smooth function.

One can assume that each noise factor follows a statistical distribution. The distributions of noise factors in turn determine the distribution of the response characteristic. This is an added complexity that is not discussed in this paper. We assume that all values of a noise factor are equally likely; that is, each noise factor follows a uniform distribution over its possible values.

The process of replication can be visualized as follows. The replication process selects one or more points from the noise space and then returns the values of the response at the selected points. A response value is the combined effect of the experimental factors, the conditioned factors, and the noise factors. The properties of the replicate values such as independence and representativeness are thus determined by the manner in which the noise space is sampled. For instance, the response values corresponding to the neighboring points of the noise space are likely to be correlated. This is the reason for serial autocorrelation often found in time series. Similarly, when the process of sampling replicate values is such that not all points of the noise space have the possibility of being selected, the realized sample values will not be representative of the distribution of the response characteristic. Thus to improve the properties of the replicate sample values, the experimenter must improve the manner of sampling the noise space. This assertion is valid regardless of the extent to which the noise factors and the noise space are understood.

STRATIFIED SAMPLING OF NOISE SPACE WITH ONE UNIT PER STRATUM

It is well known that a few variables (noise factors) often cause the most variation. This is known as the Pareto principle. Therefore, the first step in proper planning of replications is to identify the relevant noise factors. This requires subject-matter knowledge and experience. Cause and effect diagrams, histograms, stratification, check sheets, Pareto diagrams, scatter plots, and control charts are helpful in manufacturing situations to identify noise factors from the available data (see, for example, Ishikawa [1])

I have seen automated assembly operations of electronic and mechanical parts that produce defects at parts per million and per billion levels (see Box et al. [2], and Bowman et al. [3]). Similarly many of the National Institute of Standards and Technology (NIST) methods to measure fundamental physical constants have virtually zero replication variation (see, for example, NIST Special Publication 250 [4]). These examples indicate that the responsible engineers have identified almost all sources of variation and controlled them. Therefore I believe that a continual pursuit of the noise factors is of paramount importance, particularly in industrial applications. The knowledge of noise factors can be used to obtain more meaningful replicate values.

However, within the scope of the study it may not be possible to identify important noise factors. The availability of time and resources are real limitations. In addition, the experimenter may run into the limits of experience. In such cases it is useful to identify surrogates to noise factors. For example, it is often observed that measurements change from day to day. Factors such as day are not real noise factors. The real noise factors are things that change within a day such as temperature and humidity. We will therefore call factors such as day pseudo noise factors. Another example is discussed by Kacker and Shoemaker [5] where the real noise factors were the variations of composition and temperature in a reactor, but the spacial positions in the reactor were used as surrogate noise factors. Other examples of pseudo noise factors include operator, batch, experimental unit, spatial location on a unit, and measuring apparatus. From the replication viewpoint, it is useful to identify pseudo noise factors when it is not possible to identify real noise factors. My experience is that it is usually not too difficult to identify several pseudo noise factors that are meaningful in view of the objectives.

In many cases some part of the total variation in replicate values will have no discernible cause. The noise space can therefore be partitioned into two parts: a part associated with identified real or pseudo noise factors and the remaining part whose causes of variation are unknown. We will call the former identified and the latter unidentified noise space. The possibility of interactions between identified and unidentified noise factors is an added complexity that is not discussed here. I will first discuss replication sampling from the identified noise space and then from the unidentified noise space. For brevity, while discussing sampling from identified noise space, I define noise factors as real or pseudo noise factors and noise space as identified noise space.

The identified noise space is necessarily finite dimensional, each dimension corresponding to a noise factor. Since the values of a continuous noise factor must be rounded off to a few places after the decimal point, we can pretend that all identified noise factors are discrete. The objectives and the scope of the study define the range of the possible values of each noise factor. Thus a noise space can be defined from the replicate sampling viewpoint as a finite dimensional space where each dimension corresponds to a noise factor and the possible values of each noise factor (hypothetical population) are finite. The levels of the noise factors define the individual points of the noise space. Since the possible values of each noise factor are finite, the noise space has a finite number of points.

In relation to identified noise factors, the essential problem of replication is thus how to sample a finite noise space so that the replicate values have desirable properties such as independence and representativeness. The theory of finite (one-dimensional) population sampling and the literature of two-dimensional sampling such as geographical surveys is relevant (see, for example, Cochran [6]). The principal that is most useful in sampling noise space is stratification. Both independence and representativeness are realized when the noise space is stratified until only one point is sampled from each stratum. Thus the total number of strata will equal the number of replications.

Suppose the number of noise factors is k, and the population of the ith noise factor is stratified in n_i strata. Then the total number of strata will be $n = n_1 \times n_2 \times \ldots \times n_k$. When the replication process is designed such that one point is selected from each of the n strata, one obtains a sample of n replicate values. The sampling within the strata can be random or judicious. Thus the replication process will be stratified sampling with one unit per stratum chosen randomly or judiciously.

In the case study of Kacker and Shoemaker [5], there were two pseudo noise factors: the spacial location on a circular wafer and the position of the wafer on a seven-sided susceptor. The wafer was stratified into five sectors: top, left, right, bottom, and center. The susceptor was stratified into 14 positions corresponding to the top and the bottom positions on each of the seven sides. Thus the noise space consisted of $5 \times 14 = 70$ strata, and each stratum was a sector on a wafer. One measurement was taken from somewhere in the middle of each sector resulting in 70 measurements. Thus randomness was inherent in the process of making the measurement within each sector. Thus these 70 values could be assumed to be approximately random. Since the whole wafer and each side of the susceptor was covered by stratification, the measurements were representative of the within- and between-wafer measurements. The measurements were approximately independent because they came from different sectors.

The number of strata that should be assigned to a noise factor depends on its share of the total variation in the response characteristic. Those factors that account for larger variation in the response should have a larger number of strata and hence a greater representation in replicate sampling.

The minimum number of strata for a noise factor is two. With k noise factors, therefore, there will be $n = 2k$ strata in the noise space. But 2k can be more than the number of replications that can be done within the scope of the study. In such cases one can use the ideas of fractional-factorial plans and fixed- and mixed-level orthogonal arrays (see, for example, Kacker, Lagergren, and Filliben [7]) to reduce the number of strata.

By treating the noise factors as experimental factors and equating the strata with the levels of the factors, one can use orthogonal arrays of strength 2, 3, and 4 (resolutions 3, 4, and 5) to select a fraction of strata from all possible strata. Thus the number of replications is reduced by reducing the number of strata. The projective properties of orthogonal arrays (see, for example, Box, Hunter and Hunter [8], p. 388) preserve representativeness to some extent. Phadke [9], describes how an orthogonal array OA36(211 × 312) was used to define 36 strata covering 21 noise factors, ten with 2 strata each and eleven with 3 strata each. Phadke's example is not discussed in terms of stratifying noise space. But the real justification for the success of that approach is that the levels of noise factors represent the neighboring noise region; that is, the noise levels can be viewed as selections from strata formed by the neighboring regions of the noise space.

In the context of randomized block experiments with every treatment occurring once in each block, strata of a noise space correspond to the blocks, assuming that the variations of units within a block and inter-block variations encompass all identified noise factors. The stratification of noise space means that the blocks cover the whole population of eligible units. The replicate response values associated with a particular treatment from different blocks are likely to be independent because different noise conditions are associated with the blocks. When the blocks cover the population of eligible units, the replicate values of the response will be representative.

There are two ways to select one measurement from each stratum: randomly or judiciously. Random selection requires that all units in the stratum must be labeled and then one selected through a mechanical random process. Judicious selection may not require labeling of the stratum units, and some randomness may be built into the judicious selection. In the example of Kacker and Shoemaker [5], the measurement point was selected somewhere approximately in the middle of the sector on a wafer. Thus it was approximately random. The extra work for proper random selection was not worth the benefits.

The noise factors and their surrogates need not be controllable even for the sake of experimentation. As long as a noise factor behaves in a systematic pattern and it is possible to measure it, one can plan the replication process around the behavior of the noise factor. For example, suppose the temperature of an in-process material is a noise factor and suppose it changes systematically within a day. The experimenter can then select certain noise levels and schedule measurements when the noise factor attains the selected levels.

Let us now consider sampling from the unidentified noise space. Even when the noise factors are unknown, they may be known to be embodied in identifiable artifacts. For example the unidentified noise factors may be associated with identifiable experimental materials, animals, agricultural plots, etc. For replication purposes, the time-related unidentified noise factors can be associated with discrete time intervals. Thus unidentified noise space can be sampled by selecting the artifacts that embody the unidentified noise factors at random from the class of eligible artifacts. This is what is done in completely randomized experimental designs. In the absence of a more detailed knowledge of noise factors random allocation of experimental artifacts to replications is all one can do to increase the independence and representativeness of the replicate values. The replicate values will include the effects of unidentified noise factors if they become operative during the period of replication.

PRECISION AND INDUCTIVE BASIS

From the viewpoint of a scientist, two aspects of replicate values describe their usefulness: their precision and their inductive basis. Precision is a generic term for the degree of mutual agreement among the replicate measurements. A measure of precision is the standard deviation of the measurements—the smaller the standard deviation, the higher the precision. For a statement of precision to be unambiguous, the replicate measurements must be independent; and the conditioned factors that are kept fixed and the noise factors that are allowed to vary must be fully identified. Thus, for example, within the context of interlaboratory study of a test method, engineers speak of within-laboratory precision as repeatability, and between-laboratory precision as reproducibility.

Results of higher precision are less uncertain and hence more meaningful. Similarly in comparative experiments, higher precision of replicate values makes it easier to evaluate the effects of experimental factors. Therefore, higher precision is preferred to lower precision.

A scientist's interests are rarely limited to the results of a specific experiment or study. He is usually interested in the generalized conclusions that can be reasoned from one or more studies. The reasoning process that proceeds from the results of one or more specific experiments to a generalization is called induction. The extent to which the results of a study can be generalized depends on the background conditions in which the experiment was done. Thus we can say that the inductive basis of a set of replicate values are the background conditions to which the replicate values are intended to be applicable. For example, when the experimental units that produce the replicate values are chosen through random sampling from a well-defined set of eligible units, the set of eligible units is the inductive basis. Similarly, to increase the inductive basis of replicate values associated with a particular treatment setting in a randomized block design, the blocks are made to be as different as possible from each other. Here the blocks are surrogates of the noise factors. Thus the ranges of the levels of noise factors and their strata define the inductive basis: the wider the range of noise factors that is stratified, the larger the inductive basis. The levels at which the conditioned factors are kept fixed, however, limit the inductive basis. When a scientist is interested in the generalized conclusions that can be drawn from specific examples, a larger inductive basis is preferable to a smaller one.

It appears that high precision and a high inductive basis can be competing requirements. For example, to increase the precision of a test method, it is a common practice to identify important sources of variation and control them at certain levels. That is, such sources of variation are treated as conditioned factors whose levels are controlled rather than noise factors whose levels are allowed to vary. The increase in precision is thus obtained by reducing the inductive basis of the replicate values. Generally speaking when the noise factors are controlled, precision increases; and when the noise factors are allowed to vary, the inductive basis increases. The choice of which noise factors should be controlled and which should be allowed to vary depends upon the objectives of the study.

The approach to designing replication processes described in the previous sections guides the scientist to achieve a balance between the precision and the inductive basis. First, the importance of identifying the sources of variation and their surrogates cannot be overemphasized. Then, on the basis of the objectives and the scope of the study, these sources of variation are categorized as conditioned factors or as noise factors. The conditioned factors are controlled at fixed levels. The fixed levels of conditioned factors increase the precision of the replicate values. And the stratified sampling of the noise space increases the inductive basis. The thinking required to categorize the sources of variation leads to a balance between the precision and the inductive basis.

REPLICATION AND RANDOMIZATION

Replication is most frequently done as an integral part of a comparative experiment. Specifically, replication is done jointly with the other principles of experiment design. These principles of experimentation include replication, randomization, local control (also called blocking), balanced arrangement, controls (no treatment applied), and blindfold technique (see Eisenhart [10]). Eisenhart calls these principles canons of sound experimentation and states that "the principal function of replication is to provide an opportunity for randomization to perform its functions." Eisenhart further states that "replication in conjunction with randomization tends to reduce bias, by providing opportunity for effects of uncontrolled factors to balance out and thus provides means of increasing precision."

Cox ([11], pp. 81–82) states the following: "In many, if not most, experiments important uncontrolled variation may arise from several sources, notably in the experimental material, in the various stages of applying the treatments, and in the taking of the observations. It is important that the randomization should cover all important sources of variation connected with the experimental units and that, as far as practicable, the different experimental units receiving the same treatment should be dealt with separately and independently at all stages at which important errors may arise, one stage being in the application of treatments."

Thus the intent of replication is to achieve a balanced coverage of the full range of each noise factor. Further, for comparative experiments the replications for each treatment must be on an equal footing. In those cases where the noise factors cannot be identified except that they are known to be embodied in the experimental materials and various stages of experimentation, replication with randomization is the most appropriate technique to achieve balanced coverage of the noise space. The reason is that randomization gives an equal opportunity to the units that embody the noise factors to affect each treatment. Thus the balanced coverage is obtained in the sense of chance occurrence.

Many industrial and physical-science applications exist for which substantial knowledge of noise factors is either available or can be acquired. In many advanced technology applications such as automated assembly operations and computer-integrated design and manufacturing, most, if not all, of the important variables affecting the product characteristics are identified. This testifies that noise factors can be identified in certain situations. In such cases available information on the noise factors can be used in the design of replications to directly sample the identified part of the noise space through stratified sampling. Not using available information on noise factors is like throwing away critical information that affects the results. Of course the randomization method is still useful to sample the unidentified noise space.

SAMPLE SIZE

It will be hard to find a practicing statistician that has not faced the following question: "How many observations (replicate sample values) do I need to take for each treatment?" The textbook answer requires estimates of the population variance, the difference in treatment effects to be detected, and the probabilities of type one and type two errors. It is usually difficult to guess these quantities. In addition, the sample size numbers obtained through such arguments are often much larger than what the experimenter is able or willing to do. Furthermore, this is only a part of the problem. Even more serious issues include the following: the population to be sampled may not be well defined or may not be stable, and the replicate values may not be independent or may not be representative. Thus the estimates of the required sample size based on textbook formulas may have no meaning.

In most cases the sample size actually used is determined by external factors such as the availability of materials and other economic and time-related factors. The issue then shifts to one of choosing a replication scheme to get the most information with a minimum number of replicate values. When replication is done blindly without paying attention to the potential causes of variation, the replicate sample values may not be very informative. Thus the available information on the causes of variation or noise factors must be used in the replication process.

The approach of identifying the noise factors and stratified sampling of these noise factors increases the information content of the replicate values. Independence and representativeness are just two aspects of the information content. Further research on the notion of the information content of a set of replicate sample values is needed to better understand the advantages of stratified sampling in planning replications. The theory of relative information promoted by Jumarie [12], and the competing gnostic theory of Kovanic [13] hold promise in this regard. The well-known comparisons of stratified random sampling and simple random sampling in one-dimensional finite-population sampling show that for the same sample size, stratified sampling produces more precise estimates of the population mean [6]. Conversely, stratified sampling produces the same precision as simple random sampling with a smaller sample size.

TAGUCHI'S NOISE ARRAY

To my knowledge, Taguchi [14] is responsible for the explicit notion of noise factors and their use in comparative experiments. His approach to sampling noise space involves the following steps. First treat each noise factor as an experimental variable and select two or more test levels for each variable. The levels are chosen sufficiently far apart to adequately cover the noise space. Second use an orthogonal array to construct an experiment layout involving the noise factors. This layout is called the noise array. The rows of the noise array identify selected points from the noise space (see Kackar [15] and Phadke [9] for details). As is typical of Taguchi's work, no justification is given as to why the noise array approach to sampling noise space is better than random replications.

We can think of each row of a noise array (points in the noise space) as representing a stratum of the noise space. With this interpretation, we can think of Taguchi's noise array as a stratified sample with one unit per stratum selected judiciously. This appears to be the justification behind Taguchi's noise arrays. Random replication to cover unidentified noise factors can be done in addition to the stratified sampling for the identified noise factors.

CONCLUSIONS

The variables that affect the measured values of a characteristic can be categorized as experimental factors, conditioned factors, and noise factors. The number of noise factors is much larger than the experimental and conditioned factors because errors in these latter factors are noise. From the replication viewpoint, experimental and conditioned factors are kept fixed at their nominal levels. The replicate values represent the combined effects of noise factors at the given (nominal) conditions of experimental and conditioned factors.

Replication is most commonly used in three contexts: to estimate combined variance, to identify important components of variance and their contributions, and to reduce variance (increase precision) of replicate values. In each case it is critical to identify important noise factors, and it is advantageous to use them in the design of the replication process. When one estimates the components of variance and combined variance, stratifying the noise factors and sampling one unit from each stratum make the replicate values representative and independent. To increase precision or reduce variance, important noise factors must be identified and controlled.

The idea of stratifying noise factors is not exceptional; it is a generalization of the idea of blocking in designing experiments where blocks correspond to different levels of noise factors. When replication is done without thinking of the noise factors and their effects are not adequately represented in the replicate values, we have a case of blind replication. Such blind replications give a false sense of security.

REFERENCES

1 K. Ishikawa, Guide to Quality Control, Asian Productivity Organization, Tokyo, 1976.
2 G. E. P. Box, R. N. Kacker, V. N. Nair, M. S. Phadke, A. C. Shoemaker and J. C. F. Wu, Qual. Prog. (1988) 37–41.
3 K. O. Bowman, T. H. Hopp, R. N. Kacker and R. J. Lundegard, Statistical Quality Control Technology in Japan, Chance: New Directions for Statistics and Computing, Springer-Verlag, Inc., New York, 4, no. 3 (1991) 15–21.
4 J. D. Simmons (ed.), NIST Special Publication-250, NIST Calibration Services Users Guide 1991, National Institute of Standards and Technology, Gaithersburg, Md., U.S.A.
5 R. N. Kacker and A. C. Shoemaker, AT&T Tech. J., 65, no. 2 (1986) 39–50.
6 W. G. Cochran, Sampling Techniques, 3rd ed., John Wiley and Sons, Inc., New York, 1977.
7 R. N. Kacker, E. S. Lagergren and J. J. Filliben, J. Res. Nat. Inst. Stand. Tech., 96, no. 5 (1991) 577–591.
8 G. E. P. Box, W. G. Hunter and J. S. Hunter, Statistics for Experimenters, John Wiley and Sons, Inc., New York, 1978.
9 M. S. Phadke, Quality Engineering Using Robust Design, Prentice Hall, Englewood Cliffs, N.J., 1989.
10 C. Eisenhart, Bull. Int. Stat. Inst., 37 (1960) 339–350.
11 D. R. Cox, Planning of Experiments, John Wiley and Sons, Inc., New York, 1958.

12 G. M. Jumarie, Subjectivity, Information, Systems: Introduction to a Theory of Relativistic Cybernetics (Studies in Cybernetics, Vol. 12), Gordon and Breach Publishers, New York, 1986.
13 P. Kovanic, Prob. Control and Inf. Theory, 13, no. 4 (1984) 259–274.
14 G. Taguchi, System of Experimental Design: Vols. 1 and 2, UNIPUB, Kraus International Publications, White Plains, N.Y., 1987.
15 R. N. Kacker, J. Qual. Technol., 17, no. 4 (1985) 176–209.

Quality through Engineering Design - W. Kuo (Editor)

Variance Components Estimation—On the Status of Grubbs's Estimators

Kalyan Das,[a] Bikas K. Sinha,[b] and Bimal K. Sinha[c]

[a]Department of Statistics, Calcutta University, 35, B. C. Road, Calcutta 700 019, India

[b]Statistics and Mathematics Division, India Statistical Institute, 203, B. T. Road, Calcutta 700 035, India

[c]University of Maryland, Baltimore County, Baltimore, MD 21228, U.S.A.

Abstract

We consider the problem of estimating unequal variance components across blocks in a v × b block design with m ($\geq$ 1) observations per cell, when the underlying model is one of fixed effects. We present some results showing justification of the use of Grubbs's estimators for the variance components. Nonnegative variance estimators, with a uniformly smaller mean-square error (for each variance component) than Grubbs's estimators, are available only for small values of v.

INTRODUCTION

Grubbs [1] suggested a method of (unbiased) estimation of heterogeneous block variances in a two-way layout with one observation per cell. Considering b blocks and v varieties, let us assume the usual model:

$$y_{ij} = \mu + \alpha_i + \beta_j + e_{ij}, \quad 1 \leq i \leq v, \quad 1 \leq j \leq b \tag{1}$$

where the fixed effects μ, α_i, and β_j have their usual significance and the errors are uncorrelated with

$$V(e_{ij}) = \sigma_j^2, \quad 1 \leq i \leq v, \quad 1 \leq j \leq b. \tag{2}$$

We will adopt the notations

$$\bar{y}_{i.}, \quad \bar{y}_{.j}, \quad \text{and} \quad \bar{y}_{..}$$

with their usual meanings.

Let

$$S_j^2 = \sum_{i=1}^{v} \left(y_{ij} - \bar{y}_{i.} - \bar{y}_{.j} + \bar{y}_{..}\right)^2, \quad 1 \le j \le b \tag{3}$$

Then, Grubbs [1] proposed the estimating equations:

$$S_j^2 = E\left(S_j^2\right), \quad 1 \le j \le b \tag{4}$$

where E is the expectation wrt Eqs. (1–2).

The idea is to pretend that σ_j^2's are all equal so that the least square estimates of the model parameters are

$$\hat{\mu} = \bar{y}_{..}, \quad \hat{\alpha}_i = \bar{y}_{i.} - \bar{y}_{..}, \quad \hat{\beta}_{j.} = \bar{y}_{.j} - \bar{y}_{..}$$

Next, the within-block sum of squares (SS) S_j^2's are defined in the usual manner and their expectations are computed by using the actual model. This results in the method-of-moments estimates of the variance components.

Note 1. Grubbs considered the situation in which α_i's were random effects with mean zero, variance σ^2, and zero covariance. It was then a problem of estimating the variance components: σ^2 and $\sigma_1^2, \sigma_2^2, \ldots, \sigma_b^2$. Subsequently, Brindley and Bradley [2] examined the model with α_i's as fixed parameters. Other related references are Maloney [3], Ellenberg [4], Bradley and Brindley [5], and Bradley and Rollier [6].

In an explicit form, the equations in (4) are given by

$$S_j^2 = (v-1)\left[\left(1 - \frac{2}{b}\right)\sigma_j^2 + \frac{1}{b^2}\sum_{t=1}^{b} \sigma_t^2\right], \quad 1 \le j \le b \tag{5}$$

from which Grubbs's estimates from a v × b setup are given by

$$\hat{\sigma}_{jG}^2 = (v-1)^{-1}\left(1 - 2b^{-1}\right)^{-1}\left[S_j^2 - b^{-1}(b-1)^{-1}\sum_{t=1}^{b} S_t^2\right] \tag{6}$$

where G stands for Grubbs.

Note 2. When b = 2, then system (5) reduces to

$$S_1^2 = (v-1)\left(\frac{\sigma_1^2 + \sigma_2^2}{4}\right) = S_2^2 \tag{7}$$

and hence only two blocks, σ_1^2 and σ_2^2, cannot be unbiasedly estimated in a v × 2 setup.

Note 3. One way out of this problem is to consider a v × 2 setup with multiple observations per cell. Another possibility is to extend the setup by choosing b ≥ 3 blocks and estimating unbiasedly all the b variance components, as is given by Eq. (6). In this setup as well, we could have multiple observations per cell.

In the following, we will explore both situations.

STATUS OF GRUBBS'S ESTIMATORS: CASE OF TWO BLOCKS

We consider a v × 2 setup with m (> 1) observations per cell. The model adopted now would be given by

$$y_{ijk} = \mu + \alpha_i + \beta_j + e_{ijk} \qquad (8)$$
$$V(e_{ijk}) = \sigma_j^2, \quad \text{covariances} = 0, \quad 1 \le i \le v, \quad 1 \le j \le 2, \quad 1 \le k \le m.$$

The notations

$$\bar{y}_{...}, \quad \bar{y}_{i..}, \quad \bar{y}_{.j.}, \quad \bar{y}_{ij.}$$

will be adopted with their usual meanings. As before, let us use the within-block SS:

$$S_j^2 = \sum_i \sum_k \left(y_{ijk} - \bar{y}_{i..} - \bar{y}_{.j.} + \bar{y}_{...}\right)^2, \quad j = 1, 2 \qquad (9)$$

It is easy to note that

$$S_j^2 = \sum_i \sum_k \left(y_{ijk} - \bar{y}_{ij.}\right)^2 + m \sum_i \left(\bar{y}_{ij.} - \bar{y}_{i..} - \bar{y}_{.j.} + \bar{y}_{...}\right)^2 \qquad (10)$$
$$= V_j + W_{jj}, \quad j = 1,2$$

$$V_j \sim \sigma_j^2 \chi_\gamma^2, \quad \gamma = v(m-1); \quad j = 1, 2 \qquad (11)$$

$$W_{11} = W_{22} \equiv V_3(\text{say}) \sim \frac{\sigma_1^2 + \sigma_2^2}{4} \chi_{v-1}^2 \qquad (12)$$

Then the estimating equations for Grubbs's estimators are given by

$$E\left(S_j^2\right) = \gamma\sigma_j^2 + \frac{v-1}{4}\left(\sigma_1^2 + \sigma_2^2\right), \quad j = 1, 2 \qquad (13)$$

from which

$$\hat{\sigma}_{jG}^2 = \frac{1}{\gamma}\left[V_j + \frac{V_3}{4} - \left(\frac{\upsilon - 1}{4}\right)\frac{V_1 + V_2 + \frac{3}{2}}{\left(\gamma + \frac{\upsilon - 1}{2}\right)}\right], \quad j = 1, 2 \tag{14}$$

Observe that S_j^2 in Eq. (9) refers to within-block SS whereas the V_j's refer to error SS. Clearly, Eq. (13) results in the moment estimates given in Eq. (14). These are Grubbs's estimators.

On the other hand, we have available the natural or naive (N) estimators:

$$\hat{\sigma}_{jN}^2 = \frac{V_j}{\gamma}, \quad j = 1, 2 \tag{15}$$

We establish that Grubbs's estimators of variance components are *superior* to the naive estimators wrt squared error loss when we consider total variability as the criterion for comparison.

Theorem 1

$$\sum_{j=1}^{2} V\left(\hat{\sigma}_{jG}^2\right) < \sum_{j=1}^{2} V\left(\hat{\sigma}_{jN}^2\right)$$

$$\forall \sigma_1^2, \sigma_2^2 > 0, \quad \forall \upsilon \geq 2, \quad \forall m \geq 2$$

We will provide a proof of this result in a general $v \times b$ frame for all $b \geq 2$.

Note 4. Whereas the N estimators are nonnegative, Grubbs's estimators are likely to assume negative values as well. In Table 1, we provide some computational results to indicate that the chances for the Grubbs's estimates to assume negative values are indeed very small (less than 5%), even for small values of v and m, whenever σ_1^2 / σ_2^2 is moderate.

Table 1. $P(\hat{\sigma}_{1G}^2 > 0)$ as a function of $\rho = \sigma_1^2 \sigma_2^{-2}$

ρ	0.10	0.50	0.80	1.00	2.00	2.50	5.00	10.00
$P(\rho)$ (m = 2, v = 4)	1.00	0.9996	0.9988	0.9956	0.9844	0.9790	0.9409	0.8878
$P(\rho)$ (m = 3, v = 3)	1.00	1.00	1.00	1.00	0.9999	0.9998	0.9996	0.9993

Next we examine the status of Grubbs's estimators wrt admissibility in the class of unbiased estimators. We establish in the Appendix that for a $v \times 2$ setup with m observations per cell, the statistics (V_1, V_2, V_3) are jointly sufficient for the variance components σ_1^2 and σ_2^2. Because the V_i's are quadratic in the observations, we find that for any quadratic unbiased estimator of the variance components, conditional expectation (given V_i's) lead to an estimator that will be linear in the V_i's. We prove that in a suitable subclass of quadratic unbiased estimators generated by linear combinations of V_1, V_2, and V_3, Grubbs's estimators are indeed admissible wrt the criterion of total variability. Precisely, we claim the following:

Theorem 2 *No quadratic unbiased estimators of σ_1^2 and σ_2^2 that are linear in the V_i's (and have the same coefficient for V_3 in both the estimators) can dominate Grubbs's estimators uniformly in (σ_1^2, σ_2^2) wrt total variability.*

Proof. Let $\hat{\sigma}_{jU}^2 = C_{1j}V_1 + C_{2j}V_2 + C_{3j}V_3$ be an unbiased estimator of σ_j^2, $j = 1, 2$. It is easy to verify that $\hat{\sigma}_{jU}^2$ can be expressed as

$$\hat{\sigma}_{jU}^2 = \frac{V_j}{\gamma} + C\left(V_1 + V_2 - \frac{4\gamma V_3}{\upsilon - 1}\right), \quad j = 1, 2 \tag{16}$$

where C is an arbitrary constant. Note that $C_N = 0$ for $\hat{\sigma}_{jN}^2$, and $C_G = -\dfrac{1}{2\gamma\left(1 + \frac{2\gamma}{v-1}\right)}$ for $\hat{\sigma}_{jG}^2$. In view of this observation, we rewrite Eq. (16) as

$$\hat{\sigma}_{jU}^2 = \hat{\sigma}_{jN}^2 + C\left(V_1 + V_2 - \frac{4\gamma V_3}{\upsilon - 1}\right), \quad j = 1, 2 \tag{17}$$

From the distributional properties of V_1, V_2, and V_3 indicated in Eqs. (11–12) and from their independence, it is easy to deduce that

$$\begin{aligned}\sum_{j=1}^{2} V\left(\hat{\sigma}_{jU}^2\right) = \; & 4C^2\left[\left(\sigma_1^4 + \sigma_2^4\right)\left(\gamma + \frac{\gamma^2}{\upsilon - 1}\right) + 2\sigma_1^2\sigma_2^2\frac{\gamma^2}{\upsilon - 1}\right] \\ & + 4C\left(\sigma_1^4 + \sigma_2^4\right) + \frac{2}{\gamma}\left(\sigma_1^4 + \sigma_2^4\right)\end{aligned} \tag{18}$$

This expression is a quadratic in C. Consider the parameter points (σ_1^2, σ_2^2) with $\sigma_1^2 = \sigma_2^2 > 0$. For such points,

$$\sum_{j=1}^{2} V\left(\hat{\sigma}_{jU}^2\right) \propto 8C^2\left[\gamma + \frac{2\gamma^2}{\upsilon - 1}\right] + 8C + \frac{4}{\gamma} \tag{19}$$

and the minimum value for variation in C is attained when $C_{opt} = -\dfrac{1}{2\gamma\left(1+\frac{2\gamma}{v-1}\right)}$ and this choice coincides with C_G. It thus follows that Grubbs's estimators provide the least value of the total variability at such parameter points and they do so uniquely. Hence the claim. **QED**

Note 5. If we look back to Eq. (18) and minimize the same for variations in C, we obtain

$$C_{opt}\left(\sigma_1^2,\sigma_2^2\right) = -\frac{\sigma_1^4+\sigma_2^4}{2\left[\gamma\left(\sigma_1^4+\sigma_2^4\right)+\frac{\gamma^2}{\upsilon-1}\left(\sigma_1^2+\sigma_2^2\right)^2\right]} \tag{20}$$

It follows that for all $(\sigma_1^2, \sigma_2^2) > (0,0)$,

$$C_0 = -\frac{1}{2\gamma\left(1+\frac{\gamma}{\upsilon-1}\right)} \le C_{opt}\left(\sigma_1^2,\sigma_2^2\right) \le -\frac{1}{2\gamma\left(1+\frac{2\gamma}{\upsilon-1}\right)} = C_G \tag{21}$$

Thus all estimators of the form shown in Eq. (16) with $C_0 \le C \le C_G$ turn out to be admissible. Because $C_N = 0$ lies outside this interval, the N estimators are inadmissible wrt total variability.

Note 6. If we single out the variance components and examine the status of Grubbs's estimators separately for each one of them, it turns out that Grubbs's estimators are still admissible in the class of estimators of the form shown in Eq. (16). One can show that all such estimators with a choice of C satisfying

$$-\frac{1}{\gamma\left(1+\frac{\gamma}{\upsilon-1}\right)} \le C \le 0 \tag{22}$$

are admissible. The choice C = 0 (leading to $\hat{\sigma}_{jN}^2$) is acceptable only when the parameter space includes points with $\sigma_1^2 = 0$, $\sigma_2^2 = 0$, or both.

STATUS OF GRUBBS'S ESTIMATORS: GENERAL CASE

This time we consider a $v \times b$ setup with m observations per cell. Thus we refer to the model in Eq. (8) with $1 \le j \le b$. The notations $y_{...}$, $y_{i..}$, $y_{.j.}$, and $y_{ij.}$ all apply as usual, and Eqs. (9–10) hold for $1 \le j \le b$. We establish in the Appendix that $(V_1, V_2, \ldots, V_b, \underline{\underline{W}})$ are jointly sufficient statistics for the variance components where

$$\text{(i)}\quad V_j = \sum_i \sum_k \left(y_{ijk} - \bar{y}_{ij.}\right)^2, \quad 1 \le j \le b$$

$$\text{(ii)}\quad W_{jj} = m \sum_i \left(\bar{y}_{ij.} - \bar{y}_{i..} - \bar{y}_{.j.} + \bar{y}_{...}\right)^2 \tag{23}$$

$$\text{(iii)}\quad W_{jj'} = m \sum_i \left(\bar{y}_{ij.} - \bar{y}_{i..} - \bar{y}_{.j.} + \bar{y}_{...}\right)\left(\bar{y}_{ij'.} - \bar{y}_{i..} - \bar{y}_{.j'.} + \bar{y}_{...}\right)$$

$$1 \le j \ne j' \le b$$

(iv) V_j's are independent of $W_{jj'}$'s

As before, we define

$$S_j^2 = \sum_i \sum_k \left(y_{ijk} - \bar{y}_{i..} - \bar{y}_{.j.} + \bar{y}_{...}\right)^2 = V_j + W_{jj} \tag{24}$$

for $1 \le j \le b$. Then the moment equations for deriving Grubbs's estimators are again given by: $S_j^2 = E(S_j^2)$, $1 \le j \le b$. In an explicit form, the equations are

$$S_j^2 = \gamma\sigma_j^2 + (v-1)\left[\left(1 - \frac{2}{b}\right)\sigma_j^2 + \sum_{t=1}^{b} \sigma_t^2 / b^2\right], \quad 1 \le j \le b \tag{25}$$

yielding

$$\hat{\sigma}_{jG}^2 = \frac{1}{\left[\gamma + (v-1)\left(1 - \frac{2}{b}\right)\right]}\left\{S_j^2 - \left(\frac{v-1}{b^2}\right)\frac{\sum_{t=1}^{b} S_t^2}{\left[\gamma + (v-1)\left(1 - \frac{1}{b}\right)\right]}\right\} \tag{26}$$

We may again provide a two-fold justification for using Grubbs's estimates instead of the N estimates: $\hat{\sigma}_{jN}^2 = V_j/g$, $1 \le j \le b$. The main justification is that $\hat{\sigma}_{jG}^2$'s perform better than $\hat{\sigma}_{jN}^2$'s when total variability is regarded as the criterion for comparison. To see this, we need to compute $V(\hat{\sigma}_{jG}^2)$. We proceed through the following steps:

Step I It can be deduced [7] that

$$Cov\left(W_{jj}, W_{j'j'}\right) = 2(N-1)\delta_{jj'}^2 \tag{27}$$

where

$$\left.\begin{aligned}\delta_{jj} &= \left(1-\frac{2}{b}\right)\sigma_j^2+\frac{1}{b^2}\sum_{t=1}^{b}\sigma_t^2\\ \delta_{jj'} &= -\frac{1}{b}\left(\sigma_j^2+\sigma_{j'}^2-\frac{1}{b}\sum_{t=1}^{b}\sigma_t^2\right)\end{aligned}\right\} \tag{28}$$

Step II Writing

$$\left.\begin{aligned}\eta &= \frac{(v-1)/b^2}{\gamma+(v-1)\left(1-\frac{1}{b}\right)}\\ \theta &= (\theta_2,\theta_3,\ldots,\theta_b)'\\ \theta_j &= \sigma_j^2/\sigma_1^2, \quad j=2,3,\ldots,b\\ \theta_1 &= 1\end{aligned}\right\} \tag{29}$$

we may deduce that

$$\begin{aligned}V\left(\hat{\sigma}_{1G}^2\right) = \frac{2\sigma_1^4}{\left[\gamma+(v-1)\left(1-\frac{2}{b}\right)\right]^2}&\left[\gamma(1-\eta)^2+\gamma\eta^2\left(\sum_2^b\theta_t^2\right)\right.\\ &+(1-\eta)^2(v-1)\left(1-\frac{2}{b}+\frac{\Sigma_1^b\theta_j}{b^2}\right)^2\\ &+(v-1)\eta^2\left\{\sum_{j'=2}^{b}\left[\left(1-\frac{2}{b}\right)\theta_{j'}+\frac{\Sigma_1^b\theta_t}{b^2}\right]^2\right\}\\ &\left.-\eta(1-\eta)\left[\sum_{j'=2}^{b}\delta_{jj'}^2\ 2(v-1)\right]+\eta^2\sum\sum_{2\le j'\le j''\le b}2(v-1)\delta_{j'j''}^2\right]\end{aligned} \tag{30}$$

Step III A detailed simplification gives

$$V\left(\hat{\sigma}_{1G}^2\right) = \frac{2\sigma_1^4}{\left[\gamma+(v-1)\left(1-\frac{2}{b}\right)\right]^2}\left[\theta'\{d_1\mathbf{I}_{b-1}+d_2\mathbf{J}\}\theta+d_3\theta'\underline{1}+d_4\right] \tag{31}$$

where

$$
\begin{aligned}
d_1 &= \eta^2\left[\gamma+(v-1)\left(1-\frac{2}{b}\right)\right]-2(v-1)\eta/b^2\\
d_2 &= \frac{(v-1)}{b^2}\left(\eta+\frac{1}{b}\right)^2\\
d_3 &= \frac{2(v-1)}{b^2}\left(\eta-1+\frac{1}{b}\right)^2\\
d_4 &= \gamma(1-\eta)^2+(v-1)\left(1-\frac{1}{b}\right)^2\left(\eta-1+\frac{1}{b}\right)^2
\end{aligned}
\tag{32}
$$

Step IV Denoting

$$
\alpha=\gamma+(v-1)\left(1-\frac{2}{b}\right) \tag{33}
$$

we finally obtain the following expression:

$$
\begin{aligned}
V\left(\hat{\sigma}_{jG}^2\right)=\frac{2}{\alpha^2}&\left[d_1\sum_1^b\sigma_t^4+d_2\left(\sum_1^b\sigma_t^2\right)^2+(d_3-2d_2)\sigma_j^2\left(\sum_1^b\sigma_t^2\right)\right.\\
&\left.+(d_2+d_4-d_1-d_3)\sigma_j^4\right],\quad 1\le j\le b
\end{aligned}
\tag{34}
$$

We may now enunciate the following result:

Theorem 3 *Grubbs's estimates (of σ_j^2's) uniformly dominate the N estimates wrt total variability whenever $\sum_{j=1}^{b}\sigma_j^2>0$, for all $b\ge 2$, $v\ge 2$, $m\ge 2$.*

Proof. The expression in Eq. (34) gives

$$
\begin{aligned}
&\sum_{j=1}^{b}V\left(\hat{\sigma}_{jG}^2\right)\\
&\quad=\frac{2}{\alpha^2}\left\{\left[(b-1)d_1+d_2-d_3+d_4\right]\left(\sum_1^b\sigma_t^4\right)+\left[(b-2)d_2+d_3\right]\left(\sum_1^b\sigma_t^2\right)^2\right\}\\
&\quad\le\frac{2}{\alpha^2}\left\{\left[(b-1)d_1+d_2-d_3+d_4\right]+b\left[d_3+(b-2)d_2\right]\right\}\left(\sum_1^b\sigma_t^4\right)\\
&\quad=\frac{2}{\alpha^2}\left\{(b-1)\left[d_1+d_3+(b-1)d_2\right]+d_4\right\}\left(\sum_1^b\sigma_t^4\right)
\end{aligned}
\tag{35}
$$

Using Eq. (32) and simplifying, we obtain

$$(b-1)\left[d_1+d_3+(b-1)d_2\right]+d_4=\alpha\frac{\left[\gamma+(v-1)\left(1-\frac{1}{b}-\frac{1}{b^2}\right)\right]}{\left[\gamma+(v-1)\left(1-\frac{1}{b}\right)\right]} \tag{36}$$

Observe that

$$V\left(\hat{\sigma}_{jN}^2\right)=V\left(\frac{V_j}{\gamma}\right)=\frac{2\sigma_j^4}{\gamma},\quad 1\le j\le b \tag{37}$$

It is, therefore, enough to show that

$$\frac{2}{\alpha^2}\cdot\alpha\cdot\frac{\gamma+(v-1)\left(1-\frac{1}{b}-\frac{1}{b^2}\right)}{\gamma+(v-1)\left(1-\frac{1}{b}\right)}\le\frac{2}{\gamma}$$

$$\Leftrightarrow\quad 1-\frac{(v-1)/b^2}{\gamma+(v-1)\left(1-\frac{1}{b}\right)}\le 1+\frac{(v-1)\left(1-\frac{2}{b}\right)}{\gamma}$$

which is true for all $b \geq 2$, $v \geq 2$, $m \geq 2$. **QED**

The other justification is that even for moderate values of v and m, $P(\hat{\sigma}_{jG}^2 > 0, 1 \leq j \leq b)$ tends to be quite large whenever $\sigma_{max}^2/\sigma_{min}^2$ is moderate.

IMPROVED NONNEGATIVE VARIANCE ESTIMATORS

In this section we make a component-wise comparative study between Grubbs's estimators and suitably defined nonnegative estimators. For this, we relax the condition of unbiasedness and examine only square error loss. Note that for such a loss, the risk function is the mean square error (MSE). We consider the particular case of a $v \times 2$ setup with m observations per cell and focus our attention only on σ_1^2.

It is well known that, although $\frac{V_1}{\gamma}$ is an unbiased estimator for σ_1^2, the scaled estimator $\frac{V_1}{\gamma+2}$ is uniformly better and thereby renders $\frac{V_1}{\gamma}$ to be inadmissible. In the following, we compare the estimators $\frac{V_1}{\gamma+2}$ and $\hat{\sigma}_{1G}^2$ when $b = 2$.

Theorem 4 *The estimator* $\frac{V_1}{\gamma+2}$ *possesses a smaller MSE than* $\hat{\sigma}_{1G}^2$ *uniformly in* $\sigma_1^2 > 0$, $\sigma_2^2 > 0$ *iff* $v < 6$, *for* $m = 2$ *and* $v < 5$ *for* $m = 3$.

Proof. Recall Eq. (31). It follows that

$$V\left(\hat{\sigma}_{1G}^2\right) > \frac{2\sigma_1^4 d_4}{\left[\gamma + (v-1)\left(1-\frac{2}{b}\right)\right]^2} \tag{38}$$

because

$$d_1 + d_2 = \gamma\eta^2 + (v-1)\left[\eta\left(1-\frac{1}{b}\right) - \frac{1}{b^2}\right]^2 > 0, \quad d_3 > 0, \quad d_4 > 0 \tag{39}$$

Again,

$$MSE\left(\frac{V_1}{\gamma+2}\right) = \frac{2\sigma_1^4}{(\gamma+2)} \tag{40}$$

so that

$$MSE\left(\frac{V_1}{\gamma+2}\right) \Big/ V\left(\hat{\sigma}_{1G}^2\right) < 1$$

uniformly in $(\sigma_1^2, \sigma_2^2) > (0,0)$ iff

$$\frac{2}{(\gamma+2)} < \frac{2d_4}{\left[\gamma + (v-1)\left(1-\frac{2}{b}\right)\right]^2}$$

$\Leftrightarrow$

$$MSE\left[\gamma + (v-1)\left(1-\frac{2}{b}\right)\right]^2 < (\gamma+2)d_4 \tag{41}$$

Recalling the formula for d_4 in Eq. (32), we obtain when $b = 2$

$$d_4 = \gamma(1-\eta)^2 + \left(\frac{v-1}{4}\right)\left(\eta - \frac{1}{2}\right)^2$$

where from Eq. (29)

$$\eta = \frac{(v-1)/b^2}{\gamma + (v-1)\left(1-\frac{1}{b}\right)} = \frac{v-1}{4\gamma + 2(v-1)} \text{ with } b = 2.$$

Further, condition (41) reduces to

$$
\begin{aligned}
& \gamma^2 < (\gamma+2) d_4 \\
\Leftrightarrow\ & \gamma^2 < (\gamma+2)\left[\gamma(1-\eta)^2 + \frac{(v-1)}{4}\left(\eta - \frac{1}{2}\right)^2\right] \\
\Leftrightarrow\ & \gamma^2 < (\gamma+2)\left\{\gamma \cdot \left[\frac{4\gamma+(v-1)}{4\gamma+2(v-1)}\right]^2 + \frac{(v-1)\gamma^2}{[4\gamma+2(v-1)]^2}\right. \\
\Leftrightarrow\ & \gamma < (\gamma+2)\left\{\frac{[4\gamma+(v-1)]^2 + (v-1)\gamma}{[4\gamma+2(v-1)]^2}\right\} \qquad (42) \\
\Leftrightarrow\ & \gamma[4\gamma+2(v-1)]^2 < (\gamma+2)\left\{[4\gamma+(v-1)]^2 + \gamma(v-1)\right\} \\
\Leftrightarrow\ & \gamma(v-1)^2 + 2\gamma(v-1)[4\gamma+(v-1)] < \gamma^2(v-1) \\
& +2\left\{[4\gamma+(v-1)]^2 + \gamma(v-1)\right\} \\
\Leftrightarrow\ & 2(4\gamma+v-1)[(v-1)(\gamma-1)-4\gamma] < \gamma(v-1)(\gamma+3-v)
\end{aligned}
$$

Note that $\gamma = v\,(m-1)$. It is now readily verified that Eq. (42) fails to hold for $v \geq 7$, for any $m > 2$. Further, for $m \geq 3$, Eq. (42) fails to hold when $v \geq 6$. This proves the theorem. **QED**

Note 7. For $b > 2$, condition (41) is seen to hold only when $v \leq 3$, regardless of the value of $m \geq 2$. It is thus observed that the best-scaled estimator only rarely dominates over Grubbs's estimator. It would be instructive to compare total variability of Grubbs's estimators with the component-wise best-scaled estimators.

Note 8. It is evident that the N estimators cannot dominate Grubbs's estimators component-wise for any value of v.

CONCLUSIONS

In the present article, we have highlighted the importance of Grubbs's estimators by analyzing their properties. In the process, we have been able to show their dominance (in terms of total variability) over the available N estimators. When component-wise variability (MSE) is considered, the performance of Grubbs's estimators is also quite good compared with the N estimators because only rarely can each component of the former individually be dominated by the best-scaled N estimator. The inadmissibility of Grubbs's estimator, of course, can be shown by developing some nonnegative nonquadratic estimators as in Stein [8].

ACKNOWLEDGMENTS

Research of Kalyan Das is partly supported by UGC Grant # MTH 90687. The work was completed during sabbatical leave of Bimal K. Sinha at the Indian Statistical Institute.

REFERENCES

1 F. E. Grubbs, J. Am. Stat. Assoc., 43 (1948) 243–264.
2 D. A. Brindley and R. A. Bradley, J. Am. Stat. Assoc., 80 (1985) 711–714.
3 C. J. Maloney, Grubbs' estimators to date, in: Proc. 18th Conf. on Design of Experiments in Army Research Development and Testing, Part 2, 1973, pp. 523–547.
4 J. H. Ellenberg, J. Am. Stat. Assoc., 72 (1977) 407–411.
5 R. A. Bradley and D. A. Brindley, Commun. Stat.-Simul., 15 (1986) 27–34.
6 R. A. Bradley and D. A. Rollier, The extension of Grubbs estimators for classes of incomplete block designs. in: Probability and Statistics, Essays in Honor of F. A. Graybill, J. N. Srivastava (ed.), Elsevier Scientific Publishers (North-Holland), Amsterdam, Netherlands, 1988.
7 T. W. Anderson, An Introduction to Multivariate Statistical Analysis, John Wiley and Sons, Inc., New York, 1984.
8 C. Stein, Ann., Inst. Stat. Math., 16 (1964) 155–160.

APPENDIX

Reduction to Canonical Form:

From our model in Eq. (1), we define, for each i = 1, . . . , v, j = 1, . . . , b, the residuals

$$R_{ij} = \sqrt{m}\left(\bar{y}_{ij.} - \bar{y}_{i..} - \bar{y}_{.j.} + \bar{y}_{...}\right)$$

and

$$V_j = \sum_{i=1}^{v}\sum_{k=1}^{m}\left(y_{ijk} - \bar{y}_{ij.}\right)^2$$

where

$$\bar{y}_{ij.} = \frac{1}{m}\sum_{k=1}^{m} y_{ijk}$$

$$\bar{y}_{i..} = \frac{1}{bm}\sum_{j=1}^{b}\sum_{k=1}^{m} y_{ijk}$$

$$\bar{y}_{.j.} = \frac{1}{vm}\sum_{i=1}^{v}\sum_{k=1}^{m} y_{ijk}$$

$$\bar{y}_{...} = \frac{1}{bvm}\sum_{j=1}^{b}\sum_{i=1}^{v}\sum_{k=1}^{m} y_{ijk}$$

Clearly, each R_{ij} is orthogonal to V_j, and the V_j's are all independently distributed as σ_j^2, χ_γ^2, $\gamma = v\,(m-1)$, $j = 1, \ldots, b$.

Let us first find out the joint distribution of R_{ij}'s. This seems important because the sufficiency can then be used to reduce the dimension of R_{ij}'s and also to examine what functions of the R_{ij}'s emerge as vital statistics for estimation of variance components.

In view of the assumption made in model Eq. (8), we have, for fixed j,

$$\mathbf{R}_{.j} = \left(R_{1j}, \ldots, \mathbf{R}_{vj}\right)' \sim N_v\left(0, \Sigma_{\mathbf{jj}}\right) \tag{A.1}$$

where

$$\begin{aligned} \Sigma_{iijj} &= \operatorname{var}\left(R_{ij}\right) = \left(1-\frac{1}{v}\right)\left(1-\frac{2}{b}\right)\sigma_j^2 + \left(1-\frac{1}{v}\right)\frac{1}{b^2}\sum_{l=1}^{b}\sigma_l^2 \\ &\qquad \text{(same for all } i\text{)} \\ \Sigma_{ii'jj} &= \operatorname{cov}\left(R_{ij}, R_{i'j}\right) = -\frac{1}{v}\left(1-\frac{2}{b}\right)\sigma_j^2 - \frac{1}{vb^2}\sum_{l=1}^{b}\sigma_l^2 \\ &\qquad \text{(same for all } i \neq i'\text{)} \end{aligned} \tag{A.2}$$

To obtain the joint distribution of all the R_{ij}'s, we compute

$$\Sigma_{jj'} = \operatorname{cov}\left(\mathbf{R}_{.j}, \mathbf{R}_{.j'}\right) = \left[\left(\Sigma_{ii'jj'}\right)\right], \quad j \neq j'$$

where

$$\Sigma_{iijj'} = -\left(1-\frac{1}{v}\right)\frac{1}{b}\left(\sigma_j^2 + \sigma_{j'}^2 - \frac{1}{b}\sum_{l=1}^{b}\sigma_l^2\right) \quad \text{(same for all } i\text{)}$$
$$\Sigma_{ii'jj'} = \frac{1}{vb}\left(\sigma_j^2 + \sigma_{j'}^2 - \frac{1}{b}\sum_{l=1}^{b}\sigma_l^2\right) \quad \text{(same for all } i \neq i'\text{)} \tag{A.3}$$

We then have

$$\begin{pmatrix} \mathbf{R}_{.1} \\ \vdots \\ \mathbf{R}_{.b} \end{pmatrix} \sim N_{vb}[0, \Sigma] \tag{A.4}$$

where

$$\Sigma = \left[\left(\Sigma_{jj'}\right)\right], \quad j, j' = 1, 2, \ldots, b$$

Note that each $\Sigma_{jj'}$ has its own intraclass structure. Further, each Σ_{jj} is a $(v \times v)$ positive semidefinite matrix of rank $v-1$ with all the $v-1$ positive eigenvalues the same (e. g., δ_{jj}), and each $\Sigma_{jj'}$ is a $(v \times v)$ singular matrix with the same nonzero eigenvalue (e. g., $\delta_{jj'}$) with multiplicity $(v-1)$. We obtain

$$\delta_{jj} = \left(1-\frac{2}{b}\right)\sigma_j^2 + \frac{1}{b^2}\sum_{l=1}^{b}\sigma_l^2$$
$$\delta_{jj'} = -\frac{1}{b}\left(\sigma_j^2 + \sigma_{j'}^2 - \frac{1}{b}\sum_{l=1}^{b}\sigma_l^2\right) \tag{A.5}$$

Obviously, it is possible to find a single orthogonal matrix $\Gamma\,(v \times v)$, such that its first row is proportional to $\mathbf{1}\ [= (1, 1, \ldots, 1)']$. From Eq. (A.4), we then have the transformed vector

$$\mathbf{Z} = \left(I_b \otimes \Gamma\right)\begin{pmatrix} \mathbf{R}_{.1} \\ \vdots \\ \mathbf{R}_{.b} \end{pmatrix} \sim N_{vb}\left(0, \Sigma^*\right) \tag{A.6}$$

where

$$\Sigma^* = \left[\left(\Sigma^*_{jj'}\right)\right], \quad \Sigma^*_{jj'} = \begin{pmatrix} 0 & 0 \\ 0 & \delta_{jj'}\mathbf{I}_{v-1} \end{pmatrix}, \quad j, j' = 1, \ldots, b$$

Writing

$$\mathbf{Z} = \left(\mathbf{Z}'_1, \ldots, \mathbf{Z}'_b\right), \quad \mathbf{Z}_j = \left(Z_{0j}, \mathbf{Z}^*_{.j}\right)', \quad \mathbf{Z}^*_{.j} = \left(Z_{1j}, \ldots, Z_{v-1,j}\right)'$$

and noting that $Z_{0j} = 0$ (with probability 1) for all j, we get

$$\begin{pmatrix} \mathbf{Z}^*_{.1} \\ \vdots \\ \mathbf{Z}^*_{.b} \end{pmatrix} \sim N_{b(v-1)}\left(\mathbf{0}, \mathbf{I}_{v-1} \otimes \Lambda\right), \quad \Lambda = \left[\left(\delta_{jj'}\right)\right] \quad j, j' = 1, \ldots, b \tag{A.7}$$

We thus obtain (v – 1) independent groups of random b-dimensional vectors

$$\mathbf{Z}_{i.} = \left(Z_{i1}, \ldots, Z_{ib}\right)' \sim N_b(0, \Lambda), \quad i = 1, \ldots, v-1 \tag{A.8}$$

It follows that

$$\underline{\underline{\mathbf{W}}} = \sum_{i=1}^{v-1} \mathbf{Z}_{i.}\mathbf{Z}'_{i.} \sim \text{Singular } W_b(v-1, \Lambda) \tag{A.9}$$

Finally, the canonical reduction results in (V_1, V_2, . . . , V_b, $\underline{\underline{W}}$) as jointly sufficient statistics for the variance components $\sigma_1^2, \sigma_2^2, \ldots, \sigma_b^2$.

In the particular case when b = 2, $\Lambda = [(\delta_{jj'})]$ with elements defined in Eq. (A.5) reduces to

$$\Lambda = \frac{\sigma_1^2 + \sigma_2^2}{4}\begin{pmatrix} 1 & -1 \\ -1 & 1 \end{pmatrix}, \quad \text{rank}(\Lambda) = 1 \tag{A.10}$$

Thus, for each i = 1, 2, . . . , v – 1, we have from Eq. (A.8),

$$\left.\begin{aligned} &Z_{i1} + Z_{i2} = 0 \quad \text{with probability 1} \\ &Z_{i1} - Z_{i2} \sim N\left(0, \sigma_1^2 + \sigma_2^2\right) \end{aligned}\right\} \tag{A.11}$$

Clearly, $Z_{i1} - Z_{i2} = 2Z_{i1} = -2Z_{i2}$, $\forall$ i.
Therefore,

$$\begin{aligned} V_3 &= \sum_{i=1}^{v-1} \left(Z_{i1} - Z_{i2}\right)^2 \\ &= m\sum_{i=1}^{v}\left[\left(\bar{y}_{i1.} - \bar{y}_{.1.}\right) - \left(\bar{y}_{i2.} - \bar{y}_{.2.}\right)\right]^2 \\ &= 4\mathbf{R}'_{.1}\mathbf{R}_{.1} = 4\mathbf{R}'_{.2}\mathbf{R}_{.2} \end{aligned} \tag{A.12}$$

where $\mathbf{R}_{.j}$'s have been defined in Eq. (A.1). Note that, in view of Eq. (A.11),

$$V_3 \sim \left(\sigma_1^2 + \sigma_2^2\right)\chi_{v-1}^2$$

so that, for each j = 1, 2,

$$W_{jj} = \mathbf{R}'_{.j}\mathbf{R}_{.j} \sim \frac{\sigma_1^2 + \sigma_2^2}{4}\chi_{v-1}^2 \tag{A.13}$$

Clearly,

$$V_3 = 4W_{11} = 4W_{22} \tag{A.14}$$

Quality through Engineering Design - W. Kuo (Editor)

Performance of β Adjustment Factor in Industrial Designs

Ratan Dasgupta

Statistics and Mathematics Division, Indian Statistical Institute, 203 Barrackpore Trunk Road, Calcutta, 700 035 India

Abstract

The method of β correction for testing, estimating, and analyzing industrial experiments is assessed. Its relative merit is judged with Taguchi's method. The theoretical results are supplemented by computer simulation for a relatively small sample size.

INTRODUCTION

The β adjustment factor has been successfully used to analyze the results of various process and control designs in industrial experiments. To get an idea of the technique, consider a production process where μ_o is the target value of some characteristic. Let μ be the (unknown) actual value of the characteristic and $\hat{\mu}$ be an unbiased estimate of μ, i.e., $E\hat{\mu} = \mu$. Also suppose that $\text{var}(\hat{\mu}) = m_2$ is known.

The aim is to select a correction factor β such that the amount of correction $-\beta(\hat{\mu} - \mu_o)$ when incorporated in the process parameter μ gives the corrected value $\mu - \beta(\hat{\mu} - \mu_o)$ with the smallest L_2 distance, on average, from the target value μ_o. That is, we want to select β such that

$$E[\mu - \beta(\hat{\mu} - \mu_o) - \mu_o]^2 \text{ is a minimum, i.e.,} \tag{1}$$

$$E[-\beta(\hat{\mu} - \mu) + (1-\beta)(\mu - \mu_o)]^2 = \beta^2 m_2 + (1-\beta)^2(\mu - \mu_o)^2 \tag{2}$$

is a minimum.

Here we consider a particular time t where $\mu = \mu_t$ is fixed. The treatment of the general case with μ_t on a time interval will be considered in detail elsewhere.

This technique of Dr. Taguchi has generally been used for adjusting the production process instead of traditional 3σ limits and for analyzing results of industrial experiments. See Taguchi [1] for an example where an orthogonal array L_{16} has been analyzed with two levels of 10 factors.

In this paper, we assess the performance of the technique and propose a modified procedure, taking into account the cost of implementation. The relative merits of both methods are judged for a large sample and then supplemented by small sample simulations. These simulations suggest that the proposed procedures are better. The results are presented in the next section followed by a comparison of the techniques. Next, a few modifications and extensions are indicated. Finally the results are summarized in the discussion section. The graphs indicate that the proposed procedures are superior to Taguchi's existing procedure when the mean square error is the criterion and the population is normal. The graphs are explained in detail in the discussion section.

RESULTS

To obtain the minimum of Eq. (1), differentiate Eq. (2) w.r.t. β and equate it to zero to get

$$\beta = 1 - m_2 / \left[m_2 + (\mu - \mu_o)^2\right] \quad \in (0,1) \tag{3}$$

The minimum value of Eq. (1) is $\beta m_2 = m_2(\mu - \mu_o)^2/[m_2 + (\mu - \mu_o)^2]$ and the second derivative is positive.

Observe also that $\beta m_2 < m_2$ when $\beta < 1$ and

$$\beta m_2 < (\mu - \mu_o)^2 \tag{4}$$

the L_2 distance from the target μ_o if no correction is made.

From (3) and (4),

$$\beta m_2 < min\left[m_2, (\mu - \mu_o)^2\right] \tag{5}$$

The above inequality may not be satisfied if β is replaced by its estimate $\hat{\beta}$.

If $\hat{\mu}$ is the maximum likelihood estimate (MLE) of μ, then the MLE of β is the following:

$$\hat{\beta} = 1 - m_2 \Big/ \left[m_2 + (\hat{\mu} - \mu_o)^2\right] \quad \in (0,1) \tag{6}$$

In the existing procedure, i.e., for Taguchi's procedure,

$$\begin{aligned} \hat{\beta} &= 1 - m_2 / (\hat{\mu} - \mu_o)^2, \quad if\ (\hat{\mu} - \mu_o)^2 > m_2 \\ &= 0, \quad if\ (\hat{\mu} - \mu_o)^2 \le m_2 \end{aligned} \tag{7}$$

There, i.e., for Taguchi's procedure, $1/[m_2 + (\mu - \mu_o)^2]$ is approximated by $1/(\hat{\mu} - \mu_o)^2$ as $E(\hat{\mu} - \mu_o)^2 = m_2 + (\mu - \mu_o)^2$.

But $E(\hat{\mu}-\mu_o)^{-2} \neq [m_2+(\mu-\mu_o)^2]^{-1}$. Later this is shown to be a poor approximation.

The estimated gain in the mean square error (MSE) when the correction is incorporated [see Eq. (4)] is

$$(\hat{\mu}-\mu_o)^2-\hat{\beta}m_2=(\hat{\mu}-\mu_o)^4\Big/\left[m_2+(\hat{\mu}-\mu_o)^2\right]\simeq(\hat{\mu}-\mu_o)^2 \tag{8}$$

if m_2 is relatively small.

For example, if $\hat{\mu}$ is based on n observations, then in general $m_2 = \mathrm{Var}(\hat{\mu}) = 0(n^{-1})$.

There is a general cost, for example, machine-hour cost, in terms of equivalent unit in which MSE is measured, for stopping the production to implement the correction, especially when the process is not really out of control. Let this amount be C.

Hence estimated net gain for relatively small m_2 is

$$(\hat{\mu}-\mu_o)^4\left[m_2+(\hat{\mu}-\mu_o)^2\right]^{-1}-C\simeq(\hat{\mu}-\mu_o)^2-C \tag{9}$$

The correction in Eq. (6) should be incorporated if Eq. (9) is nonnegative, i.e.

$$\begin{aligned}\hat{\beta}&=1-m_2\Big/\left[(\hat{\mu}-\mu_o)^2+m_2\right], && \textit{if } (\hat{\mu}-\mu_o)^2\geq C\\ &=0, && \textit{otherwise}\end{aligned} \tag{10}$$

The existing method does not take the cost factor into consideration. Equation (10) may be compared with existing method with C = 0.

COMPARISON OF PROPOSED PROCEDURE WITH EXISTING PROCEDURE

For simplicity let $\mu_o = 0$ and suppose that $\hat{\mu}$ is based on n observations. Take $\mathrm{Var}(\hat{\mu}) = 1/n$. Then $1-\beta = (1/n)/(u^2+1/n) = f(\mu)$. The existing procedure approximates this value by $g(\hat{\mu}) = (n^{-1}\hat{\mu}^{-2} \vee 1)$, whereas the proposed procedure estimates it by $f(\hat{\mu}) = (1/n)/(\hat{\mu}^2+1/n)$. We shall show that $f(\hat{\mu})$ is a better estimator for $f(\mu)$ than is $g(\hat{\mu})$. Here $x \vee y = \max(x,y)$.

By Taylor's expansion,

$$f(\hat{\mu})-f(\mu)=(\hat{\mu}-\mu)\,f'(\mu)\,[1+o(1)] \tag{11}$$

Thus the MSE of

$$f(\hat{\mu})=[f'(\mu)]^2\,[1+o(1)]/n \tag{12}$$

Similarly the MSE of $g(\hat{\mu})$, as an estimate of $f(\mu)$, is the following:

$$\begin{aligned} g(\hat{\mu}) &= [g'(\mu)]^2[1+o(1)]/n+[g(\mu)-f(\mu)]^2, \mu^2 \neq 1/n \\ &= 4[1+o(1)]/[n+(1/4)]; \ \mu^2 = 1/n \end{aligned} \tag{13}$$

Since the last terms of Eq. (13) are $0(1)$, $g(\hat{\mu})$ is an "inferior" estimator of $f(\mu)$ compared with $f(\hat{\mu})$; in fact, $g(\hat{\mu})$ is not even consistent for $f(\mu)$ for all values of μ, e.g., $\mu^2 = 1/n$.

A similar picture is revealed by computer simulation even for small samples, n = 1, 5, 10. For 10^5 simulations from $N(\mu,1)$, $\mu \in [0,3]$ for n = 5,10, $\mu \in [0,6]$ for n = 1, the estimated MSE in Eq. (1) is plotted for both procedures with $\mu_o = 0$. Except for a small neighborhood of zero, which shrinks as n increases, the proposed procedure is better than the existing procedure. Even after truncation at the same point as that of the existing procedure (i.e., taking $C = m_2 = \mathrm{Var}(\hat{\mu})$ so that two procedures become more comparable), the situation remains the same. See Figs. 4–6, 10–12. Since the curves always remain below the graph $y = \mu^2$ except near the origin, this suggests that correction is almost always preferred. In a particular zone of μ, curves cross the line $y = 1/n$, the variance of $\overline{X}$. This is because $\overline{X}$ is an admissible estimator of μ. Observe that when $\beta = 1$, i.e., we correct the full amount, then we get $\overline{X}$ as the amount to be corrected. Near the end of the curves for large values of m, both the curves are near the line $y = 1/n$ from above, indicating that $\beta \doteq 1$; that is, almost full correction is necessary.

These procedures can be examined from a Bayesian point of view as well. Assume $X \sim N(\mu,1)$ and consider a prior distribution, $\mu \sim N(0,\tau^2)$. Then $EX = EE(X|\mu) = E\mu = 0$, $\mathrm{Cov}(\mu,X) = E(\mu X) = E\mu^2 = V(\mu) = \tau^2, V(x) = EX^2 = EE(X^2|\mu) = E(\mu^2+1) = 1+\tau^2$. Thus, $E(\mu|X) = X\, Cov(\mu, X)/V(X) = X\tau^2/(1+\tau^2)$.

An empirical Bayes estimate of μ is then obtained by maximizing the marginal likelihood, i.e., by taking $X^2 = 1 + \tau^2$. Then the empirical Bayes estimator of μ is

$$\begin{aligned} \hat{\mu} &= X(X^2-1)/X^2, && \text{if } X^2 > 1 \\ &= 0, && \text{otherwise} \end{aligned}$$

This refers to Taguchi's estimator.

However, this is not an admissible estimator of μ because it is not analytic in X. If one takes the estimate of τ^2 from the distribution of μ, then $\hat{\tau}^2 = \mu^2$; and if μ^2 is approximated by X^2 from the conditional distribution of X, then $E(\mu|X) = X\,X^2/(1+X^2)$ is the proposed estimator $\hat{\beta}X$ [see Eq. (6) with $m_2 = 1$, $\hat{\mu} = X$. Although this is analytic in X, it is not clear whether this estimator is admissible. Also note that for any fixed $\beta \in (0,1)$, βX is an admissible estimator for μ. Hence any fixed correction is admissible. The point is that

β has to be selected in a logical way. Both the procedures discussed here have their own merits.

Some further modifications and extensions

Apart from linear correction, one may introduce a quadratic correction term in squared error loss; i.e., one can select β and γ in such a way that

$$E\left[\mu - \beta(\hat{\mu} - \mu_o) - \gamma(\hat{\mu} - \mu_o)^2 - \mu_o\right]^2 \tag{14}$$

is a minimum. Straightforward computation involving differentiation with respect to β and γ gives

$$\begin{aligned} \beta &= \left[(\mu - \mu_o)^2 - \gamma\, m_3'\right] / m_2' \\ \gamma &= (\mu - \mu_o)\left[(\mu - \mu_o)m_3' - m_2'^2\right] / \left(m_3'^2 - m_4'\, m_2'\right) \end{aligned} \tag{15}$$

where $m_r' = E\,\hat{\mu}^r$. Estimates of β and γ are obtained by replacing μ by $\hat{\mu}$ in Eq. (15). Here one must pay close attention because the quadratic effect may be unnecessarily magnified if wrongly estimated. For a normal distribution, a precise estimate of m_2 suffices. Then one can proceed to minimize Eq. (14) with $\mu = 0$. Thus

$$\beta = (\mu - \mu_o)^2 / m_2, \quad \gamma = m_2(\mu - \mu_o)/m_4 = (\mu - \mu_o)/3m_2 \tag{16}$$

Therefore,

$$\hat{\beta} = (\hat{\mu} - \mu_o)^2 / m_2, \quad \hat{\gamma} = (\hat{\mu} - \mu_o)/3m_2 \tag{17}$$

The following two-step procedure may be a simple alternative to the above selection of β and γ. Denote $Y = (\mu - \mu_o)$, $x = (\hat{\mu} - \mu_o)$. After selecting $\hat{\beta}$ [$= \beta_o$, for example as in Eq. (6)], one may be interested to select γ in such a way that

$$E\left[Y - \beta_o\, x - \gamma\, x^2\right]^2 \tag{18}$$

is a minimum with respect to γ. Differentiating w.r.t. γ, one gets

$$Ex^2\left[Y - \beta_o x = \gamma x^2\right] = 0 \quad i.e., \quad \gamma = \left(Ym_2' - \beta_o m_3'\right) / m_4' \tag{19}$$

Here $\beta_o = \hat{\beta}$ of Eq. (6). Therefore,

$$\begin{aligned} \hat{\gamma} &= \left[(\mu - \mu_o)m_2' - \beta_o m_3'\right] / m_4' \\ &= (\mu - \mu_o)/3m_2, \quad \text{for normal distribution} \end{aligned} \tag{20}$$

To see the gain in MSE after introducing the quadratic term, rewrite Eq. (18) as

$$E(Y-\beta_o x)^2 - 2Y\gamma m_2' + 2\beta_o \gamma m_3' + \gamma^2 m_4' \tag{21}$$

should be a minimum w.r.t. γ. Differentiation of Eq. (21) w.r.t. γ leads to

$$\begin{aligned} -2Ym_2' + 2\beta_o m_3' + 2\gamma m_4' &= 0 \\ i.e., \quad -2Y\gamma m_2' + 2\beta_o \gamma m_3' + 2\gamma^2 m_4' &= 0 \end{aligned} \tag{22}$$

Hence, from Eq. (22) the minimum value of Eq. (21) is the following

$$E(Y-\beta_o x)^2 - \gamma^2 m_4' \tag{23}$$

Therefore, the gain in MSE is $\gamma^2 m_4'$.

Estimated gain in MSE due to the introduction of a quadratic term is $\hat{\gamma}^2 m_4' = (\hat{\mu}-\mu_o)^2/3$ for normal distribution. In general one needs to estimate $m_r' - E\,\hat{\mu}^r$ from past data. The correction should be incorporated in the machine only when $(\hat{\mu}-\mu_o)^2 \geq C$, C being the cost of implementation.

One may also consider the absolute value of the loss, instead of the squared error loss. Then the correction factor β has to be selected such that

$$E\left|\mu - \beta(\hat{\mu}-\mu_o) - \mu_o\right| \tag{24}$$

is a minimum w.r.t. β. After some straightforward computation to minimize the above, we get

$$\beta = \left[1 + \{med(\hat{\mu}) - \mu\}/(\mu-\mu_o)\right]^{-1} \tag{25}$$

Note that β = 1, if $med(\hat{\mu}) = \mu$. From Eq. (25) we obtain the following estimate of β.

$$\hat{\beta} = \left[1 + \hat{\mu}^*/(\hat{\mu}-\mu_o)\right]^{-1} \tag{26}$$

where $\hat{\mu}^*$ is an estimate of $med(\hat{\mu}) - \mu$.

DISCUSSION

The role of the β adjustment factor in the analysis of industrial design, estimation, and testing of experimental data and process control has been theoretically assessed for large samples. This study has been supplemented by computer simulation for small samples: n = 1, 5, 10 from the $N(\mu,1)$ population,

$\mu \geq 0$. The number of simulations is taken as 10^5 in order to get a precise estimate of the resultant MSE. The existing β technique is compared with some proposed procedures that arise naturally as a consequence of estimating a parametric function and considering cost. We observe that except for the small neighborhood of origin, the proposed procedure is uniformly better than the former one. Where the β technique dominates the proposed technique, the neighborhood shrinks as n increases from 1 to 10. This observation supplements the theoretical result that for large n the proposed technique is better than the former β correction method, see Eqs. (12–13). For large value of μ and in all the values of n, the MSE of both the procedures is nearly the MSE of $\overline{x}$, indicating that for large μ almost full correction $\beta \simeq 1$ is necessary. Simulation results also indicate that if the cost of implementation is nil and if one can correct the process continuously, then a small correction is almost always better. (For example, it is now possible to vary continuously the pinion at the final head of drawing in a spinning mill much like a continuous speed regulator in a ceiling fan.) Note that although the correction proposed, e.g., for n = 1, $X^2/(1 + X^2)$ is larger in magnitude than Taguchi's correction $(X^2 - 1)/X^2$ when the latter is nonzero for $X^2 > 1$, the former is quite small for X in the small neighborhood of origin.

With due caution towards the precision of the estimate γ, one can introduce a quadratic term in the correction. In the case of normal distribution, the estimate uses the value of $m_2 = \mathrm{Var}(\hat{\mu})$, obtained from past observations. This value of m_2 is also required for estimating β in linear terms. The gain due to such introduction of new terms in MSE can also be estimated. For normal distribution this takes a simple form, $(\hat{\mu} - \mu_o)^2 / 3$. This consideration of the cost of implementing C can be taken care of by requiring that both β and γ are zeros if $(\hat{\mu} - \mu_o)^2 < C$.

For distributions with a heavy tail, the absolute value of the loss function may be more appropriate than the squared-error loss function. In such a situation, β is expressible in terms of the median of $\hat{\mu}$ [see Eq. (26)].

In Figs. 1–12, μ_o is taken to be zero. Equation (1) then reduces to $E(\mu - \beta\,\hat{\mu})^2$, a minimum w.r.t. β, indicating that this is an estimation problem. Observe also that for two estimates $\hat{\beta}_1$ and $\hat{\beta}_2$ for optimal $\beta = \beta_o(\mu)$, the one with the higher MSE will cause a higher value of Eq. (1) because $E(\mu - \beta\hat{\mu})^2$ is increasing in $|\beta - \beta_o|$. Simulation results then supplement the large sample findings that Taguchi's $\hat{\beta}$ of Eq. (7) has a higher MSE than the proposed $\hat{\beta}$ of Eqs. (6) or (10) with $C = m_2$ for almost the entire parametric space of μ. We use the abbreviation " β prop(un) truncated " for (un) truncated β in the proposed method in Eqs. (6) or (10) with $C = m_2$.

Figures 1–3 (for n = 1, 5, and 10) compare to the MSE in Eq. (1) for the proposed method with $\hat{\beta}$ in Eq. (6) with the MSE of Taguchi's method with $\hat{\beta}$ in Eq. 7 for different values of μ. These are obtained by 10^5 simulations from the $N(\mu,1)$ population.

Figures 4–6 (for n = 1, 5, and 10) refer to the same comparison of proposed method with Taguchi's method, but this time Eq. (6) is replaced by a truncated version [Eq. (10)] with the same truncation point $C = m_2$ as that of Taguchi's method for $(\hat{\mu} - \mu_o)^2$. Here $m_2 = 1/n$, n = 1, 5, 10. These figures clearly show that both Eqs. (6) and (10) are superior to Taguchi's technique in Eq. (7) over a large region of the parametric space. Furthermore, the untruncated proposed method in Eq. (6) is superior to the truncated proposed method in Eq. (10). The values of μ in a small neighborhood of zero where Taguchi's method performs slightly better shrink as n increases from 1 to 10. In the figures, the dotted line refers to the MSE of proposed method whereas the broken line with the higher peak refers to the MSE of Taguchi's method for different values of μ.

The counterparts of Figs. 1–6 are Figs. 7–12; here the MSE (2) is computed by pretending that $\hat{\beta}$ is constant, although $\hat{\beta}$ is a function of X_i, $i = 1, \ldots, n$. Observe that the main features of Figs. 7–12 remain unchanged although their values differ at all the points from Figs. 1–6.

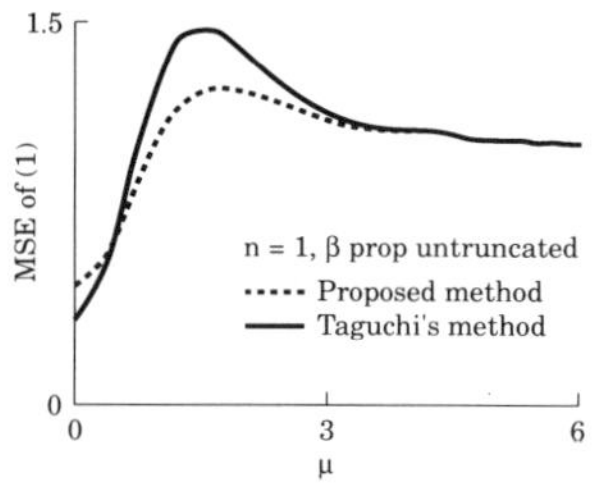

Fig. 1. MSE (1) of procedure (6) and Taguchi's procedure (7) with one observation.

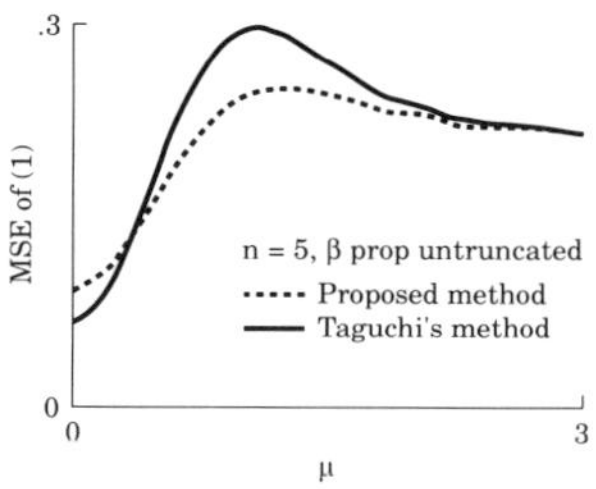

Fig. 2. MSE (1) of procedure (6) and Taguchi's procedure (7) with five observations.

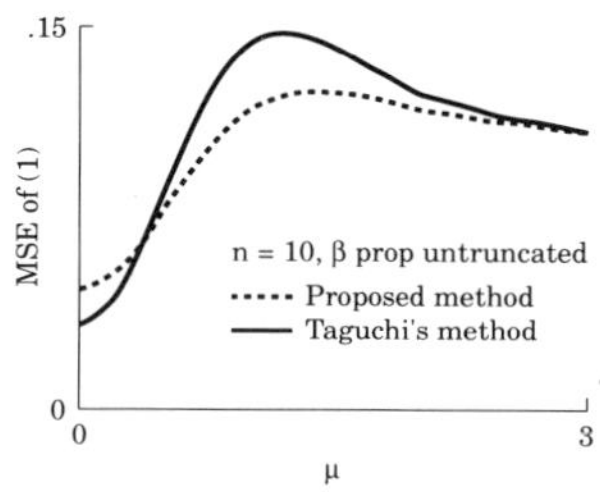

Fig. 3. MSE (1) of procedure (6) and Taguchi's procedure (7) with ten observations.

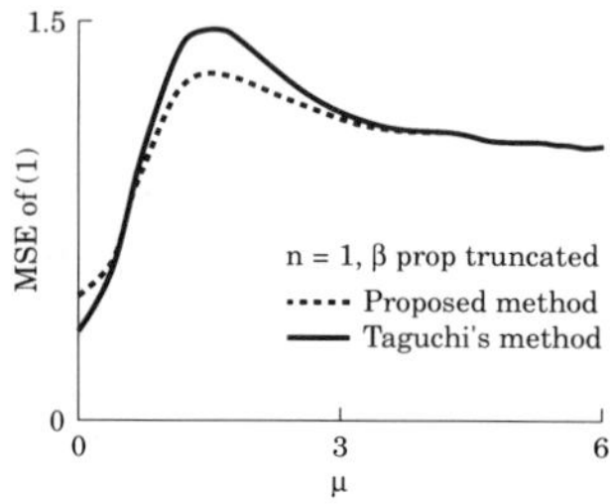

Fig. 4. MSE (1) of procedure (10) and Taguchi's procedure (7) with one observation.

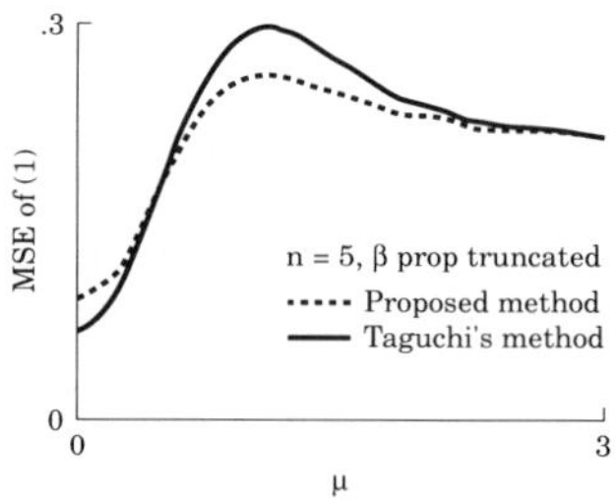

Fig. 5. MSE (1) of procedure (10) and Taguchi's procedure (7) with five observations.

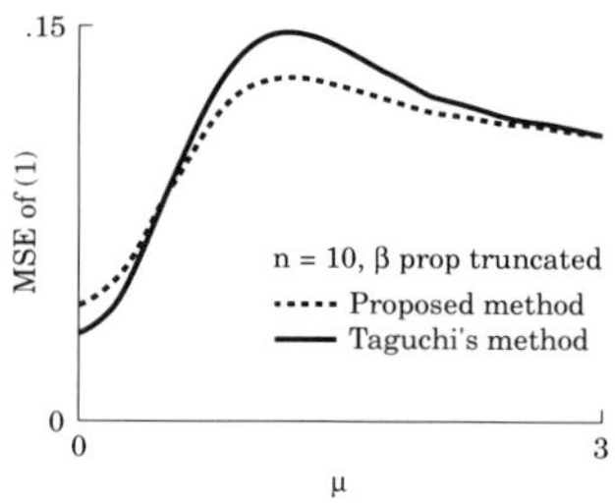

Fig. 6. MSE (1) of procedure (10) and Taguchi's procedure (7) with five observations.

Figures 7–12 also indicate that one may not expect that the MSE should be $\leq m_2$, $m_2 = 1/n$, $n = 1, 5, 10$, even if one pretends that $\hat{\beta}$ is a constant as in Eq. (2).

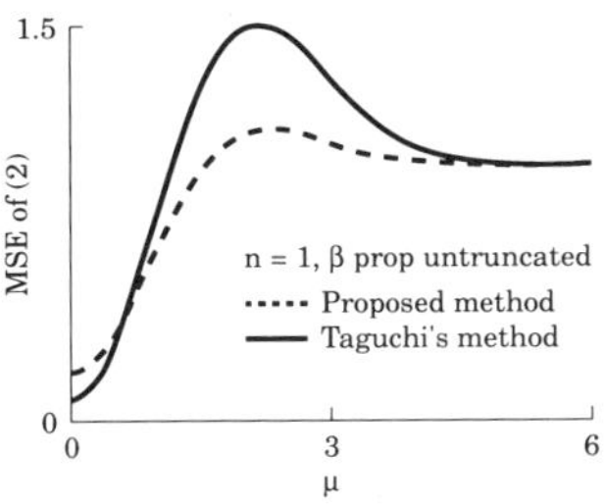

Fig. 7. MSE (2) of procedure (6) and Taguchi's procedure (7) with one observation.

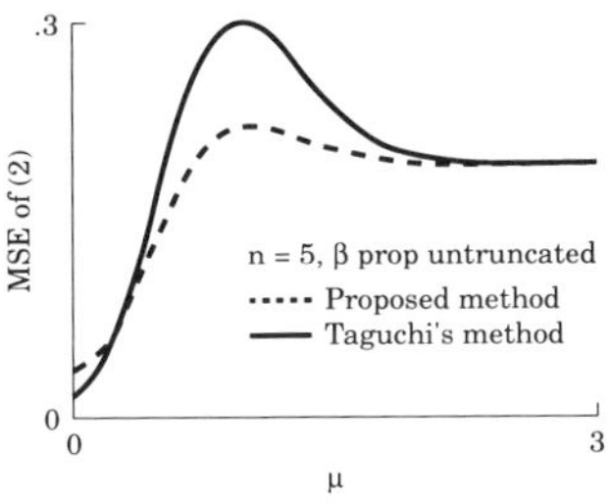

Fig. 8. MSE (2) of procedure (6) and Taguchi's procedure (7) with five observations

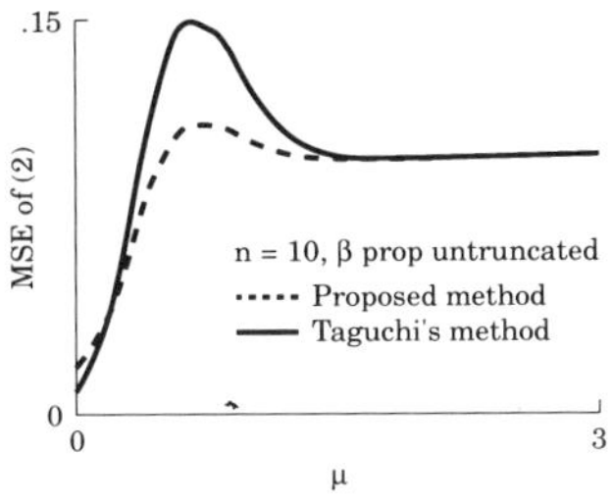

Fig. 9. MSE (2) of procedure (6) and Taguchi's procedure (7) with ten observations.

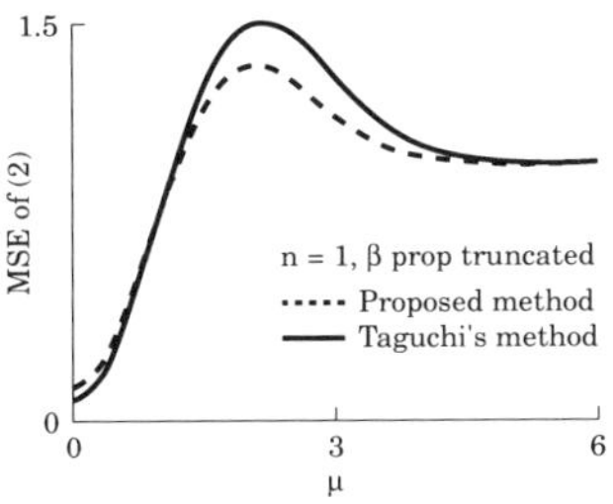

Fig. 10. MSE (2) of procedure (10) and Taguchi's procedure (7) with one observation.

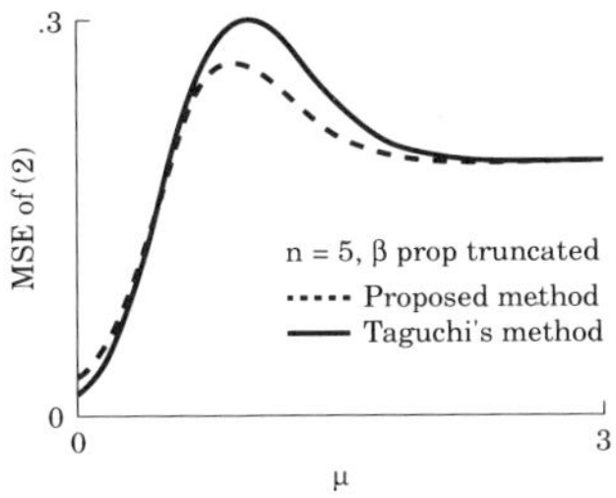

Fig. 11. MSE (2) of procedure (10) and Taguchi's procedure (7) with five observations.

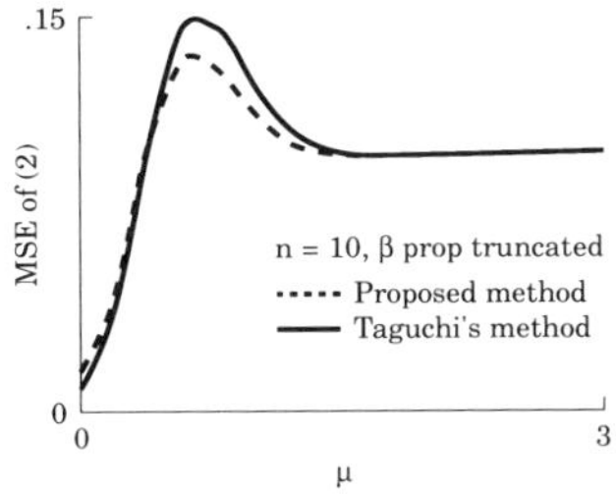

Fig. 12. MSE (2) of procedure (10) and Taguchi's procedure (7) with ten observations.

Both the proposed procedure and that of Taguchi are conservative shrinkers for $\mu = \mu_o$ (= 0 here). As a result, at zero the MSEs of these procedures are drastically lower than the MSE of the usual estimator $\overline{X}$ (corresponding to $\beta = 1$ with full correction). For the procedures under investigation, $\beta = \beta(\overline{X}) < 1$, and these behave like shrinkage-type estimators. The gain near zero is, in some sense, balanced by a higher MSE than that of $\overline{X}$ for μ a little away from zero. Observe that the peaks of the MSE for Taguchi's procedure are almost 50% more than $1/n = m_2 = \text{Var}(\overline{X}) = \text{MSE}$ [Eq. (1) when β equals one. From Eq. (2) and for a fixed β, $0 < \beta < 1$, the MSE $\to \infty$ as $\mu \to \infty$, although $\beta\overline{X}$ is an admissible estimator for μ. The procedures discussed here are superior to above correction $\beta\overline{X}$, with $\beta \in (0,1)$ fixed, because both the MSEs of proposed procedure and that of Taguchi converge to m_2 as $\mu \to \infty$.

The computer simulations are done on a VAX 8650 system using IMSL subroutine *rnset, rnnof* with 10^5 simulation from $N(\mu,1)$ for each fixed value of μ.

ACKNOWLEDGMENTS

The investigation started from a query made by Mr. N. T. V. Ranga Rao of Hyderabad ISI to the author. Professor J. K. Ghosh made various useful suggestions of preparation of the manuscript. Dr. Debasis Sengupta helped in computer simulation.

REFERENCE

1 G. Taguchi, System of Experimental Design: Engineering Methods to Optimize Quality and Minimize Costs, UNIPUB, Kraus International, White Plains, N.Y., 1987.

Quality through Engineering Design - W. Kuo (Editor)

Statistical Properties of Hadamard Matrices of Order 16

X. D. Sun and C. F. J. Wu

Department of Statistics and Actuarial Science, University of Waterloo, Waterloo, Ontario, Canada N2L 3GI

Abstract

Hadamard matrices of order 16 due to M. Hall provide a unique opportunity for studying the statistical implications of nonregular designs because they are the smallest designs in which both full aliasing and partial aliasing occur. The full-aliasing relations can be defined by a group called the incomplete defining contrast subgroup. Implications on the estimation of interactions are also considered.

INTRODUCTION

Hall [1] discovered that there are exactly five nonisomorphic Hadamard matrices of order 16. (A Hadamard matrix is an orthogonal matrix with ±1 as its entries. Two Hadamard matrices are isomorphic if one can be obtained from the other by row changes, column changes, or sign changes within any column.) Interesting as it may be from the combinatorial point of view, its statistical implications have not been explored. The main purpose of this paper is to outline some properties of these matrices as statistical designs.

Hall [1] called his five matrices Classes I, II, III, IV, and V, which we call H16.I to H16.V. The first matrix H16.I is the well-know 16-run fractional factorial design, whose columns form a group under entry-wise multiplication of columns. The last four matrices do not have this property, and some interesting opportunities arise as a consequence. We call the first type of designs regular and the second type nonregular. Nonregular designs have traditionally been used for studying main effects only. Recently Hamada and Wu [2] pointed out the potential of these designs for entertaining and studying interactions and proposed an analysis strategy. Wang and Wu [3] gave a design-theoretical explanation for the success of the Hamada-Wu strategy. What distinguishes the two types of designs is the presence of partial aliasing of effects. For regular designs any two effects are either orthogonal or fully aliased. For nonregular designs some effects are partially aliased, i.e., the aliasing coefficient is between 1 and –1. The best known is the 12-run Plackett-Burman [4] design, in which the aliasing coefficient between any main effect and any two-factor interaction (2fi) is either 1/3 or –1/3. It is also true that only partial aliasing occurs for Plackett-Burman designs of order 20, 24, and 28. What makes the present study interesting is the observation that for each of the H16.II-V designs, both full aliasing and partial aliasing occur. The full aliasing part of the relations can be defined by a group, which we call the incomplete defining contrast subgroup. In the next section we spell out these groups for H16.II-V and study their design implications. Then we study the problem of factor assignment for these four designs, and finally, by using graph representations, we demonstrate that nonregular design like H16.II-V can be more powerful in entertaining and estimating interactions than regular designs like H16.I.

Some basic definitions for regular two-level fractional factorial designs are given below. A 2^k factorial design has k factors, each at two levels coded by ± 1 conventionally. All possible combinations of the $\pm$ signs of k factors form 2^k runs. A $2^{k-p}(p < k)$ fractional factorial design only employs a 2^{-p} fraction of the 2^k runs and is defined by a set of p independent defining words or generators. All 2^p defining words generated from the p generators form a group called the *defining contrast subgroup*. The number of factors (or letters) in a word is called *word-length*. The vector $W = (w_3, w_4, \ldots)$ is called the word-length pattern, where w_i denotes the number of words of length i in the defining contrast subgroup. Resolution is defined as the smallest r with $w_r \geq 1$. For two designs d_1 and d_2 with r begin the smallest value such that $w_r(d_1) \neq w_r(d_2)$, we say that d_1 has less aberration than d_2 if $w_r(d_1) < w_r(d_2)$. If there is no design with less aberration than d_1, then d_1 has minimum aberration. The aberration criterion can further distinguish designs with the same resolution. It is thus a useful and convenient criterion for selecting designs.

HADAMARD MATRICES OF ORDER 16 DUE TO M. HALL

Among Hall's five matrices, H16.I can be completely described by its defining contrast subgroup and the related properties such as resolution and aberration. The last four matrices have, however, received little attention in the statistical design literature. Their properties as statistical designs are rather unknown, let alone their applications in practical experiments. In this section, we present some new properties of these four designs through the notion of the incomplete defining contrast subgroup.

We observe that for each of the four designs there are certain algebraic relations among its columns, i.e., some factors are fully aliased with the interactions of other factors. For example, in H16.II, the relation I = 2345 holds, i.e., the product of columns 2 to 5 is the identity column I, which consists of all + signs. This relationship implies aliasings among the effects of factors 2, 3, 4, and 5, i.e., the main effect 2 is fully aliased with 345, the interaction 23 is aliased with 45, etc. The aliasing pattern provides information about the effects that cannot be simultaneously estimated. Here we use factor i and column i interchangeably.

We note that some defining words of length 3 were presented by Hall [1], which he called triples. However, Hall did not point out any further column relations that seemed irrelevant to the construction of the matrices. By close investigation of H16.II-V, we observe that there are more defining relations among the columns than what can be expressed as Hall's triples. For example, in H16.III, the defining relation $I = 8\, t_0\, t_2\, t_4$ (we use $t_0, t_1, t_2, t_3, t_4, t_5$ to indicate columns 10, 11, 12, 13, 14, 15) cannot be derived from the triples of H16.III. All possible defining relations among the columns of these four designs have been found through a complete search. There are a total of 1,023, 511, 255, and 255 defining words for H16.II-V in comparison with 2,047 defining words for H16.I.

For each design, we list a set of independent defining words (generators) as follows, where the words inside the parentheses were not found in Hall [1].

H16.I $\quad I = 1\,2\,3 = 1\,4\,5 = 1\,6\,7 = 1\,8\,9 = 1\, t_0\, t_1 = 1\, t_2\, t_3$
$\quad = 1\, t_4\, t_5 = 2\,4\,6 = 2\,8\, t_0 = 2\, t_2\, t_4 = 4\,8\, t_2$

H16.II $\quad I = 1\,2\,3 = 1\,4\,5 = 1\,6\,7 = 1\,8\,9 = 1\, t_0\, t_1 = 1\, t_2\, t_3$
$\quad = 1\, t_4\, t_5 = 2\,4\,6 = 2\,8\, t_0 = 2\, t_2\, t_4$

H16.III $\quad I = 1\,2\,3 = 1\,4\,5 = 1\,6\,7 = 1\,8\,9 = 1\, t_0\, t_1 = 1\, t_2\, t_3$
$\quad = 1\, t_4\, t_5 = 2\,4\,6 = (8\, t_0\, t_2\, t_4)$

H16.IV $\quad I = 1\,2\,3 = 1\,4\,5 = 1\,6\,7 = 1\,8\,9 = 1\, t_0\, t_1 = 1\, t_2\, t_3$
$\quad = 1\, t_4\, t_5 = (2\,4\,6\,8\, t_0\, t_2\, t_4)$

H16.V $\quad I = 1\,2\,3 = 1\,4\,5 = 1\,6\,7 = 2\,4\,6$
$\quad = (8\,9\, t_0\, t_1 = 8\,9\, t_2\, t_3 = 8\,9\, t_4\, t_5 = 8\, t_0\, t_2\, t_4)$

We call the group generated from these defining relations the *incomplete defining contrast subgroup.* The word incomplete emphasizes the fact that this group does not completely determine all the relations among the columns. In addition to the fully aliased effects implied by the group, there are some pairs of effects that are partially aliased, which contribute to the complexity of the aliasing structure. For example, if we assign factors A, B, C, D, E, F, and G to columns 1, 2, 3, 4, 5, 8, and 9 of H16.V, the contrast l_1 for column 1 estimates the sum of many effects: $l_1 \rightarrow A + BC + DE + \frac{1}{2}FG$. Thus A is fully aliased with BC and DE and partially aliased with FG with the coefficient $\frac{1}{2}$.

The word-length pattern of the incomplete defining contrast group is defined in the same way as in the first section. Table 1 shows this pattern.

Table 1. Word-length pattern for H16.I-V.

	Word-length												
	3	4	5	6	7	8	9	10	11	12	13	14	15
H16.1	35	105	168	280	435	435	280	168	105	35	0	0	1
H16.II	19	57	72	120	243	243	120	72	57	19	0	0	1
H16.III	11	33	24	40	147	147	40	24	33	11	0	0	1
H16.IV	7	21	0	0	99	99	0	0	21	7	0	0	1
H16.V	7	21	0	0	99	99	0	0	21	7	0	0	1

H16.I has the largest number of defining words of any length and the number of defining words are in descending order for H16.I-V. Interestingly, H16.IV and H16.V have the same word-length patterns.

Another piece of information obtainable from the incomplete defining contrast group is the structure of three different types of 2fi's described as

1. Type A: fully aliased with main effect
2. Type B: fully aliased with other 2fi's but not with main effects
3. Type C: not fully aliased with any main effects or other 2fi's

These three types of 2fi's are analogous to the notions given in Wu and Chen [5]: ineligible 2fi, eligible but not clear 2fi, and clear 2fi.

Type-A 2fi's appear in length-3 words; type-B 2fi's appear in length-4 but not length-3 words, type-C 2fis do not appear in any length-3 or length-4 words. Type-B 2fi's impose the constraint that no two aliased 2fi's can be included in the same model. Type-C 2fi's impose fewer constraints on the estimable models but cause the model to have D-efficiency less than one since they are not orthogonal to other effects. We use the following D efficiency to measure design efficiency:

$$D = \frac{1}{n}|X'X|^{1/p}$$

where the columns of $X_{n\times p}$ correspond to the effects in the model including the grand mean, the main effects, and the specified interactions. From this discussion it is clear that having a large number of type-C 2fi's and a small number of type-A 2fi's is desirable (see Table 2).

Table 2. Summary of the numbers of 2fi's of different types for each design.

Design	2fi Type	No. of 2fi's	Comments
H16.I	A	105	
	B	0	
	C	0	
H16.II	A	57	
	B	48	each aliased with 3 other 2fi's
	C	0	
H16.III	A	33	
	B	72	48 aliased with another 2fi 24 aliased with 4 other 2fi's
	C	0	
H16.IV	A	21	
	B	84	each aliased with another 2fi
	C	0	
H16.V	A	21	
	B	28	each aliased with 3 other 2fi's
	C	56	

H16.V is the only design that has 2fi's of type C. The reason is that any interaction ij between columns i and j, where i = 1, . . . , 7 and j = 8, . . . , 15, never appears in any defining word of length 3 or 4. Altogether there are 56 (that is,

7 × 8) such 2fi's. Among the five designs, H16.V is considered to be the best, which is followed by H16.IV, H16.III, H16.II, and H16.I in descending order.

EXPERIMENTAL PLANS WITH FEWER THAN 15 FACTORS

So far we have assumed that all the columns are used up for factor assignment. If there are fewer than 15 factors, how should they be assigned to the columns? To use Hall's 16-run designs in practical experimentation, we are confronted with the problem of assigning factors to the columns of the matrices. This task is relatively easy in fractional factorial designs like H16.I, where resolution and aberration criteria can be used. For nonregular designs like H16.II-V, a simple approach adopted in this paper is to apply the same resolution and aberration criteria to the incomplete defining contrast subgroup. Since partial aliasings of effects are not captured by the incomplete defining contrast subgroup, the full capacity for estimating nonregular designs is not measured by these criteria. The authors have recently developed a more general criterion called estimation capacity, which works for fractional factorial as well as nonregular design and allows a more comprehensive comparison of designs such as H16.I-V. These results will appear soon.

For a given subset of columns, it is straightforward to obtain the corresponding incomplete defining contrast subgroup from the original incomplete defining contrast subgroup by eliminating the defining words that involve column labels not present in the subset. Minimum aberration designs of five to eight factors based on the word-length pattern of the incomplete defining contrast subgroup are given in Table 3. The designs indicated by * are "no aberration" designs in the sense that there is no fully aliased relation among the columns.

Table 3. Minimum aberration designs of five to eight factors for H16.I-V.

No. of factors	Design	Columns	Word-length Pattern 3456789
5	H16.I	1 2 4 8 15	0 0 1
	H16.II	* 1 2 4 8 12	0 0 0
	H16.III	* 1 2 4 8 10	0 0 0
	H16.IV	* 1 2 4 6 8	0 0 0
	H16.V	* 1 2 4 8 9	0 0 0
6	H16.I	1 2 4 7 8 11	0 3 0 0
	H16.II	4 5 8 10 12 15	0 0 0 1
	H16.III	* 1 2 4 8 10 12	0 0 0 0
	H16.IV	* 1 2 4 6 8 10	0 0 0 0
	H16.V	* 1 2 4 8 9 10	0 0 0 0
7	H16.I	1 2 4 7 8 11 13	0 7 0 0 0
	H16.II	1 4 6 8 10 12 15	0 1 2 0 0
	H16.III	2 4 7 8 10 12 15	0 0 0 0 1
	H16.IV	* 1 2 4 6 8 10 12	0 0 0 0 0
	H16.V	*1 2 4 8 9 10 12	0 0 0 0 0

Table 3. Continued.

No. of factors	Design	Columns	Word-length Pattern 3456789
8	H16.I	1 2 4 7 8 11 13 14	0 14 0 0 0 1
	H16.II	4 5 6 7 8 9 12 14	0 5 0 2 0 0
	H16.III	2 3 4 5 8 10 12 15	0 1 0 2 0 0
	H16.IV	1 2 4 6 8 10 12 15	0 0 0 0 0 1
	H16.V	1 2 4 7 8 9 10 12	0 1 0 0 0 0
9	H16.I	1 2 3 4 5 8 9 14 15	4 14 8 0 4 1 0
	H16.II	4 5 6 8 9 10 12 13 14	0 9 0 6 0 0 0
	H16.III	2 4 7 8 9 10 11 12 14	0 3 2 0 2 0 0
	H16.IV	2 3 4 5 6 8 10 12 14	0 1 0 0 2 0 0
	H16.V	1 2 4 7 8 9 10 11 12	0 2 0 0 0 1 0

Note that there is no length-3 defining word for designs with nine factors obtained from H16.II-V, i.e., they all have resolution IV. For H16.I, the best design with nine factors has only resolution III. In this sense, H16.II-V are superior to H16.I.

ACCOMMODATION OF SPECIFIC INTERACTIONS

In this section we show how H16.II-V can provide better experimental plans than H16.I in case special patterns of interactions are deemed important and should be estimated clear of each other and of the main effects.

Consider an experiment consisting of six two-level factors A, B, C, D, E, and F in two groups. A design is required to estimate all the main effects and factor interactions within each group; the inter-group factor interactions are assumed to be negligible. Also, as required by most experimenters, the run size should be as small as possible.

Two ways of grouping are considered: {AB} ∪ {C, D, E, F} and {A, B, C} ∪ {D, E, F}. The model estimation requirement in these two situations can be expressed by the interaction graphs

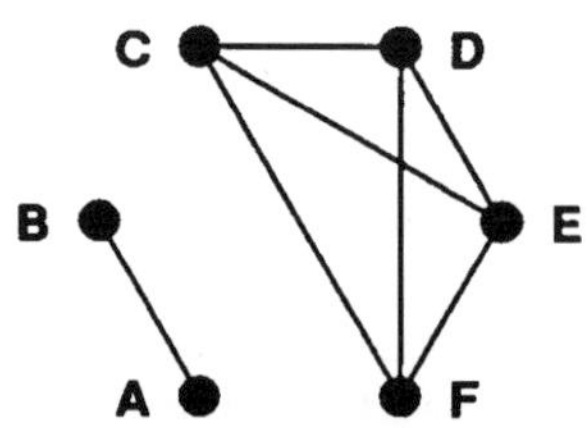

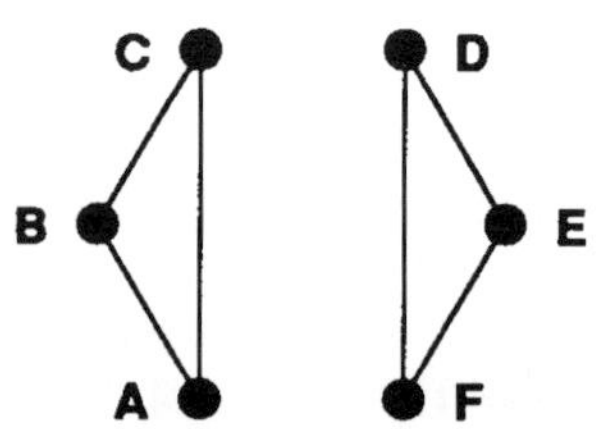

where dots represent factors and lines represent interactions. There are respectively 14 and 13 numbers of parameters to be estimated (grand mean plus main effects plus required interactions). Sixteen-run fractional factorial design (2^{6-2}) seems to be appropriate for this task. However, among the four types of 2^{6-2} designs, none of them provides a solution to this problem as indicated by the 11 interaction graphs in Wu and Chen [5, pages 171–172]. These 11 graphs are all subgraphs of the four graphs in Appendix B (i).

Instead, we can use the columns 1, 2, 8, 10, 11, and 13 of H16.V as factors A, B, C, D, E, and F. Main effects are obviously all estimable. For the remaining 9 (that is, 15–6) degrees of freedom, we consider all *feasible interaction graphs*, which represent all the sets of 2fi's that can be simultaneously estimated, assuming all other 2fi's and higher-order interactions are negligible. A complete collection of graphs for this assignment is given in Appendix B (v). The problem in the example can be easily solved by simply picking graphs no. 6 and no. 7. However, we should point out that for these designs, the 2fi's are not necessarily orthogonal to the main effects or to other 2fi's and hence result in some loss of estimation efficiency. The D-efficiency is given below each design in Appendix B.

For H16.I-IV, we also give their interaction graphs for six factors in Appendix B(i–v). In terms of the number of nonisomorphic graphs, H16.I is inferior to H16.II-V. Among H16.II-V, no design is uniformly better than the others. Why can H16.II-V allow more flexible estimation of interactions than H16.I? First, from the defining relations in the incomplete defining contrast subgroup, some interactions do not appear in length-3 or length-4 words and therefore are estimable. But this does not give a full explanation since partial aliasings are not captured by the group. Second, Wang and Wu [3] use the notion of *hidden projection* (as opposed to *geometric projection* employed by Box and Hunter [6]) for fractional factorial designs to explain why interactions are often estimable in nonregular designs. The hidden projection properties of these designs are bestowed by the complexity of the partial aliasing relations.

ACKNOWLEDGMENTS

The research was supported by the Natural Sciences and Engineering Research Council of Canada and the Manufacturing Research Corporation of Ontario.

REFERENCES

1 M. Hall, Jr., Research Summary, No. 36-10, Jet Propulsion Laboratory, Pasadena, Calif., 1 (1961) 21–26.
2 M. Hamada and C. F. J. Wu, J. Qual. Technol., 24 (1992) 130–137.
3 J. C. Wang and C. F. J. Wu, A Hidden Projection Property of Plackett-Burman and Other Designs, Technical Report, University of Waterloo, Waterloo, Ontario, Canada, 1992.
4 R. L. Plackett and J. P. Burman, Biometrika, 33 (1946) 305–325.

5 C. F. J. Wu and Y. Chen, Technometrics, 34 (1992) 162–175.
6 G. E. P. Box and J. S. Hunter, Technometrics, 3 (1961) 311–351.

APPENDIX A

Hall's 16-run non-regular orthogonal designs

H16.II

0	1	2	3	4	5	6	7	8	9	10	11	12	13	14	15
+	+	+	+	+	+	+	+	+	+	+	+	+	+	+	+
+	+	+	+	+	+	+	+	-	-	-	-	-	-	-	-
+	+	+	+	-	-	-	-	+	+	+	+	-	-	-	-
+	+	+	+	-	-	-	-	-	-	-	-	+	+	+	+
+	+	-	-	+	+	-	-	+	+	-	-	+	+	-	-
+	+	-	-	+	+	-	-	-	-	+	+	-	-	+	+
+	+	-	-	-	-	+	+	+	+	-	-	-	-	+	+
+	+	-	-	-	-	+	+	-	-	+	+	+	+	-	-
+	-	+	-	+	-	+	-	+	-	+	-	+	-	+	-
+	-	+	-	+	-	+	-	-	+	-	+	-	+	-	+
+	-	+	-	-	+	-	+	+	-	+	-	-	+	-	+
+	-	+	-	-	+	-	+	-	+	-	+	+	-	+	-
+	-	-	+	+	-	-	+	+	-	-	+	-	+	+	-
+	-	-	+	+	-	-	+	-	+	+	-	+	-	-	+
+	-	-	+	-	+	+	-	+	-	-	+	+	-	-	+
+	-	-	+	-	+	+	-	-	+	+	-	-	+	+	-

H16.III

0	1	2	3	4	5	6	7	8	9	10	11	12	13	14	15
+	+	+	+	+	+	+	+	+	+	+	+	+	+	+	+
+	+	+	+	+	+	+	+	-	-	-	-	-	-	-	-
+	+	+	+	-	-	-	-	+	+	+	+	-	-	-	-
+	+	+	+	-	-	-	-	-	-	-	-	+	+	+	+
+	+	-	-	+	+	-	-	+	+	-	-	+	+	-	-
+	+	-	-	+	+	-	-	-	-	+	+	-	-	+	+
+	+	-	-	-	-	+	+	+	+	-	-	-	-	+	+
+	+	-	-	-	-	+	+	-	-	+	+	+	+	-	-
+	-	+	-	+	-	+	-	+	-	+	-	+	-	+	-
+	-	+	-	+	-	+	-	-	+	-	+	-	+	-	+
+	-	+	-	-	+	-	+	+	-	-	+	+	-	-	+
+	-	+	-	-	+	-	+	-	+	+	-	-	+	+	-
+	-	-	+	+	-	-	+	+	-	-	+	-	+	+	-
+	-	-	+	+	-	-	+	-	+	+	-	+	-	-	+
+	-	-	+	-	+	+	-	+	-	+	-	-	+	-	+
+	-	-	+	-	+	+	-	-	+	-	+	+	-	+	-

H16.IV

0	1	2	3	4	5	6	7	8	9	10	11	12	13	14	15
+	+	+	+	+	+	+	+	+	+	+	+	+	+	+	+
+	+	+	+	+	+	+	+	-	-	-	-	-	-	-	-
+	+	+	+	-	-	-	-	+	+	+	+	-	-	-	-
+	+	+	+	-	-	-	-	-	-	-	-	+	+	+	+
+	+	-	-	+	+	-	-	+	+	-	-	+	+	-	-
+	+	-	-	+	+	-	-	-	-	+	+	-	-	+	+
+	+	-	-	-	-	+	+	+	+	-	-	-	-	+	+
+	+	-	-	-	-	+	+	-	-	+	+	+	+	-	-
+	-	+	-	+	-	+	-	+	-	+	-	+	-	+	-
+	-	+	-	+	-	-	+	+	-	-	+	-	+	-	+
+	-	+	-	-	+	+	-	-	+	-	+	+	-	-	+
+	-	+	-	-	+	-	+	-	+	+	-	-	+	+	-
+	-	-	+	+	-	+	-	-	+	-	+	-	+	+	-
+	-	-	+	+	-	-	+	-	+	+	-	+	-	-	+
+	-	-	+	-	+	+	-	+	-	+	-	-	+	-	+
+	-	-	+	-	+	-	+	+	-	-	+	+	-	+	-

H16.V

0	1	2	3	4	5	6	7	8	9	10	11	12	13	14	15
+	+	+	+	+	+	+	+	+	+	+	+	+	+	+	+
+	+	+	+	+	+	+	+	-	-	-	-	-	-	-	-
+	+	+	+	-	-	-	-	+	+	+	+	-	-	-	-
+	+	+	+	-	-	-	-	-	-	-	-	+	+	+	+
+	+	-	-	+	+	-	-	+	+	-	-	+	+	-	-
+	+	-	-	+	+	-	-	-	-	+	+	-	-	+	+
+	+	-	-	-	-	+	+	+	-	+	-	+	-	+	-
+	+	-	-	-	-	+	+	-	+	-	+	-	+	-	+
+	-	+	-	+	-	+	-	+	+	-	-	-	-	+	+
+	-	+	-	+	-	+	-	-	-	+	+	+	+	-	-
+	-	+	-	-	+	-	+	+	-	-	+	-	+	+	-
+	-	+	-	-	+	-	+	-	+	+	-	+	-	-	+
+	-	-	+	+	-	-	+	+	-	+	-	-	+	-	+
+	-	-	+	+	-	-	+	-	+	-	+	+	-	+	-
+	-	-	+	-	+	+	-	+	-	-	+	+	-	-	+
+	-	-	+	-	+	+	-	-	+	+	-	-	+	+	-

APPENDIX B

Some interaction graphs for H16. I-V designs

(i) H16.I

factors	A	B	C	D	E	F
column no.	1	2	3	4	8	13

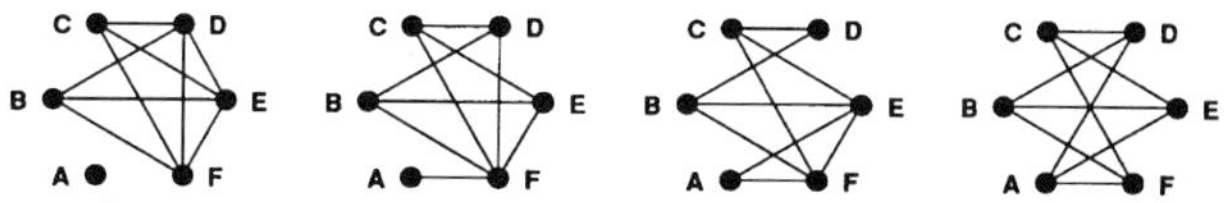

D=1.00 D=1.00 D=1.00 D=1.00

(ii) (II) H16.II

factors	A	B	C	D	E	F
column no.	1	4	6	8	11	12

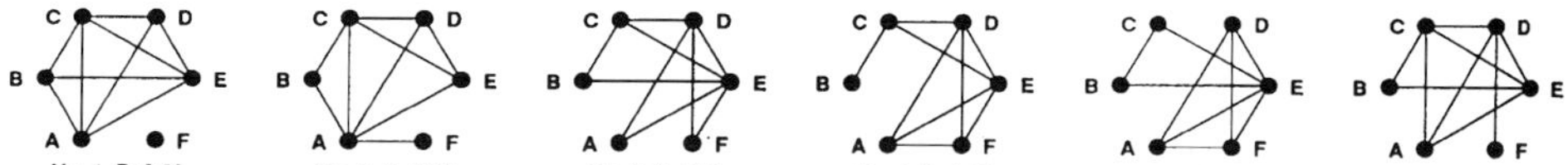

No. 1, D=0.92 No. 2, D=0.92 No. 3, D=0.84 No. 4, D=0.84 No. 5, D=0.84 No. 6, D=0.84

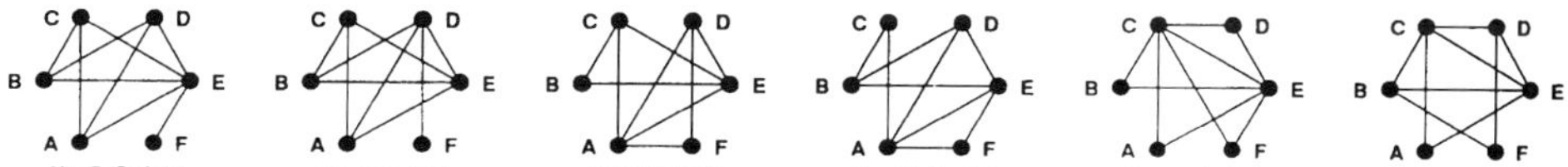

No. 7, D=0.84 No. 8, D=0.84 No. 9, D=0.84 No. 10, D=0.84 No. 11, D=0.77 No. 12, D=0.77

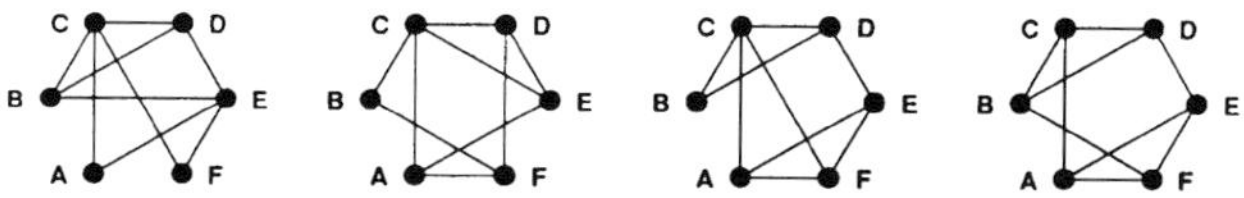

No. 13, D=0.77 No. 14, D=0.77 No. 15, D=0.77 No. 16, D=0.77

(iii) H16.III

factors	A	B	C	D	E	F
column no.	2	6	8	10	12	13

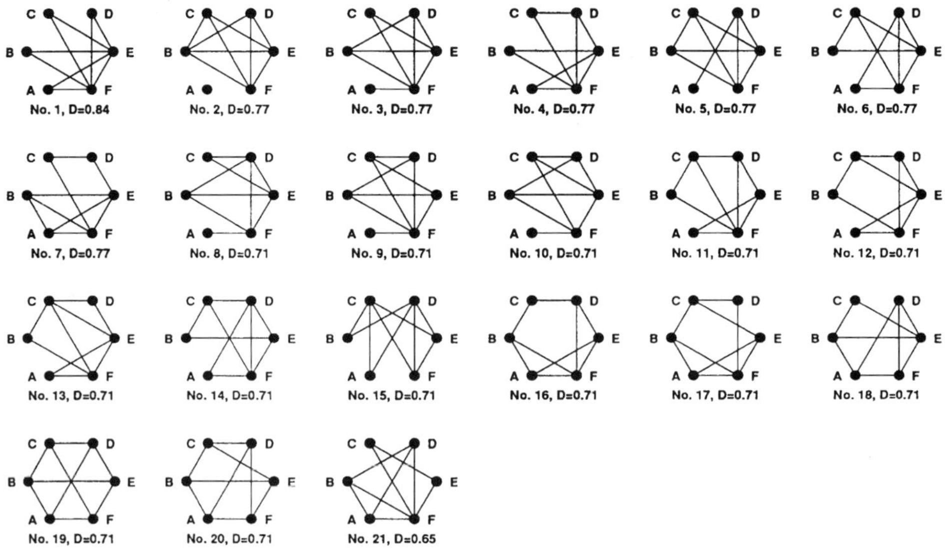

(iv) H16.IV

factors	A	B	C	D	E	F
column no.	2	4	5	8	12	14

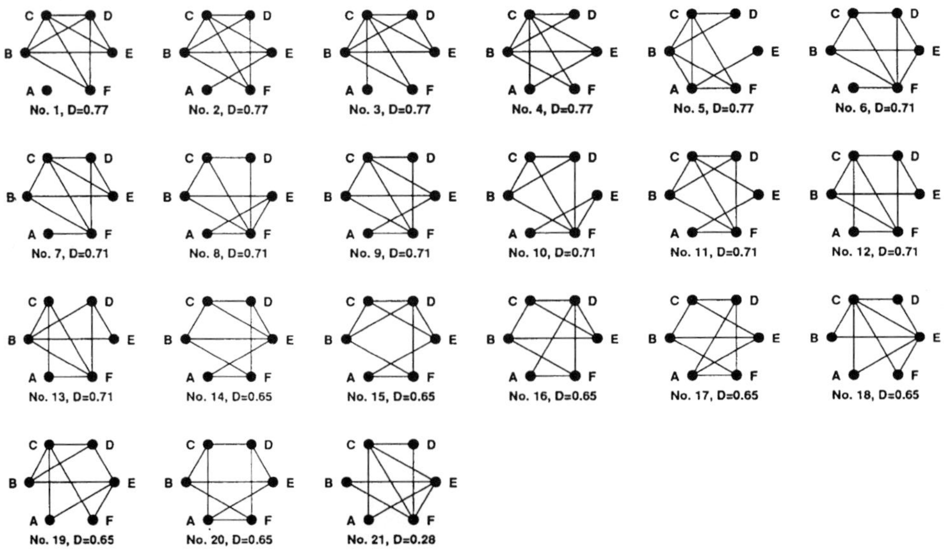

(v) H16.V

factors	A	B	C	D	E	F
column no.	1	2	8	10	11	13

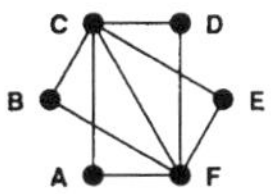

No. 1, D=0.84

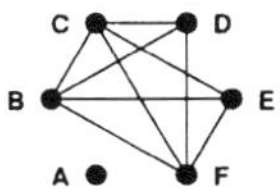

No. 2, D=0.77

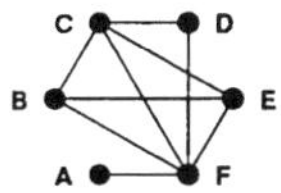

No. 3, D=0.77

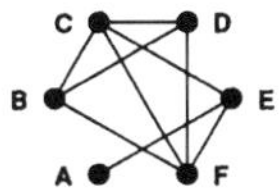

No. 4, D=0.77

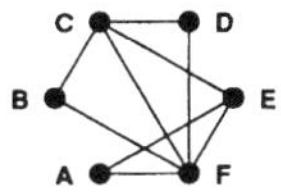

No. 5, D=0.77

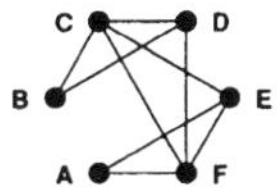

No. 6, D=0.77

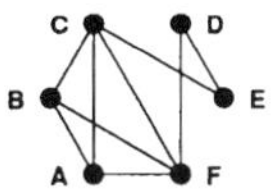

No. 7, D=0.77

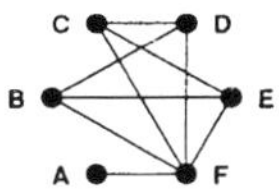

No. 8, D=0.71

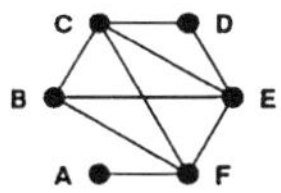

No. 9, D=0.71

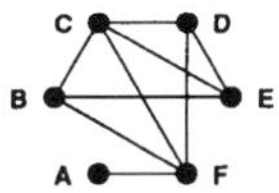

No. 10, D=0.71

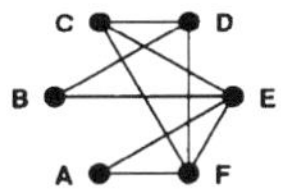

No. 11, D=0.71

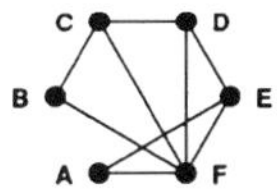

No. 12, D=0.71

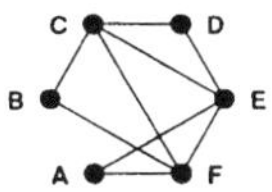

No. 13, D=0.71

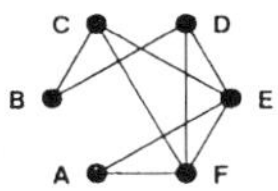

No. 14, D=0.71

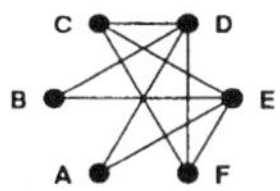

No. 15, D=0.71

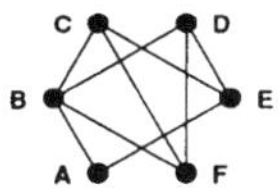

No. 16, D=0.71

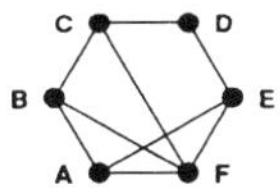

No. 17, D=0.71

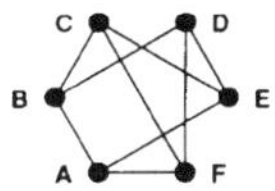

No. 18, D=0.71

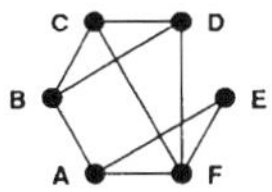

No. 19, D=0.71

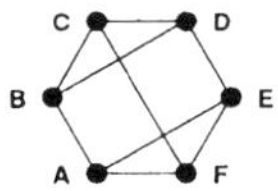

No. 20, D=0.71

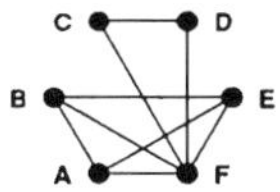

No. 21, D=0.65

Quality through Engineering Design - W. Kuo (Editor)

Multivariate Generalization of Taguchi's Signal-to-Noise Ratio for Industrial Experiments

Jagdish S. Rustagi,[a] Vijendra P. Singh,[a] Sakti P. Ghosh,[b] Kwan Y. Wong[b]

[a]ADSTAR, An IBM Company, Advanced Manufacturing Process Control, 5600 Cottle Road, San Jose, Calif. 95193, U.S.A.

[b]IBM Research Division, Almaden Research Center, 650 Harry Road, San Jose, Calif. 95120-6099, U.S.A.

Abstract

To improve the quality of industrial products through the design of experiments, G. Taguchi has used a signal-to-noise ratio for optimization. This notion is generalized for the multivariate case. For example, if the objective is "nominal is the best," Taguchi has suggested the ratio $10 \log_{10} \bar{y}^2/s^2$ for optimization, where $\bar{y}$ and s are the sample average and sample standard deviation in the univariate case. It can be generalized to $10 \log_{10} \bar{\underset{\sim}{Y}}^T \underset{\sim}{S}^{-1} \bar{\underset{\sim}{Y}}$, where $\bar{\underset{\sim}{Y}}$ is the vector of multiresponse averages and $\underset{\sim}{S}$ is the covariance of the sample. This ratio is justified on intuitive grounds as well as on consideration of some optimality criteria for designing experiments. Other ratios are also suggested. For example, using the D-optimality criterion gives the generalized ratio

$$10 \log_{10} \frac{\bar{y}_1^2 \bar{y}_2^2 \ldots \bar{y}_k^2}{\left|\underset{\sim}{S}\right|} = 10\left(\sum_{i=1}^{k} \log_{10} \bar{y}_i^2 - \log_{10}\left|\underset{\sim}{S}\right|\right)$$

where $\left|\underset{\sim}{S}\right|$ is the determinant of the sample covariance matrix and $\bar{y}_1, \ldots, \bar{y}_k$ are the averages of the k-responses.

Generalizations in case of other performance objectives are also proposed. Examples using our ADOE (automated design of experiments) software are given to illustrate the results of using generalized signal-to-noise ratios in industrial experiments.

INTRODUCTION

Use of the signal-to-noise (S/N) ratio is extensively discussed in the literature on quality popularized by Taguchi [1], and it is especially useful in choosing optimal settings of factors in experiments. Since the general concept of quality is multivariate, the optimal settings based on univariate responses in a multiresponse experiment do not lead to consistent answers. The generalized S/N ratio provides a single set of optimal levels for factors involved in the experiment. Let μ_0 be the target of a measurement y, then the quality loss defined by Taguchi is

$$Q = E(y - \mu_0)^2$$

Note that

$$Q = (\mu_0 - \mu)^2 + \sigma^2 = (bias)^2 + \text{variance}$$

where $\mu = E(y)$ and $\sigma^2 = Var(y) = E(y - \mu)^2$

Using the multivariate generalization of mean square error as $E(\underset{\sim}{y} - \underset{\sim}{\mu}_0)^T(\underset{\sim}{y} - \underset{\sim}{\mu}_0)$ where

$$\underset{\sim}{y} = \begin{pmatrix} y_1 \\ y_2 \\ \vdots \\ y_k \end{pmatrix} \qquad \underset{\sim}{\mu}_0 = \begin{pmatrix} \mu_{10} \\ \mu_{20} \\ \vdots \\ \mu_{k0} \end{pmatrix}$$

we have

$$E\left(\underset{\sim}{y} - \underset{\sim}{\mu}^T\right)\left(\underset{\sim}{y} - \underset{\sim}{\mu}\right) = \left(\underset{\sim}{\mu} - \underset{\sim}{\mu}_0\right)^T\left(\underset{\sim}{\mu} - \underset{\sim}{\mu}_0\right) + tr\,\underset{\sim}{\Sigma}$$

where

$$\underset{\sim}{\mu} = E\left(\underset{\sim}{y}\right), \quad Cov\left(\underset{\sim}{y}\right) = \underset{\sim}{\Sigma}$$

In order to get the observed characteristic on target in the univariate case, we need to multiply y by $(\mu_0)/\mu$. Then the quality loss after adjustment is

$$Q_a = k\frac{\mu_0^2}{\mu^2}\sigma^2$$

$$= k\mu_0^2\left(\frac{\sigma^2}{\mu^2}\right)$$

for some k. We assume, without loss of generality, that $k\mu_0^2 = 1$. The purpose in quality engineering is to minimize Q_a or simply to

$$\text{minimize}(10\log_{10} Q_a)$$

or

$$\text{minimize}(10\log_{10})\left(\frac{\mu^2}{\sigma^2}\right)$$

that is, to maximize Taguchi's S/N ratio.

MULTIVARIATE GENERALIZATION OF TAGUCHI'S S/N RATIO

Multivariate generalization is considered as follows. We adjust $\underset{\sim}{y}$ by scale factors given by the vector

$$\begin{bmatrix} \frac{\mu_{01}}{\mu_1} \\ \frac{\mu_{02}}{\mu_2} \\ \vdots \\ \frac{\mu_{0k}}{\mu_k} \end{bmatrix}$$

Then the adjusted $\underset{\sim}{y}$ is $\underset{\sim}{Z}$ where

$$\underset{\sim}{Z} = \begin{bmatrix} \frac{\mu_{01}}{\mu_1} y_1 \\ \frac{\mu_{02}}{\mu_2} y_2 \\ \vdots \\ \frac{\mu_{0k}}{\mu_k} y_k \end{bmatrix}$$

The covariance of $\underset{\sim}{Z}$, the adjusted vector, is $\underset{\sim}{\Lambda}$, where

$$\underset{\sim}{\Lambda} = \begin{bmatrix} \frac{\mu_{01}^2}{\mu_1^2}\sigma_{11} & \frac{\mu_{01}\mu_{02}}{\mu_1\mu_2}\sigma_{12} & \cdots & \frac{\mu_{01}\mu_{0k}}{\mu_1\mu_k}\sigma_{1k} \\ \frac{\mu_{01}\mu_{02}}{\mu_1\mu_2}\sigma_{21} & \frac{\mu_{02}^2}{\mu_2^2}\sigma_{22} & \cdots & \frac{\mu_{02}\mu_{0k}}{\mu_2\mu_k}\sigma_{2k} \\ \vdots & \vdots & & \vdots \\ \frac{\mu_{01}\mu_{0k}}{\mu_1\mu_k}\sigma_{k1} & \cdots & & \frac{\mu_{0k}^2}{\mu_k^2}\sigma_{kk} \end{bmatrix}$$

with

$$\underset{\sim}{\Sigma} = \begin{bmatrix} \sigma_{11} & \sigma_{12} & \cdots & \sigma_{1k} \\ \sigma_{21} & \sigma_{22} & \cdots & \sigma_{2k} \\ \vdots & \vdots & & \vdots \\ \sigma_{k1} & \sigma_{k2} & \cdots & \sigma_{kk} \end{bmatrix}$$

$$\sigma_{ij} = E(y_i - \mu_i)(y_j - \mu_j), \quad i, j = 1, 2, \ldots, k$$

Note that $\underset{\sim}{\Lambda}$ can be represented as

$$\underset{\sim}{\Lambda} = \underset{\sim}{M}^{-1}\,\underset{\sim}{\Sigma}\,\underset{\sim}{M}^{-1}$$

where

$$\underset{\sim}{M} = \begin{bmatrix} \frac{\mu_1}{\mu_{10}} & 0 & 0 & \cdots & 0 \\ 0 & \frac{\mu_2}{\mu_{20}} & 0 & \cdots & 0 \\ \vdots & & & & \vdots \\ 0 & \cdots & 0 & & \frac{\mu_k}{\mu_{k0}} \end{bmatrix}$$

or as

$$\underset{\sim}{\Lambda}^{-1} = \underset{\sim}{M}\,\underset{\sim}{\Sigma}^{-1}\,\underset{\sim}{M}$$

The adjusted quality loss is proportional to $\underset{\sim}{\Lambda}$. Assume for simplicity that

$$\underset{\sim}{\mu}_0 = \begin{pmatrix} \mu_{01} \\ \vdots \\ \mu_{0k} \end{pmatrix} = \underset{\sim}{1}$$

the vector having all components equal to one. Then

$$\underset{\sim}{\Lambda}^{-1} = \underset{\sim 0}{M}\ \underset{\sim}{\Sigma}^{-1}\ \underset{\sim 0}{M}$$

where

$$\underset{\sim 0}{M} = \begin{bmatrix} \mu_2 & 0 & \cdots & 0 \\ 0 & \mu_2 & \cdots & 0 \\ \vdots & \vdots & & \vdots \\ 0 & 0 & \cdots & \mu_k \end{bmatrix}$$

It is natural to maximize some function of $\underset{\sim}{\Lambda}$ such as its determinant or its trace. The trace criterion reduces to maximizing

$$\underset{\sim}{\mu}^T\ \underset{\sim}{\Sigma}^{-1}\ \underset{\sim}{\mu} \qquad \text{(i)}$$

since trace $\left(\underset{\sim}{\mathrm{M}}_0\ \underset{\sim}{\Sigma}^{-1}\ \underset{\sim}{\mathrm{M}}_0\right) = \underset{\sim}{\mu}^{\mathrm{T}}\ \Sigma^{-1}\ \underset{\sim}{\mu}$. However, if we use the determinant criterion, then we maximize

$$\det\left(\underset{\sim 0}{M}\ \underset{\sim}{\Sigma}^{-1}\ \underset{\sim}{M}\right) = \left|\underset{\sim 0}{M}\ \underset{\sim}{\Sigma}^{-1}\ \underset{\sim}{M}\right| = \mu_1^2\mu_2^2\ldots\mu_k^2\left|\underset{\sim}{\Sigma}\right|^{-1}$$

That is, the D-optimality criterion leads to maximizing the quantity

$$\mu_1^2\mu_2^2\ldots\mu_k^2\left|\underset{\sim}{\Sigma}\right|^{-1} \qquad \text{(ii)}$$

Hence, two interesting multivariate generalizations of the S/N ratio under the criterion “nominal is the best” are

(a) $\underset{\sim}{\mu}^T\ \underset{\sim}{\Sigma}^{-1}\ \underset{\sim}{\mu}$

(b) $\left(\prod_{i=1}^{k} \mu_i^2\right)\left|\underset{\sim}{\Sigma}\right|^{-1}$

Their relative importance can be judged by simulation experiments.

In case $\underset{\sim}{\Sigma}$ is singular, we can use the generalized inverse for computing the quadratic form in (a). The determinant in (b) can be replaced by the product of nonzero characteristic roots of $\underset{\sim}{\Sigma}$.

Motivated by the above discussion, we propose the following S/N generalizations.

(a) *Nominal is the best* performance criterion. For this criterion, we use

$$10 \log_{10}\left(\underset{\sim}{\bar{y}}^T \underset{\sim}{S}^{-1} \underset{\sim}{\bar{y}} \right)$$

where $\underset{\sim}{\bar{y}}$ is the average and $\underset{\sim}{S}$ is the sample covariance matrix. If $\underset{\sim}{S}$ is singular, we use the generalized inverse S^- of S. We define the ratio in that case as 10 log10$\underset{\sim}{\bar{y}}^T \underset{\sim}{S}^- \underset{\sim}{\bar{y}}$.

(b) *Smallest is the best.* For this criterion, we use

$$-10 \log_{10} \left| \underset{\sim}{S}_0 \right|$$

where

$$\underset{\sim}{S}_0 = \frac{1}{n} \begin{bmatrix} \sum_i y_{1i}^2 & \sum_i y_{1i} y_{2i} & \cdots & \sum_i y_{1i} y_{ki} \\ \sum_i y_{1i} y_{2i} & \sum_i y_{2i}^2 & \cdots & \sum_i y_{2i} y_{ki} \\ \vdots & \vdots & & \vdots \\ \sum_i y_{1i} y_{ki} & \sum_i y_{2i} y_{ki} & \cdots & \sum_i y_{ki}^2 \end{bmatrix}$$

Here $\underset{\sim}{S}_0$ is the matrix of the average sum of squares and cross products of the data. Note that the elements of the matrix are not adjusted for the averages. In case $\underset{\sim}{S}_0$ is singular, we use the product of the nonzero characteristic roots of $\underset{\sim}{S}_0$. That is, if $\lambda_1, \lambda_2, \ldots, \lambda_\ell$, $1 < k$ are positive roots of $\underset{\sim}{S}_0$, and the ratio is defined as

$$-10 \sum_{i-1}^{\ell} \log_{10} \lambda_i$$

(c) *Largest is the best.* We convert the criterion of largest is the best to smallest is the best by taking the reciprocal of the data and using the same criterion as above. That is, we use

$$-10\log_{10}\left|\underset{\sim}{T}_0\right|$$

where

$$\underset{\sim}{T}_0 = \frac{1}{n}\begin{bmatrix} \sum_i y_{1i}^{-2} & \sum_i y_{1i}^{-1}y_{2i}^{-1} & \cdots & \sum_i y_{1i}^{-1}y_{ki}^{-1} \\ \sum_i y_{1i}^{-1}y_{2i}^{-1} & \sum_i y_{2i}^{-2} & \cdots & \sum_i y_{2i}^{-1}y_{ki}^{-1} \\ \vdots & \vdots & & \vdots \\ \sum_i y_{1i}^{-1}y_{ki}^{-1} & \sum_i y_{2i}^{-1}y_{ki}^{-1} & \cdots & \sum_i y_{ki}^{-2} \end{bmatrix}$$

In case $\underset{\sim}{T}_0$ is close to being singular, we use the product of nonzero characteristic roots of $\underset{\sim}{T}_0$.

(d) *Signed target is the best.* For this criterion, we use the sample variance-covariance matrix $\underset{\sim}{S}$ by using the generalized ratio

$$-10\log_{10}\left|\underset{\sim}{S}\right|$$

In case $\left|\underset{\sim}{S}\right|$ is near zero, we use instead the product of nonzero characteristic roots of $\underset{\sim}{S}$.

Remark

Note that the sample covariance matrix, or the matrix of sums of squares and the crossproduct of data, is in general positive semidefinite. Hence the characteristic roots of the matrices used in calculating the generalized S/N ratios will be nonnegative. In case some of them are zero, we use only the positive roots in computing the ratios.

Example. Bonding experiment. We conducted an experiment with five factors—three at two levels each, one at three levels, and one at four levels—by using our developed ADOE software to determine the optimum levels of the factors with multiresponses

y_1 = increase in pitch with bonding
y_2 = increase in roll with bonding
y_3 = peel strength

in the manufacture of computers. The replications for various experiments were not the same. The above $2^3 \times 3 \times 4$ experiment was performed by using the L_{16} orthogonal array given in Table 1.

Table 1. L_{16}—array.

Experiment	Factor A	B	C	D	E
1	0	0	0	0	0
2	1	0	1	0	1
3	0	1	1	0	2
4	1	1	0	0	3
5	0	0	0	1	1
6	1	0	1	1	0
7	0	1	1	1	3
8	1	1	0	1	2
9	0	0	0	2	2
10	1	0	1	2	3
11	0	1	1	2	0
12	1	1	0	2	1
13	0	0	0	0	3
14	1	0	1	0	2
15	0	1	1	0	1
16	1	1	0	0	0

The multivariate analysis of variance can be performed to test the hypotheses that the main effects and interactions are zero.

Optimal Levels

In order to determine optimal level settings for the factors, we assume the same performance objective for all factors; that is, nominal is the best. In this case, the multivariate generalization of the S/N ratio is

$$10 \log_{10}\left(\underset{\sim}{\bar{y}}^T \underset{\sim}{S}^{-1} \underset{\sim}{\bar{y}} \right)$$

The results are given in Table 2.

Table 2. Generalized signal-to-noise ratios.

Factors	Levels			
	1	2	3	4
A	22.8560	20.7772	—	—
B	22.8808	20.4435	—	—
C	20.245	23.0652	—	—
D	20.833	20.6248	26.0788	—
E	20.6865	21.4467	22.9011	21.3914

Comparing the S/N ratio for each level computed from the average vector and covariance matrix for that level, the final settings are given in Table 3. Note that the S/N ratios are so defined that the larger number gives the better setting.

Table 3. Decision table based on S/N ratios.

Factor	Optimum Level
A	1
B	1
C	2
D	3
E	3

The results of the multivariate analysis of variance (MANOVA) may not agree with the analysis of the S/N ratios. In using MANOVA, we assume that the covariance matrix is the same for all factors, whereas for computing S/N ratios this assumption is not made.

CONCLUSIONS

In this paper, we propose a multivariate generalization of Taguchi's S/N ratio as an objective function for optimizing industrial experiments. The ADOE software has incorporated both uniresponse and multiresponse experimental design and analysis by using Taguchi's S/N ratio as the objective function. ADOE uses orthogonal arrays for experimental designs and also has many data analysis techniques, e.g., analysis of variance, analysis of means, and graphical plots for both univariate and multivariate. The S/N criterion for optimization used by Taguchi may not be valid under certain situations, e.g., for single replicate or when the determinant of the covariance matrix is close to zero. ADOE attempts to check these conditions and provides warning messages.

REFERENCE

1 G. Taguchi, System of Experimental Design, Volumes I and II, UNIPUB, Kraus International Publishers, White Plains, N.Y., 1987.

Quality through Engineering Design - W. Kuo (Editor)

Analyzing Data from Robust Parameter Design Experiments

Vijayan N. Nair

AT&T Bell Laboratories, Murray Hill, NJ 07974, U.S.A.

Abstract

The robust parameter design approach proposed by Taguchi is a very effective tool for reducing variation in products and processes and thereby improving quality. This paper reviews some methods for analyzing data from robust parameter design experiments, discusses some new developments, and outlines their advantages.

INTRODUCTION

Experimental design techniques have traditionally focused on modeling the average levels of a response variable or a quality characteristic. Taguchi's work on parameter design has demonstrated clearly the importance of variation reduction and how experimental design techniques can be used to accomplish this. In addition, his work has led to a better appreciation among quality practitioners of the usefulness of experimental design techniques for quality improvement.

Taguchi's parameter design ideas have also stimulated considerable research efforts to understand and improve his statistical methodology. This paper reviews some methods for analyzing data from robust parameter design experiments and discusses some recent developments. See also the panel discussion in Ref. [1] where several researchers and practitioners discuss the important ideas underlying parameter design and offer their perspectives on various issues. The panel discussion also provides an overview of research efforts and includes an extensive bibliography.

THE FRAMEWORK

Consider the situation where we have a product or process whose performance is measured by the quality characteristic Y. There is a fixed target value τ for this characteristic, and the goal is to have the system perform as closely as possible to the target value τ with little variation.

The output Y can be viewed as a function of (at least) two types of factors: (a) control factors $\mathbf{x}$—factors whose settings can be controlled or manipulated by the engineers, and (b) noise factors **z**—factors such as operating or customer-use conditions that cannot be controlled or are expensive to control. Variation in the noise variables causes variation in the quality characteristic. The novel idea behind parameter design is to minimize the effect of the variation in the noise variables by choosing the settings of the control factors **x** judiciously to exploit the interactions between control and noise factors.

When the input-output relationship $Y = f(\mathbf{x};\mathbf{z})$ is unknown (or complex), one can conduct physical (or computer) experiments and approximate the functional form of $f(\cdot;\cdot)$ empirically. Parameter design experiments are typically conducted in off-line environments where the important noise variables **z** are identified a priori, their settings are controlled, and their relationship with the control factors **x** studied systematically. Taguchi [2] recommends using highly fractional experiments where the control factors are varied according to an orthogonal array **X** and the noise factors are also typically varied according to another orthogonal array **Z**. The overall design is a product array obtained by crossing **X** and **Z**. If n and m denote the sizes of the control and noise arrays respectively, the run size of the product array is $n \times m$.

Let Y_{ij}, $j = 1, \ldots, m$ and $i = 1, \ldots, n$ denote the observations. Here i corresponds to the ith row of the control array (settings of the control factors $\mathbf{x}_i$) and j corresponds to the jth row of the noise array (settings of the noise factors $\mathbf{z}_j$). The data Y_{ij} are analyzed to determine the relationships between the control and noise factors and to select the appropriate settings of the control factors. In particular, the settings of **x** are chosen so that the effects of the variations in **z** on the output quality characteristic Y are as small as possible.

SIGNAL-TO-NOISE RATIOS VERSUS DATA TRANSFORMATIONS

For studying dispersion effects in the fixed-target problem, Taguchi recommends the use of the signal-to-noise (SN) ratio

$$SN_i = \frac{\overline{Y}_i^2}{S_{Y,i}^2} \tag{1}$$

where $\overline{Y}_i$ is the sample mean of the Y_{ij} at $\mathbf{x}_i$, the ith run of the control array, and S_i^2 is the corresponding sample variance. His analysis proceeds by fitting a log-linear model to the SN_i's as a function of the control factors and identifying the important (dispersion) effects. Those factors with large effects are considered important in reducing variation, and their settings are chosen suitably to minimize variation.

It is shown in Ref. [3] (see also Refs. [4] and [5]) that the use of Taguchi's SN ratio (SNR) is justified only under some special situations. To see this, let $\mu_Y(\mathbf{x})$ and $\sigma_Y^2(\mathbf{x})$ denote, respectively, the mean and variance of Y at the control factor setting **x**. To understand what we mean by important dispersion effects, we first have to know the extent to which the variance of σ_Y^2 depends on the mean μ_Y. We can write the variance as

$$\sigma_Y^2(\mathbf{x}) = v(\mu_Y(\mathbf{x}))\phi(\mathbf{x}) \tag{2}$$

where the first component is the part of the variance that depends on the mean. This component is completely determined by the constraint that the mean has to be close to the target value τ. Only the second component $\phi(\mathbf{x})$ can be reduced by appropriately choosing the control factors. In other words, the important dispersion effects are only those that affect $\phi(\mathbf{x}) = \sigma_Y^2 / v(\mu_Y)$. We can now see that Taguchi's SNR in Eq. (1) corresponds to a special case of Eq. (2) with $v(m) = m^2$. There is no reason, however, that this special case will hold in all situations.

The following, more general, data analysis strategy is recommended in Ref. [3] (see also Ref. [4]).

1. Use graphical techniques to first identify the appropriate functional form of $v(\cdot)$ in Eq. (2).
2. Transform the data from Y to $Y' = \gamma(Y)$ where $\frac{d}{dy}\gamma(y) = v^{-\frac{1}{2}}(y)$. The variance of Y' is then approximately the same as ϕ in Eq. (2). The transformed data follow, approximately, the location-scale model

$$Y'_{ij} = \mu(\mathbf{x}_i) + \phi(\mathbf{x}_i)\,\varepsilon_{ij} \tag{3}$$

   where $\mu(\mathbf{x}_i)$ denotes the location parameter and $\phi(\mathbf{x}_i)$ denotes the scale parameter.
3. Analyze the sample means and variances of the transformed data to identify the important location and dispersion effects. In particular, if the S_i^2's denote the sample variance of the transformed data, fit the log-linear model

$$\log\ S_i^2 = \mathbf{x}_i'\phi + v_i \tag{4}$$

   and estimate the dispersion effects ϕ_k by least squares.

It is easily shown that Taguchi's SNR-based analysis is approximately equivalent to applying a logarithmic transformation to the data and analyzing the sample variances of the transformed data. This is a special case of the approach outlined above. See also Ref. [6] for an alternative approach based on generalized linear models.

EFFICIENCY OF DIFFERENT METHODS OF ANALYSIS

One of the important goals in parameter design is to make inferences about the variability in the quality characteristic. Several authors have noted that it is inherently more difficult to make inferences about dispersion than about location and that experiments with the required run sizes are not economically feasible in industrial settings. For example, Gunter [7] remarks, "The statistical consequences of Taguchi's engineering insight is that . . . modeling the response variance—not just the mean—may be vital to achieve high quality and reduced cost. . . . Appearances can be deceiving, however. Although such an approach may appear to have promise, there are fundamental practical constraints that make the Taguchi strategy difficult or impossible to apply . . . one needs a lot more information to see a change in spread than to see a change in location. The crucial question is, Can we generally afford experimentation that gives us this kind of information. My answer is that I doubt it . . . ." He also presents some numerical evidence to support the argument that one needs and excessive number of experimental runs to study dispersion effects.

The problem with this conclusion is that it is based on an analysis that does not take into account the inherent structure in parameter design experiments. Nair [8] shows that more appropriate analyses that exploit the structure of parameter design experiments require considerably fewer runs. We provide here a summary of the conclusions.

Consider the models in Eqs. (3–4), and suppose we want to estimate the dispersion effects in Eq. (4). The usual methods of analyses are based on the assumption that the noise variables vary freely from run to run, i.e., the ε_{ij}'s in Eq. (3) are independent and identically distributed. This assumption is implicit in Gunter's comments [7] as well as his run-size calculations. These, however, completely ignore the basic structure of the control and noise arrays in parameter design experiments.

To see this, let us write the output (response) as

$$Y_{ij} = f(\mathbf{x}_i; \mathbf{z}_{ij}) \tag{5}$$

where $\mathbf{x}_i$ denotes the settings of the control factors at the ith run and $\mathbf{z}_{ij}$ represents all the noise variables at the ij^{th} run of the product-array. A key feature of parameter design experiments is to identify the important noise variables and vary them systematically during the experiment. Hence, the $\mathbf{z}_{ij}$'s do not vary freely from run to run. More specifically, we actually have the situation

$$Y_{ij} = f(\mathbf{x}_i; \mathbf{z}_j, \delta_{ij}) \tag{6}$$

where the $\mathbf{z}_j$ refers to the important noise variables that have been identified and whose settings are varied systematically through the noise array and the δ_{ij}'s denote the effects of the other (unidentified/uncontrolled) noise variables. Note that the values of the $\mathbf{z}_j$'s are the same across the control array runs. Thus, we should take this structure into account in the inference stage.

To see the effect of this, consider the special case of the location-scale model with a single noise variable z_j. Then, instead of Eq. (3) we actually have

$$Y_{ij} = \mu(\mathbf{x}_i) + \phi(\mathbf{x}_i)[z_j + \delta_{ij}] \tag{7}$$

Let

$$\rho = \frac{\sigma_z^2}{\sigma_z^2 + \sigma_\delta^2} \tag{8}$$

where σ_z^2 denotes the variation of the controlled noise variables and σ_δ^2 denotes the variation of the uncontrolled noise variables. Then, for the analysis based on the logarithm of the S_i^2's discussed in the last section, Nair [8] shows that the actual variances of the dispersion effects is $(1 - \rho^2)$ times the usual variances computed under the assumption that the ε_{ij}'s are independent and identically distributed. This holds both when the z_j's are fixed as well as when their values are chosen randomly from the noise distribution, but the same settings are repeated across the control runs.

Two consequences of this are immediately apparent. First, one does not need as many experimental runs as indicated in Gunter [7], especially if a considerable amount of the noise variation is controlled during the experiment. Note also that in many experimental situations, it is relatively easy or cheap to get replications; thus the size of the noise array may not be as much of a concern as the size of the control array. See also the comments by Wu in the panel discussion [1, p. 134]. Second, this has important implications for designing parameter design experiments. We can see that the efficiency of parameter design experiments will be increased if we can make r close to one. To do this, we must identify the nature of the noise variations and incorporate as many of the important noise variables as possible into the parameter design experiment.

The discussion thus far has considered the method of estimation based on analyzing the log of the S_i^2's. If the z_j's in Eq. (7) are indeed fixed, one obvious alternative is to first estimate $\phi(\mathbf{x}_i)$'s directly by using regression analysis and then model the estimated $\hat{\phi}(\mathbf{x}_i)$'s as a function of the design factors. How efficient is this compared to the method discussed previously? The model in Eq. (7) is a heteroscedastic regression model and therefore, strictly speaking, one should do weighted least squares to take into account the unequal variances. (See the next section for the use of iteratively reweighted least squares estimation in such situations.) Suppose, for now, that we use just the usual least squares estimation method to estimate $\phi(\mathbf{x}_i)$'s and then fit a log-linear model to the $\hat{\phi}(\mathbf{x}_i)$'s as a function of the control factors. Nair [8] shows that the variance of the dispersion effects estimated using this scheme is approximately $\sigma_\delta^2/(nm\sigma_z^2) = (1-\rho)/(nm\rho)$. Note that this variance also depends on ρ and decreases with σ_δ^2, i.e., as more of the noise variation is identified and controlled during the experiment.

When the noise factors are fixed, many authors have proposed modeling the response directly as a function of both the control and noise factors (see [1], Section 6). In the case of a single noise variable, the model they consider can be written as

$$Y_{ij} = \mu(\mathbf{x}_i) + \phi(\mathbf{x}_i) z_j + \delta_{ij} \tag{9}$$

Note that this model assumes the variance associated with the δ_{ij}'s is constant, i.e., does not depend on the control factors. The next section presents a more general formulation. If we estimate the $f(\mathbf{x}_i)$'s in Eq. (9) by least squares and fit a linear model to the $\hat{\phi}(\mathbf{x}_i)'s$ to estimate the dispersion effects, the variance of the estimated dispersion effects turns out to be $\sigma_\delta^2 / (nm\sigma_z^2) = (1-\rho)/(nm\rho)$, the same variance as the situation discussed in the last paragraph.

We can now compare the variances of the dispersion effects estimated under various models and under the assumption that the underlying random variables are all normally distributed. Under the model in Eq. (3) and the assumption that the ε_{ij}'s are all i.i.d. normal, the variance equals $2/[n(m-1)]$. As we saw, this analysis ignores the fact that in parameter design situations we, in fact, have the model in Eq. (7). If the δ_{ij}'s in Eq. (7) are i.i.d. normal and the z_j's are fixed or random, the variance of the estimated dispersion effects is $2(1-\rho^2)/[n(m-1)]$. Thus, the true variance is a factor $(1-\rho^2)$ times smaller than the perceived variance. If we used one of the regression analyses discussed in the last two paragraphs, then the variance of the estimated dispersion effects is $(1-\rho)/(nm\rho)$. All the analyses that take into account the structure of the parameter design experiments depend on ρ and are considerably more efficient than the one that ignores the structure. See Ref. [8] for more details and additional comparisons. See also Ref. [9] for some power comparisons.

A GENERAL FORMULATION

Thus far, we have discussed estimation methods under different models. A general formulation that includes all of these as special cases is

$$Y_{ij} = \mu(\mathbf{x}_i) + \phi_1(\mathbf{x}_i)z_{1j} + \ldots + \phi_K(\mathbf{x}_i)z_{Kj} + \theta(\mathbf{x}_i)\delta_{ij} \tag{10}$$

Here, $\mu(\mathbf{x}_i)$ measures the location effects, $\phi_r(\mathbf{x}_i)$ measures the dispersion effects associated with the noise variable z_r, and $\theta(\mathbf{x}_i)$ measures the dispersion effects of the uncontrolled and unknown noise variables. The model implicitly assumed when one uses Taguchi's SNR analysis or the sample variances S_i^2's is the one given by Eq. (7). We see from Eq. (10), however, that this model is reasonable only when all the ϕ_r's are the same and also equal θ. This is of course a rather big restriction. Note also that the response surface model considered by many authors (see Ref. [1], Section 6) assumes that θ is constant.

We now consider the general formulation in Eq. (10) and describe a data-analysis strategy for estimating the various location and dispersion effects for the case when the z_{rj}'s are fixed. See Ref. [10] for more details.

0. Initially set $\theta(\mathbf{x}_i)$ to be constant.
1. Let $h_0(\cdot), \ldots, h_K(\cdot)$ be link functions that satisfy (approximately)

$$h_0[\mu(\mathbf{x}_i)] = \mathbf{x}_i'\boldsymbol{\alpha} \tag{11}$$

and

$$h_r[\phi_r(\mathbf{x}_i)] = \mathbf{x}_i' \phi_r, r = 1, \ldots, K \tag{12}$$

Thus, the α_k's measure the location effects associated with the control factors, and the ϕ_{rk}'s measure the dispersion effects corresponding to the rth noise variable. In practice, the link functions in Eqs. (11–12) will be unknown and have to be determined a priori or empirically. Substitute Eqs. (11–12) into Eq. (10) to get

$$Y_{ij} = h_0^{-1}(\mathbf{x}'_i \boldsymbol{\alpha}) + h_1^{-1}(\mathbf{x}'_i \boldsymbol{\phi}_1) z_{1j} + \ldots + h_K^{-1}(\mathbf{x}'_i \boldsymbol{\phi}_K) z_{Kj} + \theta\delta_{ij} \tag{13}$$

Estimate the parameters α and ϕ_r's by using nonlinear or linear least squares, respectively.

2. Use appropriate model selection techniques to identify the active effects and hence an appropriate submodel.
3. To assess if $\theta = \theta(\mathbf{x}_i)$, compute "studentized" residuals r_{ij} from the fitted submodel in Step 2. Model the logarithm of $\hat{\sigma}_i^2 = \sum_{j=1}^{K} r_{ij}^2$ as a function of the design factors (as discussed in the Framework Section) to identify any important dispersion effects associated with the δ_{ij}'s.
4. If the last step reveals the presence of some dispersion effects, repeat Steps 2 and 3, but now use weighted least squares with the fitted value of $\hat{\theta}(\mathbf{x}_i)$'s. Otherwise, stop.
5. Use the identified models and the fitted values to arrive at improved settings of the design factors.

ANALYSIS FOR UNCONTROLLED NOISE VARIABLES

So far we have only discussed situations where the noise variables are controlled and their values are varied systematically in off-line experiments. In many situations, however, some of the important noise variables cannot be carefully controlled. This is certainly true in on-line experiments. Even in off-line situations, it may be too expensive or not practically feasible to control the noise variables. Phadke [11] discusses a parameter design experiment for tuning the performance of the computer system. One of the primary noise variables here was the system load, which could not be controlled precisely. See Ref. [10] for an experiment on thermal design of cabinets where the ambient temperature was not controlled.

If the noise variables vary but their values can be observed, the techniques discussed in the last section can be used to analyze the data. Instead of the noise variables z_{rj}'s being the same across control runs, their values will now vary freely from one control run to another. However, the overall data analysis strategy outlined in the last section can still be used to analyze the data. See Ref. [10] for additional details and discussion.

REFERENCES

1 V. N. Nair (ed.) Technometrics, 34 (1992) 127–161.
2 G. Taguchi, System of Experimental Design, Vol. I and II, UNIPUB, Kraus International Publishers, White Plains, N. Y., 1987.
3 V. N. Nair and D. Pregibon, AT&T Technical Journal, 65 (1975) 73–84.
4 G. E. P. Box, Technometrics, 30 (1988) 1–40.
5 R. Leon, A. C. Shoemaker and R. N. Kacker, Technometrics, 29 (1987) 253–285.
6 J. A. Nelder and Y. Lee, Appl. Stochastic Models and Data Analysis, 7 (1991) 107–120.
7 B. Gunter, Technometrics, 30 (1988) 1–40.
8 V. N. Nair, Exploiting the Structure in Parameter Design Experiments in the Analysis Stage, AT & T Bell Laboratories, Murray Hill, N.J., Technical Report, 1992.
9 D. M. Steinberg and D. Bursztyn, Noise Factors, Dispersion Effects, and Robust Design, Tel Aviv University Technical Report, Tel Aviv, Israel, 1991.
10 A. E. Freeny and V. N. Nair, Statistics Sinica, 2 (1992) 313–334.
11 M. S. Phadke, Quality Engineering Using Robust Design, Prentice Hall, Englewood Cliffs, N.J., 1989.

Quality through Engineering Design - W. Kuo (Editor)
1993 Elsevier Science Publishers B.V.

Exploratory Data Analysis Techniques as Applied to a High-Precision Turning Machine

James J. Filliben,[a] Sabri Cetinkunt,[b] Wen Lung Yu,[b] and Alkan Donmez[a]

[a]The National Institute of Standards and Technology, Statistical Engineering Division, Bldg. 1011 Room A337, Gaithersburg, MD 20899, U.S.A.[1]

[b]The University of Illinois, Chicago, IL 60680 U.S.A.

Abstract

Submicrometer precision machining requires very precise position and speed control of the motion of the machine tool axes. The accuracy of coordinated position control determines the profile accuracy of a part in contour machining, while the accuracy of speed control is the most significant factor in the resulting subsurface damage that may occur in contour or noncontour machining. In high-precision machining of brittle materials, such as diamond turning of mirror components, the chip-removal mechanics should be in the ductile regime of the material. This chip-removal process requires extremely low speeds and very precise feed rate control. This motion-control problem is difficult because of the high friction and the unpredictable nature of the friction at those very low speeds.

The present servocontrol systems typically are not capable of delivering the desired precision in motion control. In practice, friction must be compensated for by real-time closed-loop control algorithms, which in turn require an accurate understanding of the magnitude of friction effects. To this end, this paper describes an experiment and statistical analysis whose purpose is to investigate the friction inherent in a particular open-loop turning machine. A set of statistically designed off-line experiments is constructed and run. The data from these experiments are subjected to an exploratory data analysis approach. A series of graphical procedures is applied for assessing statistical control. Multiplots, cube plots, and DEX (design of experiment) mean plots are applied to assess the relative importance to factors. A new technique called block plots is introduced to examine interactions among various factors.

INTRODUCTION

Background

Ultraprecision machining is the key manufacturing process for producing parts used in high-technology engineering systems. As engineering systems become more sophisticated, the demand for precision parts becomes more stringent. Turbine blades, ultraprecision rolling element bearings used in aircraft engines, precision mirrors for optical systems, and biomedical subsystems (such as artificial hip joints) are all examples of high-precision machining applications [1–3]. Such applications typically require surface smoothness to be in the 0.02- to 0.050-μm range and profile accuracy to be less than 0.5 μm; such requirements place high demands on the precision capability of the machining process itself.

Subsurface damage is a major problem in high-precision machining of brittle materials. These defects severely limit the use of the resulting parts. Although subsurface defects can be partially mitigated by post-polishing, this is a costly and time-consuming process; it is more desirable and cost-beneficial to avoid the problem altogether at the machining stage rather than correcting it with after-the-fact polishing. Subsurface damage is caused by impact-mode chip removal. Procedurally, the best way to avoid impact chip removal is to machine in the ductile regime. Operationally, ductile regime machining is achieved by very slow feed rates under very precise feed rate control. The ideal is thus to operate at an ultralow velocity and to maintain that velocity with high precision.

Unfortunately, such low-speed machining accentuates the effect of stiction (static) friction forces existent in the machine tool guideways (especially for the commonly used plain type, double-V guideways) [4–5]. Heretofore, few experiments have focused on stiction friction effects and so relatively little is known about the essentials of this phenomenon [6]. For such low-velocity systems, stiction friction must be measured, characterized, and mathematically modeled. Stiction friction modeling plays a major role in the ultimate goal—the development of real time feedback servocontroller algorithms with friction compensation [7–8].

Although actuators and sensors are now quite adequate and may be constructed with standard (off-the-shelf), components, the standard controller systems are not adequate. The current state of affairs is that the software for the usual proportional integral derivative (PID)-type controller is not at all capable of providing the desired speed control for low-velocity, high-friction environments [9–11].

To solve this low-velocity friction problem, we must move beyond the standard PIDs and develop more sophisticated controller algorithms. Recent work in a different setting has been carried out to this end [12–14]. This paper describes our collaborative approach to the above problem, consisting of executing statistically designed off-line experiments and applying a variety of exploratory data-analysis techniques.

Experimental goals

The ultraprecision part demands of high-technology applications require ultraprecision machine tool performance. The long-range motion control goals of such systems are

1. low target velocity (in the 0.4-μm/s range; current velocities are typically in the 40-μm/s range)
2. constant velocity regulation (to within ±1% of target velocity)

To achieve these goals requires algorithmic friction compensation by the motion controller. To arrive at a real-time friction compensation algorithm is beyond the scope of this paper. Note also that an open-loop system (as we used) is not capable in itself of fully achieving such goals—a closed-loop system would be needed. The analysis of our open-loop system to assess its capacity is nevertheless valuable in developing closed-loop systems. With this in mind, the specific goal of this project is to determine for a particular high-precision, single-point diamond turning machine tool at the Precision Machining Research Facility at the National Institute of Standards and Technology (NIST)

1. the relative contribution of various factors to velocity, position, and friction variation
2. the interactions that exist between factors affecting velocity, position, and friction
3. the stability and predictability of the process and whether friction is repeatable
4. the way to characterize the stiction friction, and what factors affect it

To achieve these goals, this paper describes an experiment and presents a statistical analysis for investigating the friction inherent in a particular open-loop turning machine.

Machine tool description

The machine tool under study is a two-axis diamond turning machine. Only one of the axes was used in the experiments since both axes have similar servocontrol systems. Once the control problem for one axis is solved, it can then be readily duplicated on the other axis, and the general machining problem thus reduces to a motion coordination problem. The axis is driven by a brushless, frameless DC servomotor directly coupled with a highly sensitive tachometer and incremental encoder. The guides are double V-way steel on flat steel surfaces with no rollers. A directly coupled acme type nut-and-lead screw mechanism exists between the stage and the motor. The static friction varied between 3.4 to 5.6 N-m (30 to 50 lb-in.) depending on the table position and direction of motion. The backlash was about 18 μm (700 μin.).

The frameless, brushless DC servomotor has about 8.20 N-m (72.7 lb-in.) peak torque, and 2.99 N-m (26.5 lb-in.) continuous torque capacity at stall. The motor torque constant is 0.82 N-m/A (7.27 lb-in./A); the rotor inertia is 1.16 $\times 10^{-4}$ kg-m^2 (1.03 $\times 10^{-3}$ lb-in.-s^2). The incremental encoder resolution is 1,024 lines per revolution which provides 4,096 counts per revolution using a ×4 factor in the decoding circuit of the digital controller board. The servomotor is powered by a linear servoamplifier that can be used in velocity or current mode. The amplifier commutates the motor phase currents using the encoder signals. A high-sensitivity tachometer provides a high signal-to-noise ratio (SNR) at very low speeds. The tachometer has 55.7 v/(rad/s) (or 5.8328 v/rpm), with maximum terminal voltage of 118 v, which limits the use of the tachometer to within 20 rpm. If a very high-resolution position feedback system is used, the position resolution may be sufficient even at very low speeds to obviate the need for tachometer feedback.

A PC-bus-based motion-controller board using the Motorola[2] 56001 DSP (30 MHz) is used for position control as well as the velocity control loop depending on the amplifier mode. The board's servocontrol algorithm is capable of accepting only encoder inputs. Velocity measurement by the tachometer is fed back to the amplifier only when the amplifier is used in velocity mode.

Friction characterization

Friction is characterized in two different modes:

1. static (stiction) friction when there is almost no motion
2. dynamic friction when there is motion

For very low-speed machining applications, the transition from static to dynamic friction must be characterized. Heretofore, the precise nature of the friction-velocity relationship has not been well characterized [6]. It is goal of this project to advance the body of knowledge regarding this relationship.

For a given movement of a machine tool with proper position and velocity sensors, the stiction friction contribution can be determined and computed. If a machine tool movement results in a position-time trace, then the fundamental underlying equation relating motion and torque at any point t in time is

$$T_m(t) = T_a(t) + T_f(t)$$

where

$T_m(t)$ = motor torque
$T_a(t)$ = acceleration torque
$T_f(t)$ = friction torque

Thus the total torque expended by the machine tool is partitioned into two components—one dealing with acceleration and the other dealing with resistance to motion (friction).

We are interested in the nature of the $T_f(t)$ versus t curves—both fundamentally and in relation to values of other machine tool parameters. Since the motor torque and the acceleration torque can be estimated from amplifier current command and motor velocity measurements, we thus compute friction torque $T_f(t)$ via $T_f(t) = T_m(t) - T_a(t)$.

We are interested in $T_f(t)$ dependencies on various controllable machine, procedural, and environmental conditions (such as current command pulse

[2] Certain trade names and company products are mentioned in the text or identified in an illustration in order to adequately specify the experimental procedure and equipment used. In no case does such identification imply recommendation or endorsement by the National Institute of Standards and Technology, nor does it imply that the products used are necessarily the best available for the purpose.

height, pulse width, table position, velocity, direction, etc.). This friction equation is fundamental for our experiment.

In summary, the long-range friction-analysis sequence is as follows:

1. compute friction traces $T_f(t)$
2. relate $T_f(t)$ to machine tool parameters (position, velocity, direction, temperature, etc.)
3. model $T_f(t)$ relative to these parameters
4. incorporate such a model into controller system algorithms so as to predict and compensate for friction.

EXPERIMENT DESIGN

There are two distinct approaches to characterizing a machine tool—one based on theoretical considerations and the other based on empirical investigation. Ideally, the two approaches are synergistic in the sense that theoretical progress suggests new and interesting empirical investigations, while empirical investigations suggest modifications and extensions to existing theory.

For the problem at hand, theory is limited in allowing us to predict friction with the degree of accuracy needed for ultrasmooth machining. We thus attack the machine tool characterization problem with an empirical, data-driven approach. The approach used is called *design of experiment* or *experiment design*. Experiment design is a rigorous and disciplined approach following principles, tenets, and efficient experimental plans developed over the years. It is a powerful tool for attacking engineering problems and achieving engineering objectives [15–16]. For analyzing data from such experiments, we subsequently apply exploratory data analysis (EDA) [17] techniques to rapidly ferret out underlying structure and to develop data-driven models. The combination of experiment design and exploratory data analysis serves as the backbone of this paper.

Design essentials

For the problem at hand, the following are the essentials necessary to construct a proper experiment design.

Response variables (performance variables)

There are three response variables:

1. distance traveled
2. velocity
3. friction torque

Variations in distance traveled and velocity are indirect indicators of underlying friction. Distance is measured by an incremental encoder whose resolution is 4,096 counts per revolution of the lead screw (that is, 4,096 counts/2,540 μm = 4,096 counts/0.1 in.). Velocity is measured by a high-sensitivity tachometer with a range of 0 to 20 rpm. Alternatively, velocity may be derived from numerical differencing of distance measurements. Friction is derived by $T_f(t) = T_m(t) - T_a(t)$ as described in the friction characterization section. Note that all of the response variables are dependent on and functions of the controllable factors (such as direction, pulse height, etc.). It is of interest to know how each of these response variables is affected by changes in the factors.

Factors

There are many factors that affect traveled distance, velocity, and friction. In our experiments, some were controllable and measurable (such as pulse height), while others were uncontrollable but measurable (such as temperature). Relative to the machine characterization objectives of the experiment, the following five factors were considered:

1. pulse width (duration) of the current signal into the motor (0 to 270 ms)
2. pulse height of the current signal into the motor (0% to 100% of maximum motor torque)
3. nominal table position (0 to 254 m = 0 to 10 in.)
4. side (on left side or right side of the nominal table position)
5. direction (movement left or right relative to entire table)

The measured distance, velocity, and estimated friction will change as a function of these five factors.

Experimental goals

1. *Important factors*: Which of the above five factors affects a given response variable the most? Which affects it the least? What is the ordering of the five factors in terms of their effect on a response variable?
2. *Best settings:* What are the settings of the various factors that yield maximum precision positioning or maximum precision velocity?
3. *Machine tool distance-traveled variation*: Under constant input conditions, what is the variation in distance traveled?
4. *Machine tool low velocity*: Can the machine tool achieve low velocity (0.4 μm/s)?
5. *Machine tool velocity variation*: Can the machine tool hold a specified low velocity to within 1% of target?
6. *Friction characterization:* What is the nature and sensitivity of the friction variable?

A run

A run consists of noting (from the experimental design) the specified settings of the five factors: (1) pulse width, (2) pulse height, (3) nominal table position, (4) side (left/right) of nominal table position, and (5) direction (left/right) of movement; manually moving the table to the nominal table position; specifying the other four factor settings on the computer; sending the specified pulse width and height to the motor (the pulse height is maintained over the entire duration of the signal); and then recording the resulting elapsed distance (via encoder) and velocity (via tachometer). Readings are recorded at time intervals of about 1.7 ms and continue until the pulse width has expired (at most 270 ms). Additional readings are collected until table motion stops (approximately 30% beyond the pulse width termination point). The distance and velocity data are automatically collected, decoded, and recorded onto file. Although actual run time is always less than 1 s, the total run time (including setup) is about 5 min.

Total sample size n

The most important single number which dictates whether the chosen experimental plan will be of the full factorial type or of the fractional factorial type is the total sample size n (the total number of runs that we can actually afford to carry out—being pragmatically sensitive to both time and cost constraints). In this case, a rough upper limit to n is computable from

1. how much total experimentation time is available
2. how long it takes to make a single run

The total length of time permissible for all experiments was one week—five working days. A single run (including setup) takes 5 min, which implies 12 runs per hour, a maximum of 100 runs per day, and 500 runs per week. Thus n = 500 is the approximate upper bound on the sum total number of all runs available for all experiments.

The experiment designs

Given the overall goals of the experiment as described above and given the sample size constraints, we ran eight experiments (as summarized below in Fig. 1). The first two of these experiments are now described to illustrate the approach.

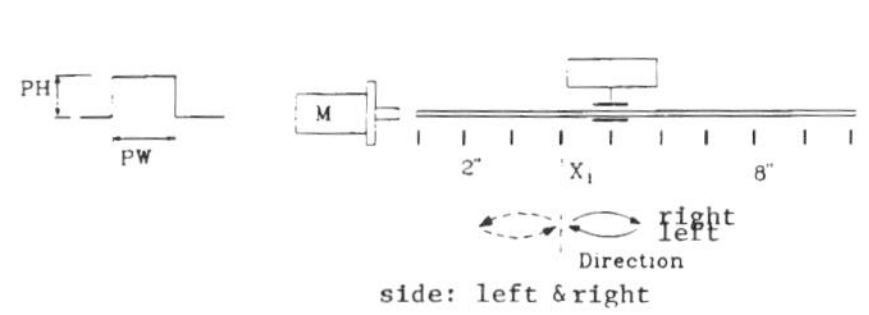

Experiment Set	Conditions of Experiment	Number of Experiments Conducted
# 1	PH{40%,100%}, PW{90,270 ms}, X_i : { 2",8" }; + −, − +; (2 × 2 × 2 × 4 + 4)	36
# 2	PH{70%}, PW{180 ms}, X_i : { 5"}; + − : Repeat 20 times	40
# 3	Repeat # 1 one day later	36
# 4	PH{70%}, PW{180 ms}; − +, + −; X_i : {5",7",2",8",4",5",6",1",9",3",5"};	44
# 5	PH{40%,100%}, PW{90,270 ms}; Repeat : 5, 5	80
# 6	PH{40%,100%}, PW{90,270 ms}; X_i : { 1", . . . , 9"}; + −, − + ; (2 × 2 × 9 + 1) × 4	148
# 7	PH{40%,55%,70%,85%,100%}, PW{90,135,180,225,270 ms}; X_i : { 2",5",8"}; + −;	90
# 8	PH{40%,70%,100%}, PW{90,180,270 ms}; X_i : {2",5",8"}; + −;	198
	Total	672

Fig. 1. Summary of experiment designs.

Experiment #1

Goal: Which of the five factors are important?

Factor settings:

1. pulse width (PW): 2 levels—90 and 270 ms
2. pulse height (PH): 2 levels—40% and 100%
3. nominal position: 2 levels—0.0508 and 0.2032 m (2 and 8 in.)
4. side: 2 levels—left (–) and right (+) of nominal position
5. direction: 2 levels—left (–) and right (+)

Design: A 2^5 full factorial design (32 runs), augmented by a "center point" (with pulse height = 70%, pulse width = 180, and nominal table position = 0.1270 m (5 in.). At this center point, a total of 4 (2×2) runs were made (2 for each side and 2 for each direction).

Total number of runs: 32 + 4 = 36

Comment: Although pulse height, pulse width, and nominal (table) position are continuous factors, it is an experiment design tenet that the significance of such factors may be ascertained by judiciously choosing a few relevant discrete values (such as pulse height at 90 ms and 270 ms. The full factorial design will tell us not only if the primary factors are important, but also if any of the interactions are important. The center-point data will allow for model verification and also allow for curvature testing.

Experiment #2

Goal: For a constant input, does the machine tool have repeatable response (distance-traveled)?

Factor Settings:

1. pulse width (PW): fixed at 180 ms
2. pulse height (PH): fixed at 70%
3. nominal position: fixed at 0.1270 m (5 in.)
4. side: fixed at right side
5. direction: 2 levels—left (–) and right (+)

Design: A completely randomized design. Go to the middle of the table. Go right and return (therefore two runs). Do we return to the original point? Repeat 20 times. Do we return to the same point?

Total number of runs: $2 \times 20 = 40$

Comment: This experiment is fundamental. We cannot model a phenomenon across factors if the phenomenon is not repeatable under fixed factor settings. This experiment tests that repeatability. If the machine tool dynamics are not sufficiently repeatable, then fundamental changes in the machine tool components need be made to assure acceptable repeatability under constant conditions.

DATA ANALYSIS

The EDA approach, pioneered by Tukey [17], is characterized by

1. keeping a completely open mind about the model
2. thinking "diagnostic" rather than "confirmatory"
3. letting the data "speak for themselves"
4. analyzing the raw data rather than analyzing gross summary statistics
5. keeping an eye to outliers and unusual subsets
6. using techniques that have few assumptions
7. testing the four basic underlying assumptions
8. emphasizing sufficiency rather than rigor
9. using graphical rather than quantitative procedures
10. applying multiple techniques to examine the data
11. letting the flow of the analysis be data driven
12. keeping summary statistics simple and robust

The remainder of this paper illustrates the use of many such EDA techniques—some of which are new—to the high-resolution turning machine problem. The EDA techniques include:

1. multitrace plots
2. multiplots
3. cube plots
4. DEX mean plots (new in part)
5. block plots (new)

We will see how the application of these techniques easily and clearly uncovers the dynamics structure of the underlying machine tool and thus serves as a powerful tool for machine tool characterization. All analyses were carried out with the public-domain graphical data analysis system DATAPLOT developed at NIST (see [18]).

Multitrace plots

In a data analysis context, the following "juxtaposition principle" is fundamental: *The essence of analysis is comparison, and the essence of comparison is juxtaposition.* A direct application of this principle is the multitrace plot which, as the name implies, is simply a plot consisting of the various response traces from different conditions or replications of an experiment. When there is no condition-to-condition effect or no replication-to-replication effect, then all such traces will be nearly identical. In practice, such is not usually the case because of systematic or random components in the equipment, procedure, measuring device, environment, etc.

Consider the following important machine-tool question regarding repeatability. Given that we start at some nominal (but fixed) table position and we supply a signal of fixed pulse width and fixed pulse height to move us away from the point and then apply the same fixed width and fixed pulse height signal to move back toward the original point, will the machine tool in fact return to the original point? Repeatability is an elementary machine tool positioning property. If a machine tool cannot reset itself by moving back to an initial point, then how can the machine tool user ever have confidence in the execution of more complex positioning sequences?

Experiment 2 addressed this repeatability problem as did Experiments 5 and 8. In Experiment 2, the pulse height was fixed (70%), the pulse width was fixed (180 ms), and the nominal table position was fixed (0.1270 m = 5 in.). The machine tool was manually moved to the specified location and four runs/legs (all with identical signal) were executed:

1. side = right; direction = right
2. side = right; direction = left
3. side = left; direction = left
4. side = left; direction = right

Ideally, after leg 2, we should have returned to the initial point; similarly, after leg 4, we should have returned to the initial point. The above manual resetting and sequencing of four runs was repeated a total of 20 times. Ideally, the positioning after any given leg should be the same across the 20 repeats. The EDA multitrace graphical analysis of Experiment 2 is given in Figs. 2–4.

In Fig. 2 the data was drawn from only the first leg (of the four)—starting from the initial point and moving out to the right side in the right direction. Ideally if the machine tool is repeatable, then we should end up with 20 nearly identical distance traces—all culminating in the same distance being covered from the 180 ms pulse-width signal.

In interpreting Fig. 2, we see that the typical distance traveled in the 180-ms burst plus carry-over is 380 μm. In terms of variation, the observed relative

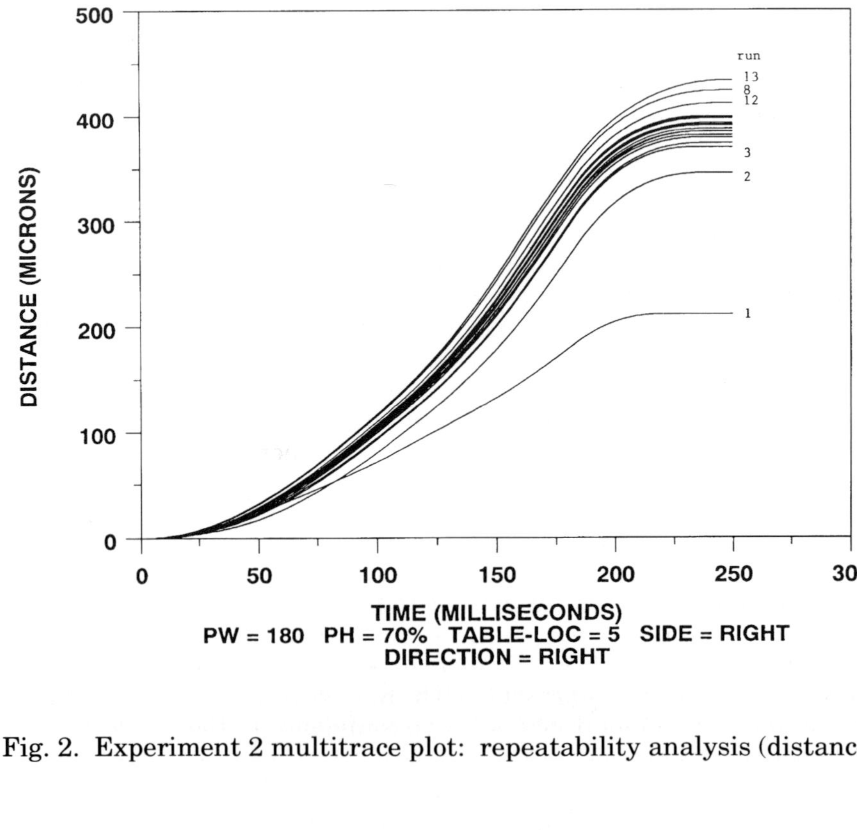

Fig. 2. Experiment 2 multitrace plot: repeatability analysis (distance).

variation exceeds 50%. This is well beyond the accepted tolerance in a high-precision engineering environment. It is clear from Fig. 2 that some runs exceeded the typical by 50 μm; other runs (e.g., the first run) were smaller than the typical by 180 μm. The range of the data was much too large—almost 200 μm. It is suspected that as the moves are repeated and the machine tool guide ways become better lubricated, the variation between runs will decrease; nevertheless, even with the first run removed, the range of the data was nearly 100 μm (about 25%)—well beyond our high-precision machining device goal. Outliers are certainly evident in run 1 where there is a startup effect. This having been identified, we must understand it to the point of predictability and then compensate for it in the controller. In addition, there is a systematic drift across the 20 runs. Initial runs travel less far; intermediate runs travel farther; final runs are more clustered at intermediate distances (to be seen more clearly in Fig. 4). Linearity is not apparent in Fig. 2. Ideally, an individual distance trace should start as quadratic (during the linear velocity ramp-up portion), continue as linear (during the constant velocity full-signal portion), continue as quadratic (during the linear velocity ramp-down portion), and finish as linear (during the 0-velocity, no-signal portion).

None of the traces in Fig. 2 have the above structure. In particular, not even the primary linear substructure is evident. The conclusion is that other forces are coming into play (for example, a large stiction friction component) that are changing the basic shape of the distance response curve. Another possibility is that the torque constant for the motor is not large enough for the encountered stiction friction.

Figure 3 is a rearrangement of the first five pairs of traces in Fig. 2. In Fig. 3 the return trip (leg 2, shown on a dotted line) has been attached to the end of the outgoing trip (leg 1, solid line). We thus see a continuous sequence for any given run; all 10 runs (out of the 40) are shown. As before, distance traveled is on the vertical axis; the horizontal axis is a coded time. If the machine tool is repeatable, then for any given run, leg 2 should return to the coded start point (0) of leg 1.

From Fig. 3 we see that a strong direction effect exists. For run 1, leg 1 ends at about 240 μm, but leg 2 does not end at 0 but at –700 μm; similarly for run 2, leg 1 ends at 300 μm, but leg 2 ends at –500 μm; etc. There is a strong left/right direction effect of roughly 500 μm.

Figure 4 is a dual-trace stiction friction analysis based on friction torque data from Experiment 2. The solid step function is motor torque; the dotted trace is friction torque. In the terminology of the friction characterization section, the horizontal axis is t, the solid step function is $T_m(t)$, and the dotted trace is $T_f(t)$. The motor torque $T_m(t)$ is a step function that is at a maximum over the entire pulse width. After the pulse width is completed, the motor torque is 0. Ideally the stiction friction torque $T_f(t)$ should be high in the first portion of the pulse width and then should drop off to 0 for the later portion of the pulse width. However, the stiction friction trace $T_f(t)$ does not drop off. It remains high near the motor torque value $T_m(t)$ over the entire pulse width. It eventually drops off after the pulse width is completed. This plot provides revealing insight into this machine tool system. It shows that over the entire pulse width all of the motor torque is "consumed" by stiction friction torque. The very important

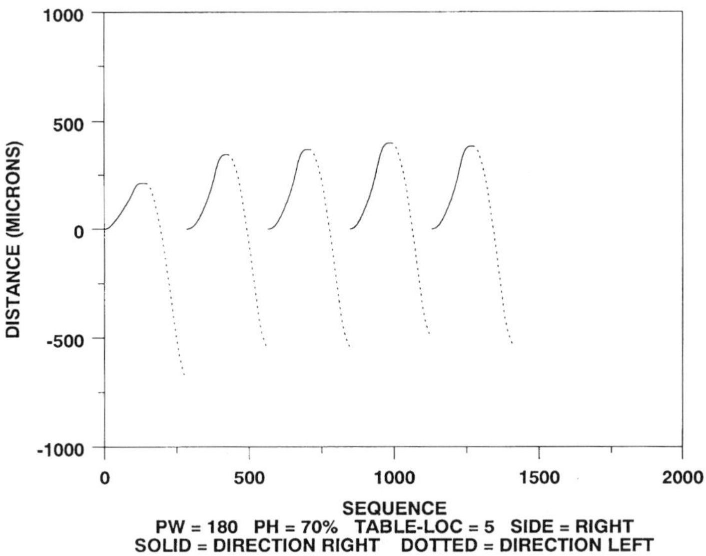

Fig. 3. Experiment 2 multitrace plot: repeatability analysis (distance).

engineering conclusion from this multitrace plot is that the motor for this machine tool system is inadequate—the friction torque should not equal the motor torque over the entire pulse width.

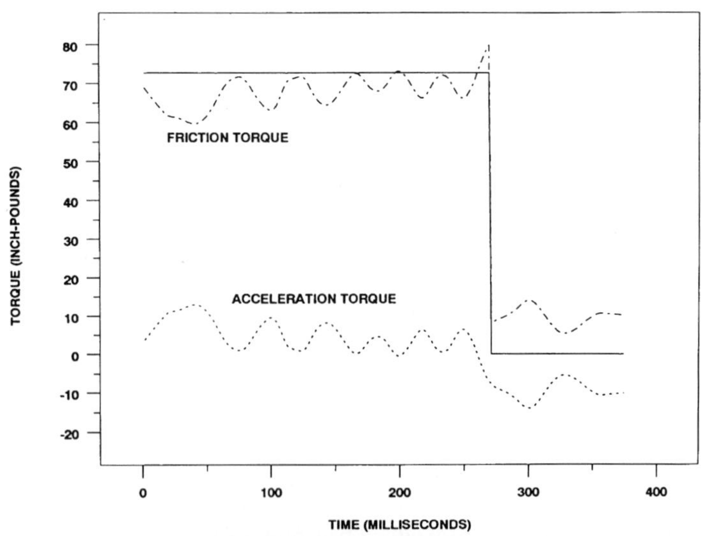

Fig. 4. Experiment 2 multitrace plot: friction analysis.

Multiplots

Another application of the juxtaposition principle is *multiplotting*—the positioning of multiple plots on the same page to enable comparisons across plots. The plots themselves may be of any relevant type, but that is less important than the overriding issue that the plots are all on the same page—within the same eye sweep—so that the analyst's natural comparative and pattern-recognition abilities come into play in the analysis.

In the previous sections we saw how the judicious use of multitrace plots uncovered important information about the repeatability of the machine tool. Figures 2 and 3 both involved distance measurements versus time. That was certainly appropriate for examining the position repeatability question. To address the low-velocity and the stable (small variation) velocity questions, we find in the experimental goals of the design essentials section two experimental goals involving velocity:

1. Can the machine tool achieve low velocity (0.4 μm/s)?
2. Can the machine tool hold a specified low velocity to within 1% of target?

Just as we generated distance-versus-time multitrace plots to examine the position repeatability questions, so too we can generate corresponding velocity-versus-time multitraces to examine velocity questions. We shall do that, but we will also invoke the multiplotting technique to collect all of this velocity information on a single page. In fact, it is useful for gaining insight into the machine tool to construct a multiplot consisting of six plots—three distance plots and also the corresponding three velocity plots. In doing so, we can not only examine distance measurements and velocity measurements but also examine how the velocity and distance measurements track with one another. Such a multiplot is shown in Fig. 5.

Figure 5 has six plots all from Experiment 2:

1. top left: distance versus time, leg 1, 20 runs (Fig. 2)
2. top middle: distance versus time, legs 1 and 2, 10 runs (an extension of Fig. 2)
3. top right: distance versus time, legs 1 and 2, 10 runs (Fig. 3)
4. bottom left: velocity versus time, leg 1, 20 runs
5. bottom middle: velocity versus time, legs 1 and 2, 10 runs
6. bottom right: velocity versus time, legs 1 and 2, 10 runs.

The velocity data could have been obtained in two different ways: (1) from tachometer measurements and (2) from numerical differencing of distance data. Both methods were used and the plots from the two methods were virtually identical. The velocity traces shown are from the second method. If the machine tool is capable of low velocity and is stable, then the velocity traces should be about 0.4 μm/s. If the machine tool has constant velocity, then the velocity traces should be within 1% of one another. If the machine tool has idealized form, then the velocity traces should (1) start up as linear ramp-up, (2) continue as constant full power, (3) continue as linear ramp-down, and (4) finish as 0 velocity when the pulse width is completed.

From any of the three velocity plots, we can see that the typical velocity is about 40 μm/ms = 40,000 μm/s. This is many orders of magnitude too fast relative to the goal of 0.4 μm/s. The goal for velocity variation was 1% variation, and with a typical velocity value of about 40 μm/ms, this translates into a

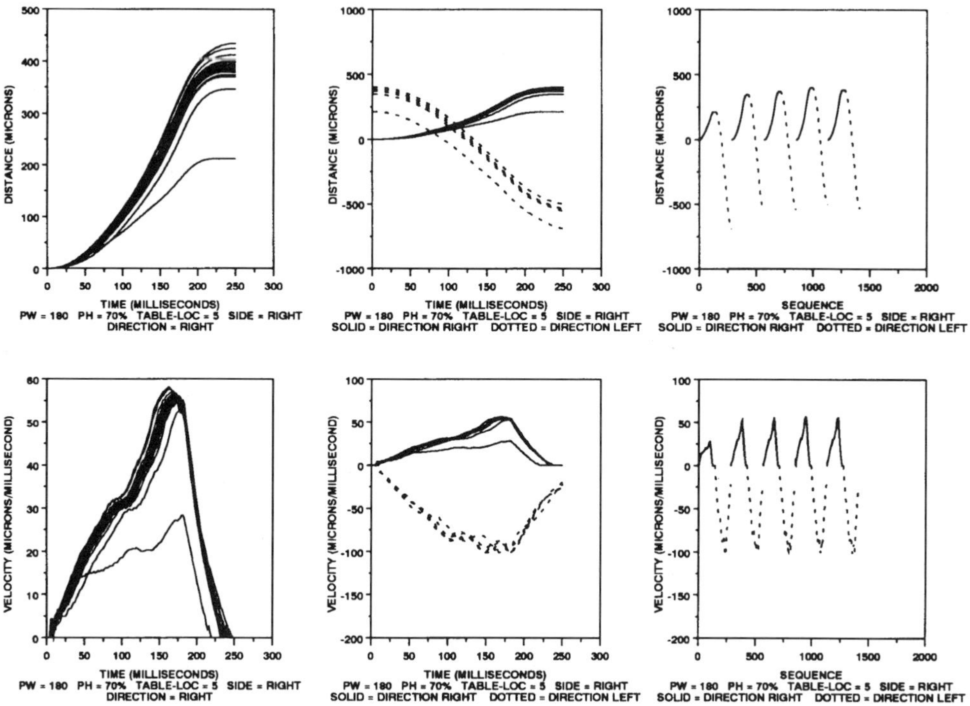

Fig. 5. Experiment 2 repeatability analysis (distance and velocity).

between-run spread of about 0.4 μm/ms. From plot 4 of Fig. 5, we see that run-to-run differences in velocity far exceed 0.4 μm/ms. Run 1 differs from typical by 20 μm/ms. Even with run 1 removed, there is about a 10 μm/ms spread across the remaining 19 runs. As before, run 1 is certainly outlying and shows a startup effect that affects velocity as well as position. As with the distance analysis, plot 4 of Fig. 5 shows there is a systematic drift in velocity across the 20 runs. There is no evidence of the ideal constant velocity portion that should dominate the middle of any single trace. Again, as with the distance analysis, either the stiction friction component is too large or the motor torque is too small (or both). Finally, in the bottom-middle and bottom-right subplots of Fig. 5, the solid velocity traces are for the trip out, the dotted velocity traces are for the return trip. Ideally, these sets of traces should be mirror images of one another; they are not. The maximum out-trip velocity has a magnitude of about 50; the maximum return-trip velocity has a magnitude of about 100. As with the distance analysis, there is a strong directional effect with velocity also.

Cube plots

There are five factors that were examined in terms of their effect on distance, velocity, and friction of the machine tool under study: (1) pulse width, (2) pulse height, (3) nominal table position, (4) side, and (5) direction.

Important fundamental questions in formulating any general model to serve as a basis for friction compensation in a controller include: Which of the above five factors affects a given response variable the most? Which affects it the least? What is the ordering of the five factors in terms of their effect on that response variable?

Experiments 1, 3, and 5 provided data to answer these questions. It is sufficient to apply an EDA graphic called a cube plot to data from Experiment 1 to show the effectiveness of the technique in providing an answer to the question of how relatively important the five factors are when compared to one another.

As described in detail in the experiment designs section, Experiment #1 was a 2^5 full factorial design (32 runs) augmented by a center point (4 runs). The primary settings for the five factors were

1. pulse width (PW): 2 levels—90 and 270 ms
2. pulse height (PH): 2 levels—40% and 100%
3. nominal position: 2 levels—0.0508 and 0.2032m (2 and 8 in.)
4. side: 2 levels—left (–) and right (+) of nominal position
5. direction: 2 levels—left(–) and right (+)

The center-point data were obtained by setting pulse height, pulse width, and nominal position at center values (70%, 180 ms, and 0.1270m [5 in.], respectively), and varying the remaining two factors (side and direction) over all four possible combinations.

Figure 6 is a collection of nine plots and was motivated by the same juxtaposition principle that motivated the general multiplot technique. The three dimensions of the cube are the following factors: (1) pulse width, (2) pulse height, and (3) nominal table position (Table Loc.). Of the 36 runs that constituted Experiment 2, Fig. 6 makes use of only nine (25%) of the runs—the runs corresponding to side: fixed at right, and direction: fixed at left. The cube plot is thus constituted by the eight vertices of the cube plus an additional plot at the center point of the cube. The cube plot is an excellent graphic when the underlying experiment design is of the 2^k full factorial type. In this case, Experiment 2 was a 2^5 full factorial design, but we are looking at a quarter-replicate and so in effect we are examining data from a 2^3 full factorial design. Each of the nine plots was generated under different pulse-width, pulse-height, and table-position settings. For example, the bottom-left-front plot was generated at pulse width = 90 ms, pulse height = 40%, and nominal table position = 0.0508m (2 in.), while the top-right-back plot was generated at pulse width = 270 ms, pulse height = 100% ms, pulse height = 100%, and nominal table position = 8 in.

Each of the nine plots in the cube plot has distance traveled (in μm) on the vertical axis, and time (in ms) on the horizontal axis. For accent, the trace has been solid filled. The horizontal axis is fixed at 0 to 400 ms. The vertical axis is fixed at 0 to 2,500 μm. Fixing both axes is crucial to make comparisons across factors. The MAX = 0. . . legend within each plot is the total distance actually traveled for that plot. Thus for the lower-left-front plot, the total distance traveled was 297 μm, while for the upper-right-back plot, the total distance traveled was 1,669 μm.

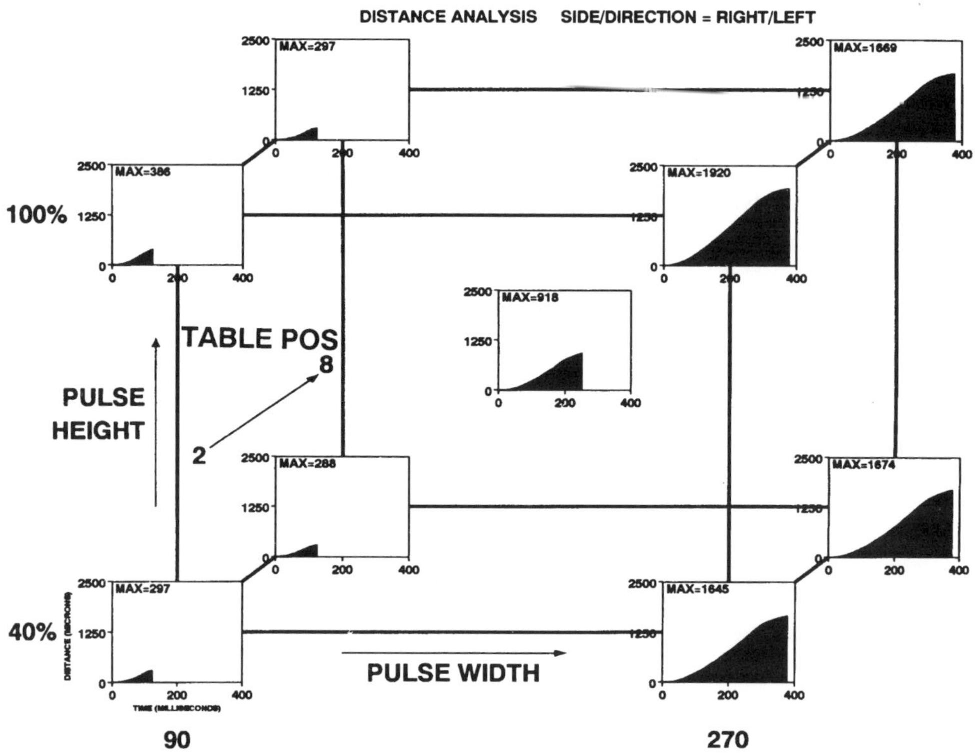

Fig. 6. Cube plot of distance traces of a single-pulse friction analysis of low-velocity turning machine in Experiment 1.

If there were no pulse height effect, then any pair (for fixed pulse width and fixed table position) of vertical plots would have nearly identical traces. If there were no pulse width effect, then any pair of horizontal plots would be nearly identical. If there were no table position effect, then any pair of front-back plots would be nearly identical. The greater the discrepancy between any pair of plots, the more significant the corresponding factor.

The pulse width is by far the most significant factor (as it should be). Comparing left to right, we see the traces are markedly different, much more so than the pulse height and table position factors. Comparing the maximum distance traveled from left to right, we see that the pulse width effect (the amount of change in distance as we change the pulse width from 90 to 270 ms) is well over 1,300 μm. The pulse height has a much smaller effect but it is basically positive. For example, a pulse height of 100% yields slightly higher distances than a pulse width of 40%. The table position also has a much smaller effect. A table position of 0.0508m (2 in.) yields slightly higher distances than table position of 0.2032m (8 in.). Thus, in order of importance, the factors as gleaned from the cube plot are pulse width, pulse height, and nominal table position. While the pulse width is the most important factor, there is also some hint of mild interactions.

DEX mean plots

The cube plots shown in Fig. 6 give us insight into the relative importance of the main effects. On the other hand, very little direct information about interactions is given by such plots. To develop a good machine tool model, we must answer the following three questions:

1. What is the relative importance of the various factors (including interactions)?
2. What is an appropriate machine tool model?
3. What is the parameter estimate for each factor effect

The visualization and estimation of the effects of these factors (including interactions) lead us to the graphical techniques of this section DEX (design of experiment) mean plots and the following section (block plots).

As before, let us examine the data from Experiment 1 restricted to side or direction = right or left. This is a 2^3 experiment (with augmented center point). What is an appropriate model for the data from this experiment? One commonly employed and very flexible model is

$$Y = c + 0.5*[B_1*X1 + B_2*X2 + B_3*X3 \\ + B_{12}*X1*X2 + B_{13}*X1*X3 + B_{23}*X2*X3 + B_{123}*X1*X2*X3]$$

where

X1 = coded pulse width	(−1 = 90 ms and +1 = 270 ms)
X2 = coded pulse height	(−1 = 40% and +1 = 100%)
X3 = coded table position	(−1 = 0.0508m [2 in.] and +1 = 0.2032m [8 in.])

and

c	= general additive constant
B_1	= pulse width effect
B_2	= pulse height effect
B_3	= table position effect
B_{12}	= pulse width-pulse height interaction effect
B_{13}	= pulse width-table position interaction effect
B_{23}	= pulse height-table position interaction effect
B_{123}	= pulse width-pulse height-table position interaction effect

One advantage of the above model is that with the substitution of the least squares estimates for the eight coefficients, the resulting prediction equation yields an exact fit at the eight vertices of the cube. These least squares estimates may be computed via any regression program; on the other hand, for two-level designs, the estimation process is arithmetically trivial [15] and the estimation is done in a fast, convenient fashion. We will arrive at estimates shortly, but first we examine the data via DEX mean plots.

Figure 7 is a matrix of DEX mean plots. This matrix is a multiplot of estimated main effects and two-term interactions from a 2^3 experiment. All six plots in Fig. 7 have mean response (that is, the mean distance traveled in µm) on the vertical axis. All six plots have two coded levels (–1 and +1) on the horizontal axis. The plots on the diagonal are main effects plots; the horizontal axes for the diagonal plots are

top left diagonal:	coded pulse width
middle diagonal:	coded pulse height
bottom right diagonal:	coded nominal table position

The off-diagonal plots are interaction mean plots. Such plots are possible because the set [–1,+1] forms a closed group under multiplication (thus, for example, the product X1*X3 also yields values of only –1 and +1). The horizontal axes for the off-diagonal plots are

top center:	coded pulse width-pulse height interaction
top left:	coded pulse width-table position interaction
right center:	coded pulse height-table position interaction

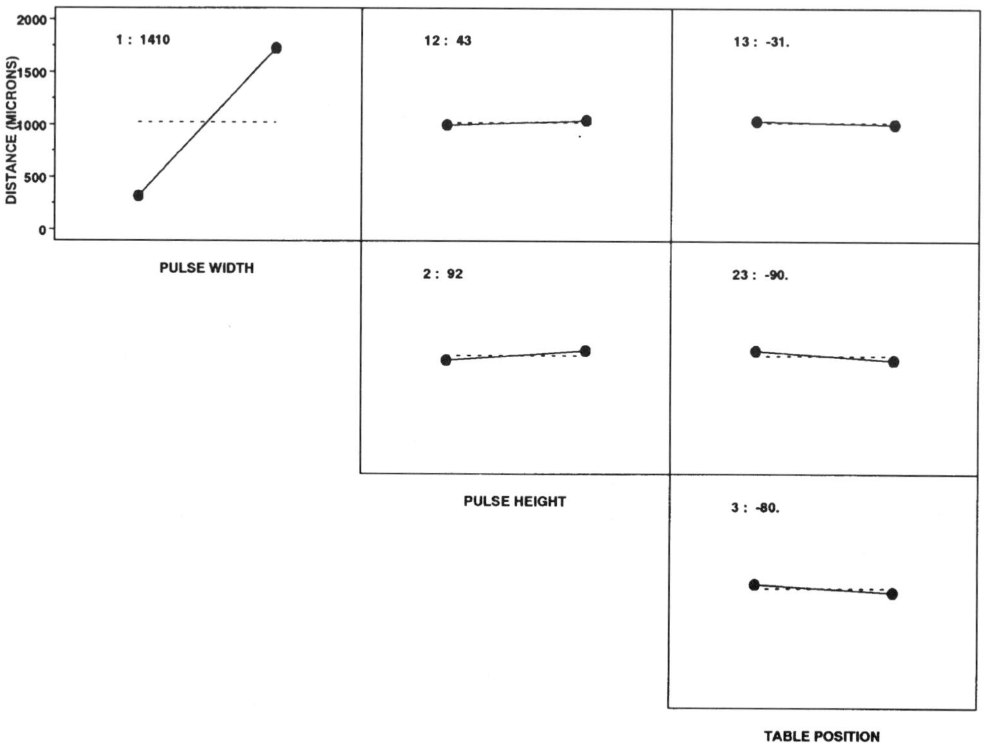

Fig. 7. Matrix of 3-factor DEX mean plots of distances for Experiment 1. Side/direction = right/left

The least squares estimated effect is given as the legend of each plot. Since all plots have identical 0- to 2,000-μm vertical axis range, the steeper the slope, the more significant the factor.

DEX mean plots form the graphical foundation for the least squares estimation of the model effects; that is, the least squares estimate for the effect for a given plot is mathematically identical to the difference in the mean response between the high setting (+1) and low setting (–1). For a factor to have an effect, the response from one setting of the factor must change significantly. For important factors, the difference in the mean response will be large; for unimportant factors, the difference in the mean response will be small. Thus steep slopes indicate important factors and shallow slopes indicate unimportant factors. By scanning the three diagonal plots, we see that the estimated

pulse width effect B_1 = 1,410
pulse height effect B_2 = 92
table location effect B_3 = –80

Similarly, the estimated

pulse width-pulse height effect B_{12} = 43
pulse width-table location effect B_{13} = –31
pulse height-table location effect B_{23} = –90

The actual estimates are given as legends within each plot. More importantly, since all six plots are given with an identical vertical axis scale (0 to 2,000 μm), a simple visual scan of the matrix of plots immediately reveals the relative importance of the three main effects and the three interactions. A scan of Fig. 7 thus yields the following ranked list:

1 pulse width (effect = 1,410)
2. pulse height (effect = 92)
3. pulse height-table position interaction (effect = 90)
4. table position (effect = –80)
5. pulse width-pulse height interaction (effect = 43)
6. pulse width-table position interaction (effect = –31)

Do the conclusions about pulse height, pulse width, table position, and interactions generalize beyond side/direction = right/left? To ascertain this, we augment the 2^3 DEX mean plots matrix as given above with similar matrices for side/direction = left/left, right/left, and right/right. Such a collection of plot matrices is given in Fig. 8. As before, steep slopes imply important factors; flat slopes imply unimportant factors. Of the 24 plots in Fig. 8, the matrix of six plots in the lower left is for side/direction = left/left, in the, lower right is for side/direction = right/left, in the upper left is for side/direction = left/right, and in the upper right is for side/direction = right/right. Within each six-plot matrix, main factors are on the diagonal and two-term interactions are off-diagonal. Note that to facilitate comparison, all 24 subplots in Fig. 8 have a common (0 to 2,000 μm) scale.

If our conclusions about the relative importance of the three main effects and three interactions is true in general, then four 6-plots should be nearly identical. If not, then this indicates an outlier condition and/or an additional dependence (or interaction) between pulse width, pulse height, or table position, with side or direction.

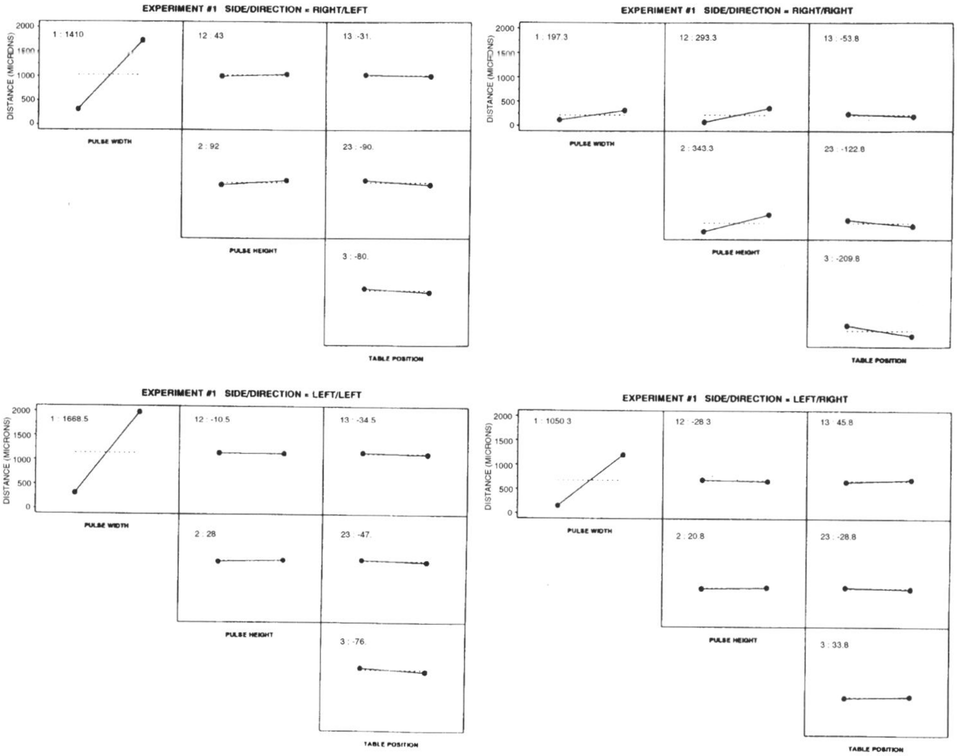

Fig. 8. Matrix of 3-factor DEX mean plots of distances for Experiment 1.

The four sets of plots are not identical; we conclude that the pulse width is by far the most important factor in three out of four cases. The fourth case, where both the side and direction are right, appears to be an outlier. This may be related to a startup effect since right/right was run first out of the four cases.

It is of interest to determine if one can arrive at even more definitive conclusions (and a model) that encompass all of the data in Experiment 1 and all five factors—including side and direction. This is done by a generalization of the three-factor mean plot matrix (Fig. 8) to a five-factor mean plot matrix given in Fig. 9.

Figure 9 consists of 15 mean plots: 5 main factor mean plots (along the diagonal) and 10 two-term interaction mean plots (off-diagonal). All plots have common vertical axis scales (0 to 2,000 μm). The plots along the diagonal reveal the relative effect of pulse width, pulse height, table position, side, and direction (from upper left to lower right). The off-diagonal mean plots provide information on the corresponding two-term interaction (there are 10 two-term interactions); thus, for example, the upper-right mean plot is for the pulse width-direction interaction. The legend is the least squares estimate of the effect.

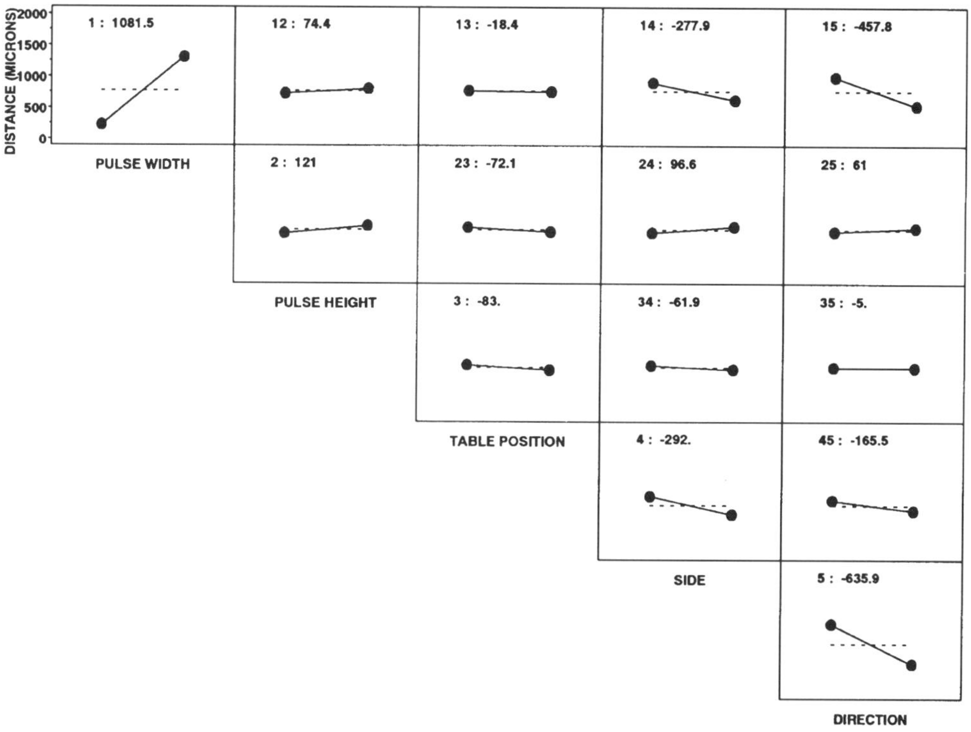

Fig. 9. Matrix of 5-factor DEX mean plots of distances for Experiment 1.

Figure 9 shows that although pulse width is the most dominant factor (as it should be), there are clearly strong contributions from direction and from some two-term interactions. The ranked list of most important factors is thus: pulse width (estimated effect = 1,081.5 μm), direction (–623.875), interaction of pulse width with direction (–457.75), side (–292), interaction of pulse width and side (–277.875), and interaction of side and direction (–165.5).

Block plots

We finish our discussion of EDA techniques by proposing a new technique, the block plot, which shows promise in ferreting out detailed main effects and interaction structure. The foundation for the block plot is the following general conclusions principle: In order to *form global conclusions about a factor, that conclusion must hold not only for all of the data, but also for all / most relevant subsets of the data.* An individual block plot focuses on one factor only (the primary factor), but it examines the behavior of that one factor across all levels of all other factors (nuisance factors). A block plot is defined as

1. A plot of the raw data with the vertical axis being the raw response. For all five block plots in Fig. 10 the vertical axis is distance traveled (in μm).

2. The plot character is the coded setting of the factor we are focusing on (that is, the primary factor). For the first block plot in Fig. 10 the primary factor is pulse width, for the second block plot, pulse height, etc.
3. The horizontal axis is a composite multifactor axis consisting of all combinations of those factors that we are not focusing on (that is the nuisance factors). For the first plot in Fig. 10 the primary factor is pulse width and thus the horizontal axis is a combination of all four remaining factors: pulse height, table position, side, and direction. In this case there will be $2 \times 2 \times 2 \times 2 = 16$ distinct combinations or blocks of these four factors.

An abbreviated guideline to interpreting block plots is as follows:

1. Consistent within-block ordering across all or most blocks implies that the primary factor is significant.
2. Systematic change in the within-block ordering with settings of some nuisance factor implies an interaction between the primary factor and that nuisance factor.
3. Systematic changes in block size with settings of some nuisance factor also imply interaction.
4. Systematic between-block location shifts across settings of some nuisance factor imply that the nuisance factor is significant.

Figure 10 contains five block plots—one each to examine each of the five factors: pulse height, pulse width, table position, side, and direction.

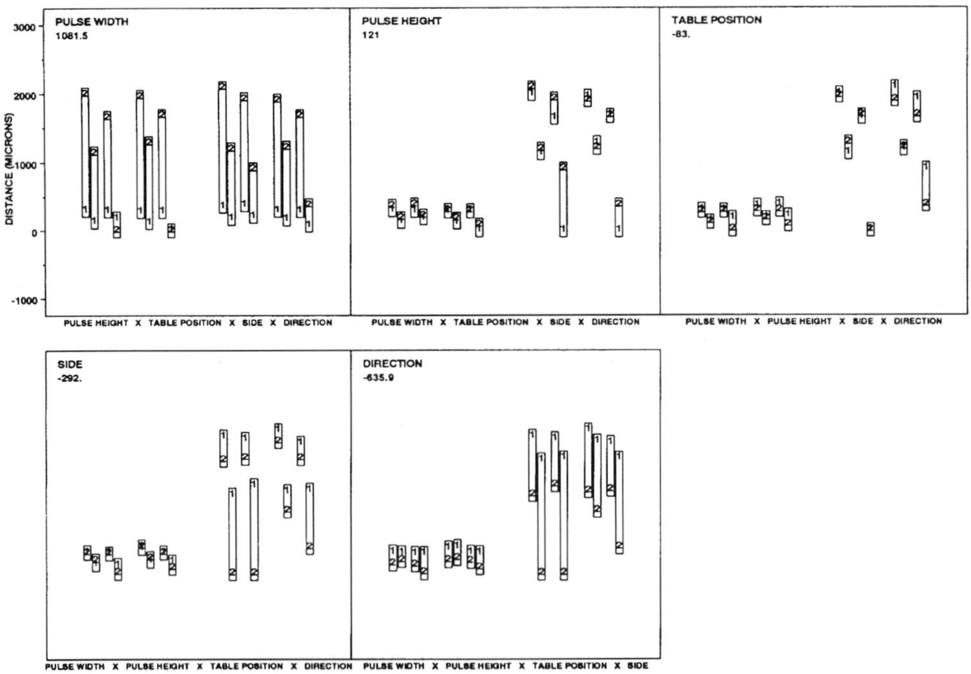

Fig. 10. Matrix of 5-factor block plots for Experiment 1.

Each of the block plots in Fig. 10 replots the same $2^5 = 32$ data points. Each block plot focuses on a different primary factor and, therefore, has different nuisance factors. For example, the top-left block plot of Fig. 10 has

1. primary factor: pulse width
2. plot points: 32 points; 16 from pulse width = 90 and 16 from pulse width = 270
3. vertical axis: distance (in μm)
4. horizontal axis: combination of all four factors other than pulse width (namely, pulse height, table position, side, and direction)
5. plot character: primary factor = pulse width

Similarly, for example, the bottom-right block plot of Fig. 10 has

1. primary factor: direction
2. plot points: 32 points; 16 from direction = left and 16 from direction = right
3. vertical axis: distance (in μm)
4. horizontal axis: combination of all four factors other than direction: pulse height, pulse width, table position, and side
5. plot character: primary factor = direction

We may conclude that the pulse width has the largest effect by far—270 ms larger than 90 ms (as it should be); pulse height has a relatively small effect with some outliers—100% larger than 40% (as it should be); table position is also a relatively small effect—2 in. higher than 8 in.; side interacts with pulse width and the right is higher than the left; and direction interacts with pulse width and the right is higher than the left.

The block plot is a powerful diagnostic technique that gives us yet another perspective into the underlying structure of the data. The plot itself can be easily augmented by some simple binomial probability calculations to carry out tests of significance without requiring homoscedasticity and normality assumptions (as typical with ANOVA F tests). For example, in examining the upper-right block plot of Fig. 10, one could pose the question: What is the probability that *direction = right* could consistently (in all 16 blocks) yield higher distances than *direction = left*—simply by chance? This is binomially equivalent to the question: What is the probability of obtaining 16 heads in 16 flips when the coin is in fact fair? The answer to both of these questions is 1/2 raised to the 16th power (≈ 0.00002), which yields a simple test of significance, that is, in this case well beyond the 5% and 1% levels.

Such significance tests are possible with the block plot, but the real worth of the plot—in the spirit of exploratory data analysis—is the added diagnostic power it puts in the hands of the analyst for detecting main effects, interactions, outliers, unusual combinations of settings, and so on.

CONCLUSIONS

The general engineering goal was to improve the performance of the control of machine tools and to attain very low speeds under large friction conditions.

The specific engineering goal was to investigate the performance of a particular high-precision single point diamond turning machine tool. The performance of the machine tool under open-loop conditions was tested and observed to be unsatisfactory for the stated friction-characterization goals of the project. In particular, we concluded the following.

1. The stiction friction component at these low speeds was extremely large. This led to a lack of repeatability of fixed-condition runs of the order of several hundred μm.
2. The current motor's maximum torque was insufficient for the encountered stiction friction. For low-velocity machining, a higher torque motor is of prime importance to achieve the required closed-loop performance.
3. A significant directional effect exists. Its interaction with pulse width is also important.

EDA conclusions

In terms of EDA, our conclusions are as follows.

1. Each EDA technique approached the data from a different perspective and so added its own contribution to understanding the system.
2. Multitrace plots were important for testing statistical control, repeatability, and predictability.
3. Multiplots, cube plots, and DEX mean plots were valuable for assessing factor effects.
4. The newly introduced block plot provided additional insight into factor effects and interactions.

ACKNOWLEDGMENTS

This work was supported in part by the American Statistical Association, the National Science Foundation, and the National Institute of Standards and Technology. The encouragement of Robert Lundegard, Chief of Statistical Engineering Division at NIST, while providing the necessary resources during the course of this project is greatly appreciated. Our thanks to NIST colleague Eric Lagergren for his many valuable suggestions in improving this manuscript.

REFERENCES

1 N. Ikawa, R. R. Donaldson, R. Komanduri, W. Konig, P. A. McKeown, T. Moriwaki and I. F. Stowers, Ann. CIRP, 39/2 (1991) 587–594.
2 T. Nakasuji, S. Kodera, S. Hara, H. Matsunaga, N. Ikawa and S. Shimada, Ann. CIRP, 39/1 (1990) 89–92.
3 K. Ueda, A. Amano, K. Ogawa, H. Takamatsu and S. Sakuta, CIRP Ann., 40/1 (1991) 554–558.
4 W. L. Nelson, J. Basic Eng. (Mar. 1961) 65–76.
5 S. Kato, N. Sato and T. Matsubayashi, Lubr. Technol., 94 (1972) 234–247.
6 J. Halling (ed.), Principles of Tribology, MacMillan Education LTD, 1989.

7 B. Armstrong-Helouvry, Stick-slip arising from Stribeck friction, in: Proceedings of the 1990 International Conference on Robotics and Automation, IEEE, Cincinnati, Ohio, May 1990, pp. 1377–1382.
8 N. E. Ehrich, An investigation of control strategies for friction compensation, M.S. thesis, University of Maryland, College Park, 1991.
9 J. W. Gilbart and G. C. Winston, Automatica, 10 (1974) 125–131.
10 E. Tung, G. Anwar and M. Tomizuka, Low velocity friction compensation and feedforward solution based on repetitive control, in: Proc. of 1991 ACC, Boston, Mass., vol. 3, pp. 2615–2620.
11 S. Cetinkunt, W. L. Yu, J. Filliben and A. Donmez, Friction characterization experiments for precision machine tool control at very low speeds, 1992 American Control Conf., Chicago, Ill., June 24–26, 1992.
12 J-S. Hu and M. Tomizuka, Adaptive asymptotic tracking of repetitive signals—A frequency domain approach, in: Proc. of 1991 ACC, Boston, Mass., vol. 3, pp. 2621–2628.
13 C. Canudas de Wit, Experimental results on adaptive friction compensation in robotic manipulators: Low velocities, in: Lecture Notes in Control and Information Sciences, V. Hayvard Khatib (ed.), Springer-Verlag, Berlin, New York, 1989, pp. 196–214.
14 S. Yang and M. Tomizuka, ASME J. of Dyn. Sys. Mea. and Control, 110 (1988) 221–227.
15 G. E. P. Box, W. G. Hunter and J. S. Hunter, Statistics for Experimenters, John Wiley and Sons, Inc., New York., 1978.
16 J. J. Filliben, R. N. Kacker and E. S. Lagergren, Experiment design for scientists and engineers, NIST Workshop for Industry, 1990.
17 J. W. Tukey, Exploratory Data Analysis, Addison-Wesley, Reading, Mass., 1977.
18 J. J. Filliben, Comput. Graph., 15, no. 3 (1981) 199–213.

Chapter 4
Statistical Approaches and Reliability

Applications of statistical approaches and reliability are introduced in the design process. Controlling statistical uncertainties is the key to producing high-quality products. Both experimental design and a tutorial on accelerated testing are provided along with an introduction of time-dependent availability for modeling performance of flexible manufacturing systems and electronic parts. Reliability techniques are implemented in design-engineering problems including those of noncoherent systems.

Quality through Engineering Design - W. Kuo (Editor)

Recent and Future Research on Practical Methods for Accelerated Testing

William Q. Meeker[a] and Luis A. Escobar[b]

[a]Department of Statistics, Iowa State University, Ames, IA 50011, U.S.A.

[b]Department of Experimental Statistics, Louisiana State University, Baton Rouge, LA 70803, U.S.A.

Abstract

Accelerated tests are used to obtain timely information on the life distribution of products. Test units are used more frequently than usual or are subjected to higher than usual levels of stress or stresses like temperature and voltage. Then the results are used to make predictions about product life at the more moderate use or design conditions. Changes in technology, calls for rapid product development, and the need to continuously improve product reliability have put new demands on the applications for these tests. In this paper, we briefly review the statistical and other ideas behind accelerated testing and give an overview of some current and planned statistical research to improve accelerated test planning and methods.

INTRODUCTION

Motivation

Today's manufacturers are facing strong pressure to develop newer, higher-technology products in record time, while improving productivity, product field reliability, and overall quality. This pressure has motivated the development of methods like concurrent engineering and has encouraged wider use of designed experiments for product and process improvement efforts. The requirements for higher reliability have increased the need for more *up-front* testing of materials, components and systems.

Estimating the time-to-failure distribution of components of high-reliability products is particularly difficult. Most modern products are designed to operate without failure for years, tens of years, or more. Thus, few units will fail in a life test of practical length at normal use conditions. For this reason, accelerated tests (ATs) are used widely in manufacturing industries, particularly to obtain timely information on the reliability of product components and materials. Generally, information from tests at high levels of stress (e.g., use rate, temperature, voltage, or pressure) is extrapolated through a physically reasonable statistical model to obtain estimates of life at lower, normal levels of stress. In some cases, stress is increased or otherwise changed during the course of a test (step-stress and progressive-stress ATs). AT results are used in the reliability-design process to assess or demonstrate component and subsystem reliability, certify components, detect failure modes, compare different manufacturers, and so forth. ATs have become increasingly important because of rapidly changing technologies, more complicated products with more components, and greater customer expectations for better reliability.

There are two fundamentally different methods of accelerating a reliability test:

- *Increase the use-rate of the product.* Consider the reliability of a toaster, which is designed for a median lifetime of 20 years, assuming a usage rate of twice each day. If, instead, we test the toaster 365 times each day, we might reduce the median lifetime to about 40 days. Also, because it is not necessary to have all units fail in a life test, useful reliability information on design can be obtained in a matter of a few days instead of months.
- *Increase the aging rate of the product.* Increasing experimental factors such as temperature or humidity can accelerate the chemical processes of certain failure modes, such as weakening of an adhesive mechanical bond or the growth of a conducting filament through an insulator. Assuming a physically reasonable statistical model relating lifetime to amount of stress, we can use the life data from the AT to estimate the lifetimes under normal factor levels.

In this paper, we will distinguish between ATs on the basis of what we are able to observe.

- *Accelerated Life Tests (ALTs):* One obtains information on time-to-failure for units that fail and lower bounds for failure time (i.e., running times) for units that do not fail.
- *Accelerated Degradation Tests (ADTs)*: One observes, at one or more points in time, the amount of degradation for a unit (perhaps with measurement error).

Literature

Nelson [1] provides an extensive and comprehensive source for background material, practical methodology, basic theory, and examples for accelerated life testing. Viertl [2], with little emphasis on applications, provides a briefer (and more technical) overview of the available statistical methods for ALTs, with more focus than Nelson [1] on a large class of statistical methods that for a variety of practical reasons seem not to have been widely used in practice. These methods include nonparametric and semiparametric statistical methods. Viertl [2] also discusses Bayesian methods for ALT test planning and analysis. Both sources cite numerous references. Mann et al. [3, Chapter 9] presents an overview of ALT planning available at the time that book was written. In this paper, we will not attempt to review material covered in these references but will instead cover some necessary background material and then describe newer directions of research.

TRADITIONAL STATISTICAL MODELS AND METHODS FOR ATS

AT models

Most parametric AT models have two components:

- A parametric model for units at a constant level of a stress factor
  1. For ALTs, this typically consists of an assumed time-to-failure distribution (or family of distributions). Although this assumption may be avoided for some applications, parametric models (e.g., Weibull and lognormal) provide important practical advantages for most applications.
  2. For ADTs, this usually involves an assumed model for degradation over time. Typically, this is a function that can be used to describe the progression toward failure over time.
- An assumed relationship(s) between one (or more) of the constant-stress model parameters and stress

These relationships express the effect on life of changing factors like temperature, voltage, humidity, and specimen size (see Nelson [1]). For example, in the lognormal accelerated-failure time regression model, it is assumed that the mean μ of the log time-to-failure-distribution is

$$\mu(TEMP) = \beta_0 + \beta_1\left[\frac{-1{,}000}{TEMP + 273.16}\right]$$

where TEMP is temperature in degrees C and that the standard deviation σ of log time does not depend on temperature. This is the Arrhenius model. The parameters β_0, β_1 and σ are to be estimated from the data. For purposes of conceptualization, this formulation has useful ties to the standard linear-regression model.

Standard data-analysis methods and software for ATs

Most AT data analyses include a combination of graphical methods (scatterplots and probability plots with possibly censored data) and analytical methods (regression analysis) based on maximum likelihood estimation (MLE). These methods differ from standard statistical methods in that typically (1) data cannot be modeled with a normal distribution, (2) data are censored, (3) interest is focused on the lower tail of the life distribution, and (4) conclusions require extrapolation both in stress and from the laboratory to the field. Nelson [1] describes in detail graphical and analytical methods for analyzing ALT data. Some other books describing statistical methods for analyzing ALT data include Lawless [4] and Cox and Oakes [5]. Bayesian methods of analysis also have been proposed (see Chapter 7 of Viertl [2] for a review).

Nelson [1] provides a detailed table that outlines the capabilities of different statistical packages that might be useful in the analysis of ALT data. This table shows that the development of commercial software typically lags the development of useful statistical methods by at least 5–10 years.

Step-stress and progressive-stress AT experiments

In some areas of reliability testing, stress on individual units is varied during the test. Nelson [1, Chapter 11] gives references and a comprehensive overview of existing models, test plans, and analysis methods for such experiments. See Viertl [2] for some other references and technical discussion.

Originally, the primary motivation for using step-stress tests, in which stress is increased or decreased on units during the test, and progressive-stress tests, in which stress is continuously increased during the test, was to turn up the stress on the surviving units to force all or most of the units to fail more quickly than with constant-stress tests. In many situations (in particular, when the failure mode and the effect of the accelerating stress are reasonably well understood and when one wants to make inferences about an approximately constant-stress environment), step-stress experiments of this kind are both overly complicated and unnecessary for the following reasons.

- Inferences from step-stress experiments require much more complicated model assumptions than those that are typically used in constant-stress ATs.
- Well-developed methods for analyzing censored data are available; it is not necessary that all units fail.
- Inferences are generally desired on the lower tail of the distribution, and waiting for all units to fail generally does not provide much additional information about the lower tail of the distribution.

Other situations, however, require step-stress tests. In particular, LuValle [6] shows how to use step-stress data to identify and make inferences about unobservable rate constants of a hypothesized kinetic model. Identifying the underlying model structure is important because this structure suggests how the product will age in an environment with nonconstant stress.

Inferences from AT experiments

It is particularly difficult to use AT results to predict actual product performance because most products tend to be complicated combinations of materials and components. AT results, however, typically provide only direct information on the life distribution of individual components and materials on the basis of carefully controlled laboratory tests. Extrapolation to make predictions on the distribution of life of a product in the field environment (which itself may be difficult to predict) is in most cases difficult or impossible to completely justify.

A typical AT is what Deming [7] would call an analytic study. When making predictions from an AT, one has to make strong assumptions about the adequacy of the AT manufacturing or testing process to describe the actual manufacturing or use process. Typically, confidence intervals based on analytic studies provide, as best, only a lower bound on the total variability and uncertainty; see Hahn and Meeker [8]. Such intervals do not account for model inaccuracy, and through extrapolation, such errors tend to be greatly amplified and potentially dominate other sources of uncertainty.

Confidence statements from nonlinear models

Confidence intervals can be used to express uncertainty from limited samples but not model-error data. Most models used in the AT experiments will require nonlinear estimation methods like maximum likelihood (ML) for which there are no exact methods for statistical inference. Instead, we rely on large sample approximations. Statistical theory provides several alternatives for setting approximate confidence intervals that can be used to assess parameter uncertainty due to sampling error. Confidence intervals based on the asymptotic normal distribution of ML estimators are commonly used in commercial software, but asymptotic approximations for coverage probabilities may be poor. Confidence intervals can also be obtained by inverting likelihood-ratio tests. Vander Wiel and Meeker [9] show that confidence intervals set by this method have actual coverage probabilities that are much closer to the nominal coverage probabilities than those computed with the standard normal-theory (Wald) method. Monte Carlo simulation (parametric bootstrap) and other resampling methods may perform even better but can require large amounts of computer time.

STATISTICAL AT MODELS BASED ON PHYSICAL/CHEMICAL MODELS

Need for special models

Inferences from ALTs typically involve extrapolation in both stress and time [10]. Models for extrapolation should have a basis in the physics and chemistry of the important failure mechanism(s). The standard accelerated failure-time regression models (based, for example, on Weibull or lognormal approximations to the failure-time distribution at a given stress) are adequate for modeling some simple chemical processes that lead to failure. In other cases, more sophisticated models are required.

Development of models for extrapolation

Extrapolation is dangerous but unavoidable in accelerated testing. Generally, the magnitude of the inevitable extrapolation error (because of the use of an approximate model) depends strongly on the degree of extrapolation. Meeker and Hahn [10] emphasize this point and give suggestions on how to design an ALT to balance the degree of extrapolation with statistical precision. The justification for extrapolation has to come from the use of physically based models that have been developed from a combination of physical and chemical theory and we would hope extensive previous experience in testing similar products and materials. In particular, some AT models assume that

- a physical/chemical process causes change in the test unit, and this change progresses over time to eventually cause failure
- the applied stress or stresses accelerate reaction rates, and this acceleration can be described by a model that is adequate over the range of testing and interest

ADT models describe the progression of the observed failure-causing process over time or the progression of some variable or variables related to failure.

ADTs have the important advantage of allowing more credible and precise reliability estimates, often with less need for extrapolation. In ALTs, one observes only the time at which failure occurs (or a running time for unfailed units).

The well-known and commonly used models for ATs (e.g., the Arrhenius model for temperature acceleration and the inverse power rule for voltage acceleration) have gained acceptance for some applications because of many successful applications and general agreement of laboratory test results with long-term field performance.

The accelerated failure-time model

The accelerated failure-time (AFT) regression model is the most widely used parametric failure-time regression model. Under this model $F_S(t) = F_0[\theta(S)*t]$ where $F_0(t)$ is the c.d.f. at some baseline stress and $F_S(t)$ is the c.d.f. at stress level S. This implies that changing stress simply scales time. Special cases include the commonly used Arrhenius, inverse power, and the Eyring models; see Nelson [1].

For a kinetic failure-causing process, the AFT model holds if

- the initial conditions in the failure-causing process (e.g., the amount of a compound available at the time that a unit is manufactured that could be transformed into a failure causing contaminant through a chemical process) and the conditions required for failure to occur (e.g., the amount of the transformed contaminant needed to cause failure) do not depend on stress
- the failure-causing kinetic process can be approximated by a single-step reaction over the whole range of stresses
- the reaction-rate constant for the failure-causing process depends on stress but is the same for all units tested at a particular stress level.

See LuValle et al. [11], Klinger [12], and Klinger et al. [13, Appendix B] for more details.

More general acceleration models

When there is more than one step in the failure-causing process and stress affects rate constants for the steps differently, then the conditions for the AFT model do not hold and different life stress relationships are needed. Conceptually, we have a general time-transformation function r such that $F_S(t) = F_0[\rho(S,t)]$.

Although attempts have been made to develop ALT inference methods for unspecified $\rho(S,t)$ (e.g., Barlow and Scheuer [14]), for most practical purposes it is necessary to specify a parametric functional form for $\rho(S,t)$, preferably derived from the chemical kinetics of the failure-causing process. LuValle et al. [15] and Martin [16] discuss the use of physical and chemical theory to derive time-transformation functions.

Meeker and LuValle kinetic model

Meeker and LuValle [17] present the results from a humidity ALT of conductive anodic-filament failures on printed circuit boards for which standard ALT models (e.g., the AFT model discussed in the previous section) are clearly inadequate. They used a hypothesized kinetic model of the failure process as a basis to derive an alternative, more general, class of parametric ALT models. They use likelihood-based methods to estimate the model parameters. The new ALT models, which are generalizations of the traditional AFT model, fit their data better than the traditional ALT models. The kinetic model also provides extrapolations to design conditions that are more consistent with actual field data.

Further work in developing models for extrapolation

Failure mechanisms involving other types of chemical reactions will lead to other models. More complicated chemical models will lead to models with more parameters. This is likely, however, to lead to numerical and identifiability problems more severe than those encountered by Meeker and LuValle [17], and methods will have to be found to deal with these issues.

Much of the previous work in this area has focused on modeling life as a function of only one accelerating stress. Developing models for two or more stresses will be important for many real applications. Given adequate statistical models and well-designed experiments, MLE theory and statistical inference will, in principle, be relatively straightforward for these more advanced models. Practical implementation in software packages, however, always lags developed theory and methods.

Software for fitting general AT statistical models

No easy-to-use general software exists for analyzing data from general statistical models. For many new models, new software or software components must be developed. In their pioneering work, Nelson and Hendrickson [18] provided useful graphical methods and allowed a user to specify a general model by writing a Fortran subroutine. Today, commercial packages like BMDP and maximum likelihood programs provide similar capabilities. Programming in high-level languages like Gauss and Splus provide other alternatives.

PLANNING ALTS AND RECENT ADVANCES

Nelson [1] reviews previous work on planning ALTs. This work involves

- developing statistically "optimum" plans that may not be suitable for real applications but may provide a baseline best plan and important insights for the planning process
- extending optimum plans to compromise plans that impose some practical constraints but generally still optimize some parts of the plan

Compromise plans sacrifice some statistical efficiency in return for improved overall properties, including improved robustness to misspecification of unknown inputs (see Ref. [10] and [19]). Because they allow for the goal of extrapolative inferences with censored data, compromise plans can provide important improvements over traditional experimental plans.

Criteria for statistically planning ALTs

To plan an ALT one needs to (1) specify the experimental region, (2) choose factor-level combinations, and (3) choose the number of test units to allocate to each factor-level combination. The appropriate criteria for choosing a test plan depend on the purpose of the experiment. From a purely theoretical point of view, one would choose the upper bound(s) on the experimental region to be as large as possible because this would increase estimation precision. As a practical matter, however, we need to restrict the upper endpoint of the experimental region so that we limit the amount of extrapolation error. Optimizing a test plan according to one criterion can result in a plan with poor properties under other criteria, and it is useful to evaluate the tradeoffs to obtain a satisfactory practical plan. In developing test plans, one can use the following figures of merit.

- Often, the primary purpose of an ALT experiment is to estimate a particular quantile in the lower tail of the life distribution at design conditions. Thus, a common primary criterion is to minimize the variance or asymptotic variance of the MLE of the target quantile.
- In other situations, one might be interested in a more general measure of estimation precision. Let F denote σ^2/n times the Fisher information matrix for the model parameters. A useful secondary criterion is $|F|$, the determinant of F. This criterion is motivated because the volume of an approximate joint-confidence region for the model parameters is inversely proportional to an estimate of $\sqrt{|\mathbf{F}|}$.
- We would like to have plans with a degree of robustness to departures from the assumed model and misspecified unknown parts. To achieve this, we may find it useful to evaluate and perhaps optimize test plans with respect to test properties under alternative and typically more general models. For example, if one is planning a single-factor experiment under a linear model, it might be useful to evaluate test properties under a quadratic model [19]. If one is planning a two-factor experiment under the assumption of a linear model with no interaction, it would be useful to evaluate test properties under a linear model with an interaction term [20].
- When there are not enough failures, ML estimates may not exist. Hamada [21] shows how to compute the probability that a test plan will result in data that will not be able to estimate all of the model parameters. As a practical matter, to have a useful amount of precision in one's estimates, one must have more than the minimum number of failures necessary for the MLEs to exist (e.g., at least four or five or more, depending on the desired precision) at each factor-level combination. Thus, Meeker [19] and Meeker and Hahn [10] report the expected number of failures at each factor-level combinations per 1,000 units in the ALT.

For most traditional models, these figures of merit depend on the asymptotic covariance matrix of the ML estimates of the model parameters. This matrix typically depends on the model parameters and on the test plan.

Basic results for planning ALTs

Previous research on planning single-factor ALTs (e.g., Nelson [1]) has provided a number of results and guidelines that are useful in planning ALTs.

- An optimum plan assuming a two-parameter life-stress relationship (usually linear) will test units at only two stress levels.
- Choosing the highest test stress to be as high as possible will increase precision and statistical efficiency. The highest stress level should not, however, be chosen to be so high that it causes new failure modes or an inadequate assumed stress/life relationship.
- The lowest test stress and the allocation to that stress can be optimized.
- Compromise test plans that use three or four levels of stress have somewhat reduced statistical efficiency but provide important practical advantages.
- Typically, one should allocate more units to lower stresses. Intuitively, there are two reasons for this: (1) because of the limited test time, there will be a smaller proportion of units failing at lower stress; and (2) we are more interested in inferences at lower stresses.
- Statistically optimum plans typically are not robust to misspecifications to the unknown inputs. A statistically optimum plan can, however, be used as a starting point to find a good plan.
- The properties of ALT plans depend on unknown inputs, such as the true model and the parameters of that model. It is important to assess the sensitivity to misspecifications in these inputs and to arrive at a compromise plan that will provide satisfactory results over the range of interest.

ALT plans with nonconstant σ

Most literature on ALT planning and analysis has developed under the AFT assumption that increased stress only scales time, which is equivalent to the assumption that stress affects only the location of the log time-to-failure distribution. However, some accelerating factors affect not only the location of the log time-to-failure distribution but also the scale parameter. Nelson [1] describes models and methods for handling such data. Meeter and Meeker [22] give references and develop methods for planning one-factor ALTs under such models. Their ideas could be extended to problems with more than one factor.

Two or more accelerating variables

ALTs sometimes use more than one accelerating or other experimental factors. Nelson [1] describes a simulation-based method of evaluating and planning a multifactor ALT and illustrates the methods on an experiment that was to be used to investigate the effects of voltage stress, insulation thickness, and type of conductor on insulation life. Escobar and Meeker [20] provide a systematic method of obtaining statistically optimum and more practical compromise test plans for two-factor ALTs with censored data.

Bayesian design of ALTs and the use of prior information

Because of the nature of the nonlinear design problem, properties of ALTs and thus the optimum plans for ALTs depend on unknown parameters. Such optimum plans are locally optimum [23]. Having to specify model parameters before an experiment is conducted strongly motivates the use of Bayesian design methods for which one needs to specify prior distributions on the model parameters.

Chaloner and Larntz [24] present a Bayesian approach to the design of ALTs with a single accelerating factor and Weibull and lognormal time-to-failure distributions. They consider the test plan problem when the interest is on one or more quantiles at design conditions. Their setting allows for prior specifications of (1) the distribution (Weibull or lognormal), (2) the assumed relationship (linear or quadratic) between the location parameter of the log-life and the accelerating factor, (3) the model parameters, and (4) the quantiles of interest. They present several examples using beta distribution priors for the model parameters and compare the Bayesian plans with the local optimum plans. Their conclusions were that the Bayesian plans are very similar to the local optimum plans. The Bayesian approach, however, allows a natural incorporation of uncertainty in the model, distributional form, and quantiles of interest.

Use of simulation in ALT planning

The optimization of ALT plans is typically based on the minimization of asymptotic (approximate) variances of ML estimators. The asymptotic variances can be expected to provide adequate approximations as long as the number failing in the subexperiments is not too small. An alternative method is to use Monte Carlo simulation to evaluate test properties. Nelson [1] uses a single simulated experiment to assess the test properties of an ALT for which analytical theory was not available. An extension of this idea is to use repeated simulation to allow an assessment of the sampling variability in the estimates and to see the range of outcomes that might be expected in the actual ALT. Such a simulation allows the analyst to obtain Monte Carlo estimates of "exact" test properties (e.g., variances of ML estimators) without having to rely on asymptotic approximations.

Software for ALT planning

The computation of test properties and the development of optimum and compromise plans requires complicated computations. Escobar and Meeker [25] provide an algorithm that allows one to compute the Fisher information matrix for proposed test plans easily under many different ALT models involving censored data. Jensen and Meeker [26] provide a computer program that allows the user to develop and compare optimum and compromise ALT plans. The program also allows the user to modify or specify plans and to evaluate their properties.

It would be helpful to extend the work of Jensen and Meeker [26] to produce an easy-to-use computer program to do the necessary computations and allow the comparison of alternative test plans. We have developed some functions in Splus that could become the basis for such a program.

ADTS, MODELING AND ANALYSIS

Previous work

Some life tests result in few or no failures. In such cases, it is difficult or impossible to assess reliability with traditional life tests that record only time-to-failure. For some components, it is possible to obtain degradation measurements over time, and these measurements may contain useful information about product reliability. Even with little or no censoring, there may be important practical advantages in analyzing degradation data, especially when some form of acceleration is to be used to get information more quickly.

- Unless there is a large amount of measurement error, there will be more useful information in degradation data than in time-to-failure data. If we can assess how close an unfailed unit is to failure, we will have a better basis for time extrapolation.
- Useful information can be obtained at lower stress levels and thus reduce the necessary degree of extrapolation and the resulting extrapolation error.
- If the degradation mechanism is observed and understood, there is a better chance of finding an accurate mechanistic model to relate degradation to stress, improving one's ability to extrapolate accurately.

Although the literature on ADT models is not abundant, important references have used degradation data to assess reliability. Nelson [1] reviews the degradation literature, surveys applications, describes basic ideas on ADT models, and by using a specific example, shows how to analyze a type of degradation data. We will review recent work not described there.

Lu and Pantula [27] use a repeated-measures model to analyze ADT data from silicon devices. Carey and Koenig [28] describe a strategy to extract reliability information from observations on the degradation of device characteristics. The strategy consists of four steps: (1) assessing the accuracy of the test equipment, (2) modeling the degradation of the individual devices as a function of aging time, (3) modeling the effect of stress on the parameters of the model used in number (2), and (4) predicting the degradation of a device at use conditions on the basis of the model given by (2) and (3). They illustrate the methodology by using degradation data from a temperature ADT for devices used to estimate the reliability of an integrated logic family that is a component of a new generation of submarine cables. Fagerstorm [29] uses a differential equation model to analyze and estimate the time-to-failure distribution. A bootstrap method is used to obtain confidence intervals. The methods are illustrated on sample paths from laser life tests. Lu and Meeker [30] use a nonlinear random-effects model and Monte Carlo methods to estimate the time-to-failure distribution from degradation data. They use a parametric bootstrap to obtain point-wise confidence intervals for the distribution. They apply the method to metal fatigue data by using the Paris-law model to describe crack growth. Their model and method could be extended to allow for acceleration.

Software for ADT data analysis

There is no ready-to-use software available for analyzing ADT data. Such software would have to do the following:

- fit nonlinear growth curve models to the sample paths
- estimate the distribution of the fixed- and random-effect model parameters
- estimate the effect of experimental factors (especially those used to accelerate the degradation tests)
- allow one to obtain estimates and confidence intervals for functions of the model parameters (e.g., percentiles of the time-to-failure distribution at the design stress).

Recent advances have been made in the development of theory, methods, and software for some of the parts that are needed. Numerous methods are available for estimating random effects from linear growth-curve models (e.g., Laird et al. [31]). More recently, computationally intensive methods for computing estimators from nonlinear growth-curve models have been suggested [32–34]. Lu and Meeker [35] describe a collection of Splus functions for degradation analysis for a single stress level, but these functions would have to be modified to allow for acceleration.

ADT planning

Important questions arise about how to design degradation experiments. Since experiments should provide the most efficient use of available resources, such planning begins with the ability to assess the precision that one can expect to achieve with a specified design. Important questions include how to choose the interval between inspections (or frequency of recordings in the cases where continuous readings can be taken), the number of units that should be tested, the factor-level combinations, and the allocations of test units.

Carey and Escobar [36] consider the test planning for a class of degradation processes when the interest is on the estimation of a given characteristic of the population in study. Their methodology is applicable to ADTs with the following characteristics: (1) each device is subjected to an elevated but constant stress level over the length of the experiment, (2) the degradation of the characteristic of interest is expected to approach a maximum asymptotic value (or plateau value) over time, and (3) the plateau values are stress dependent and they can be modeled by a simple function of stress. They provide methodology to obtain: (1) the stress levels to use in the experiment and the proportion of units to be allocated to these stresses, (2) the times to measure each device, and (3) the sample size required to estimate the population-median plateau value within a given precision.

OTHER APPLICATIONS OF OVER-STRESS TESTING

The first parts of this paper have discussed traditional ATs that are used to obtain timely information of the life distribution at design conditions by testing units at one or more higher levels of stress. The primary purpose of these tests has been to characterize product life at use conditions. There are several other important applications for over-stress reliability testing, particularly in the electronics industry. These additional applications have also been termed accelerated tests. Although statisticians have had less impact in these areas, there is a high potential for the use of statistical ideas, particularly in the planning of efficient and informative tests.

Highly accelerated life tests

When planning ALTs to make projections about life at use conditions, Meeker and Hahn [10] suggest that tests should be planned to minimize the amount of extrapolation in both stress and in time. In other applications, where inferences about life at use conditions are not needed, highly accelerated life tests (HALTs) may be useful. Confer et al. [37] discuss the use of extremely high temperature (up to 150° C) and voltage (8 times rating) to achieve acceleration ratios of up to 2,555. They suggest that such tests can be used as a means of sampling inspection for incoming component lots and as a burn-in screening test. In this application, however, there is danger that high stress will damage units that are to be put into service. Other applications include pilot tests to get information needed for planning a more extensive ALT at lower levels of stress and experiments to obtain information on the relevance of failure modes discovered in high-stress testing. When using very high levels of stress one must be on the look out for failure modes that would never occur at use conditions and have concern for the adequacy of the model.

Environmental stress testing

The pressure to quickly develop new, high-reliability products has motivated the development of new product testing methods. The purpose of these testing methods is to quickly identify and eliminate potential reliability problems early in the design stage of product development. One such testing method is known as STRIFE (stress life) testing. The basic idea of STRIFE testing is to aggressively stress and test prototype or early production units to force failures. Although it may be useful to test only one or two units, a greater degree of replication will provide important additional information on unit-to-unit variation. Typical STRIFE tests use combinations of cycling temperature and vibration. The amplitude of the cycling is increased continuously as a function of time until the end of the test. This could be days or as long as appropriate fixes for failure modes can be effected without long delays that might be needed for a complicated redesign as opposed to a simple part substitution. Bailey and Gilbert [38] report an example in which the complete STRIFE test and improvement program was successfully completed in three weeks.

Generally, failures in STRIFE testing are caused by product or process design flaws. When there is a failure in a STRIFE test, it is necessary to find and carefully study the failure's root cause. First, it is necessary to determine if the failure could occur in actual use. Knowledge plus physical and chemical modeling of the particular failure model is useful for helping to make this determination. Nelson [1] describes an example where a costly effort was made to remove a high-stress-induced failure that never would have occurred in actual use. The occurrence of such failures might indicate that the test is using some combinations of stress that are too high to be useful. When it is determined that a failure could occur in actual use, it is necessary to change the product design or manufacturing process to eliminate that cause of failure. In some cases, the fix is obvious. In other cases, additional focused research and experimentation on the component or on component interfaces may be required.

Because the results of STRIFE testing are effecting changes in the production process, it is pointless to use the resulting data to predict what will happen in normal use. Still, ideas from statistical experimental design and models relating stress to life can be useful in choosing stresses, stress ramp speed, and other aspects of the test. Nelson [1] describes STRIFE tests as "elephant tests" and describes some important issues. Again, the *Proceedings of the Institute of Environmental Sciences* often contains papers on this subject, as do various journals on reliability of electronics systems. Apparently, no statistical work has been done in this area.

Burn-in

The most common reliability problem for manufacturers and consumers of electronic equipment has been early failures, typically caused by a small number of manufacturing defects in the product population (defective components are often called freaks). Although manufacturers would prefer to build-in reliability by eliminating all manufacturing defects, it has been difficult, with rapidly changing technology, to achieve a goal of zero defects, particularly early in the production cycle. To achieve sufficiently high reliability particularly in critical applications such as space and under-sea systems, manufacturers have commonly practiced burn-in for systems and components to screen out the units that would fail early in life. For components like integrated circuits, it is common to do burn-in at high humidity and temperature. For system burn-in, it is typically impossible to use high stress without damaging sensitive components. In either case, burn-in may also involve continuous operational exercising and monitoring of the units.

A burn-in test can be viewed as a type of 100% inspection or screening of the product population to eliminate or reduce the number of defective items. Burn-in may be necessary if the output of production does not meet reliability specifications. It is important that the burn-in stress not be so high as to damage the good units. Burn-in is expensive and thus the duration of burn-in is typically limited. The decision on how long to run a burn-in can be based on the desired reliability level and the distribution of the observed failures during the screen. For a stable production process, which has been adequately characterized, this can be set in advance. Otherwise, a sequential or dynamic stopping rule may be needed. Kuo [39] provides and applies a cost-optimization model for important burn-in decisions. He also reviews previous literature on this subject.

Jensen and Petersen [40] provide an engineering approach to this subject. Also, see Nelson [1]. Statistical methods can and have been useful for helping to choose stress levels and length of burn-in and in using burn-in data to assess the state of the production process and the likely field reliability of a product going into service.

Environmental-stress screening

Environmental-stress screening (ESS) was developed as a more economical and more effective means of removing defects from a product population by testing units at the system or subsystem level. Because systems and subsystems cannot tolerate high stress, ESS uses mild but more complicated stressing. High temperature and humidity at the component level are replaced by more moderate temperature cycling, physical vibration, and perhaps stressful operational regimes such as running computer CPU chips at faster than usual clock speeds and lower than usual voltages to help identify the weak units. These tests can be thought of as generalizations of step-stress tests, but their purposes are much different. The tests are sometimes called "shake and bake" tests. Again, the goal is to screen out, as effectively and as quickly as possible, the defective items without otherwise doing harm to the system.

Numerous articles on ESS testing appear each year in the *Proceedings of the Institute of Environmental Sciences.* Tustin [41] provides a motivational description of the methodology and several references. Nelson [1] gives additional references, including military standards.

Statisticians have had little impact in the development of ESS methods. There are, however, some areas where statistical methods could have an impact on ESS. Some of these will require the development of better models to relate the effect that complicated stressing has to the life distribution of both the defects and the nondefects. For example, advanced application of the statistical principles of experimental design, modeling, and data analysis can be used to help

- choose stress conditions that are best for weeding out manufacturing defects, but minimize the chance of doing damage to good units
- design screens to provide feedback that can be used to improve product or process design to reduce the frequency of or to eliminate manufacturing defects
- assess information that would allow prediction of field reliability (The complicated stress patterns make it difficult or impossible to use ESS data directly to make predictions about field performance. Clearly, however, with enough screening data, correlated with field data, one could find relationships that would allow such predictions.)

The development of physical or statistical models to describe the effect of ESS stress patterns on product life would be useful; however, this task will not be easy. LuValle [42] reports experimental evidence indicating that varying multiple stresses in typical ESS procedures leads to complicated effects on life. There can be interactions among stresses, and there may be "memory" of past stress patterns; that is, from a particular point in time, future degradation may depend not only on the current amount of degradation but also on how (i.e., under which stress patterns) that degradation has accrued. This is in contrast to traditional step-stress failure models (see Nelson [1]).

Like burn-in, ESS is an inspection and screening scheme. In line with the modern quality precept of eliminating reliance on mass inspection, most manufacturers prefer not to do burn-in or ESS. These methods are expensive and may not be totally effective. By improving the reliability through continuous improvement of the product design and the manufacturing process, it may be possible to reduce or eliminate reliance on screening tests, except perhaps in the most critical applications.

CONCLUSIONS

Accelerated testing will continue to be an important tool for reliability improvement. Although ATs have been used for decades, the current push for rapid product development and greater reliability are placing new demands on this important technology. Careful planning and modeling are essential for successful AT programs. For the most common applications of AT, conclusions involve extrapolation and will be highly dependent on the adequacy of one's model. Proper application of AT requires detailed knowledge about the causes of failure and the relationship between these causes and stress. It is important for reliability engineers, physical scientists, and statisticians to collaborate in

- planning ATs
- developing adequate models for AT data
- developing methods of analysis and graphical display of AT models
- drawing conclusions from AT data.

ACKNOWLEDGMENTS

Gerry Hahn, Marcia Martens Pierson, and a referee provided numerous helpful comments on an earlier draft of this paper. We also thank Mike LuValle for providing helpful comments and references, especially for our section on AT models.

REFERENCES

1 W. Nelson, Accelerated Testing: Statistical Models, Test Plans, and Data Analyses, John Wiley and Sons, Inc., New York, 1990.
2 R. Viertl, Statistical Methods for Accelerated Life Testing, Vandenhoeck and Ruprecht, Göttingen, Germany, 1988.
3 N. R. Mann, R. E. Schafer and N. D. Singpurwalla, Methods for Statistical Analysis of Reliability and Life Data, John Wiley and Sons, Inc., New York, 1974.
4 J. F. Lawless, Statistical Models and Methods for Life Time Data, John Wiley & Sons, Inc., New York, 1982.
5 D. R. Cox and D. Oakes, Analysis of Survival Data, Methuen, New York, 1984.
6 M. J. LuValle, Microelectron. Reliab., 30 (1990) 591–603.
7 W. E. Deming, The Amer. Stat., 29 (1975) 146–152.
8 G. J. Hahn and W. Q. Meeker, Statistical Intervals: A Guide for Practitioners, John Wiley and Sons, Inc., New York, 1991.
9 S. A. Vander Wiel and W. Q. Meeker, IEEE Trans. on Reliab., R-39 (1990) 346–351.

10 W. Q. Meeker and G. J. Hahn, How to Plan Accelerated Life Tests: Some Practical Guidelines, Vol. 10 of the ASQC Basic References in Quality Control: Statistical Techniques, American Society for Quality Control, Milwaukee, Wis., 1985.
11 M. J. LuValle, T. L. Welsher and J. P. Mitchell, A new approach to the extrapolation of accelerated life test data, in: The Proceedings of the Fifth International Conference on Reliability and Maintainability, Biarritz, France, 1986, pp. 620–635.
12 D. J. Klinger, On the notion of activation energy in reliability: Arrhenius, Eyring, and thermodynamics, in: Proceedings of the Reliability and Maintainability Symposium, IEEE, New York, 1991, pp. 295–299.
13 D. J. Klinger, Y. Nakada and M. A. Menendez (eds.), AT&T Reliability Manual, Van Nostrand Reinhold, New York, 1990.
14 R. E. Barlow and E. M. Scheuer, Technometrics, 13 (1971) 145–159.
15 M. J. LuValle, T. L. Welsher and K. Svoboda, J. Stat. Phys., 52 (1988) 311–320.
16 J. W. Martin, Durability of Build. Mat., 1 (1982) 175–194.
17 W. Q. Meeker and M. J. LuValle, An Accelerated Life Test Model Based on Reliability Kinetics, Technical Report, Department of Statistics, Iowa State University, Ames, 1991.
18 W. Nelson and R. Hendrickson, 1972 User Manual for STATPAC–A General Purpose Program for Data Analysis and for Fitting Statistical Models to Data, General Electric CR&D Technical Report 72GEN009, 1972.
19 W. Q. Meeker, Technometrics, 26 (1984) 157–172.
20 L. A. Escobar and W. Q. Meeker, Planning Accelerated Life Tests with Two or More Experimental Factors, Department of Statistics, Iowa State University, Ames, 1992.
21 M. Hamada, Comparing Accelerated Life Test Plans by Their Non-estimability Probabilities, Department of Statistics and Actuarial Science, University of Waterloo, Waterloo, Ontario, 1990.
22 C. A. Meeter and W. Q. Meeker, Optimum Accelerated Life Tests with Nonconstant σ, Technical Report, Department of Statistics, Iowa State University, Ames, 1990.
23 H. Chernoff, Ann. of Math. Stat., 24 (1952) 586–602.
24 K. Chaloner and K. Larntz, Bayesian Design for Accelerated Life Testing. Technical Report, School of Statistics, University of Minnesota, Minneapolis, 1990.
25 L. A. Escobar and W. Q. Meeker, Fisher Information Matrix for the Extreme Value, Normal, and Logistic Distributions and Censored Data, Technical Report, Department of Statistics, Iowa State University, Ames, 1992.
26 K. L. Jensen and W. Q. Meeker, ALTPLAN: Microcomputer software for developing and evaluating accelerated life test plans, Paper presented at the Annual ASA Meetings, Anaheim, Calif., 1990.

27 J. C. Lu and S. G. Pantula, A Repeated-measurements Model for Over-stressed Degradation Data, Department of Statistics, North Carolina State University, Raleigh, 1989.

28 M. B. Carey and R. H. Koenig, IEEE Trans. on Reliab., 40 (1991) 499–506.

29 R. Fagerstorm, Stochastic differential equation modeling of laser degradation, Paper presented at the Institute of Mathematical Statistics Special Topics Meeting on Statistics in Industry, Philadelphia, Abstract in Inst. of Math. Stat. Bull., 20 (1991) 159.

30 C. J. Lu and W. Q. Meeker, Using Degradation Measures to Estimate a Time-to-Failure Distribution, Preprint No. 90-20, Department of Statistics, Iowa State University, Ames, 1992.

31 N. Laird, N. Lange and D. Stram, J. Am. Stat. Assoc., 82 (1987) 97–105.

32 A. Racine-Poon and A. F. M. Smith, Population models, in: Statistical Methodology in the Pharmaceutical Sciences, D. A. Berry (ed.), Marcel Dekker, New York, 1989.

33 M. J. Lindstrom and D. M. Bates, Biometrics, 46 (1990) 673–687.

34 M. J. Palmer, B. F. Phillips and G. T. Smith, Biometrics, 47 (1991) 623–635.

35 C. J. Lu and W. Q. Meeker, Splus Functions for Using Degradation Measures to Estimate a Time-to-Failure Distribution, Preprint No. 90-20, Department of Statistics, Iowa State University, Ames, 1992.

36 M. B. Carey and L. A. Escobar, Experimental Design for a Class of Accelerated Degradation Tests, Technical Report, Department of Experimental Statistics, Louisiana State University, Baton Rouge, 1991.

37 R. Confer, J. Canner and T. Trostle, Use of Highly Accelerated Life Test (HALT) to Determine Reliability of Multilayer Ceramic Capacitors, in: Proceedings of the Electronic Components & Technology Conference, 1991, pp. 320–322.

38 R. A. Bailey and R. A. Gilbert, STRIFE Testing for Reliability Improvements, in: Proceedings of the Institute of Environmental Sciences, 1981, pp. 119–121.

39 W. Kuo, IEEE Trans. on Reliab., R-33 (1984) 145–156.

40 F. Jensen and N. E. Petersen, Burn-in, An Engineering Approach to Design and Analysis of Burn-in Procedures, John Wiley and Sons, Inc., New York, 1982.

41 W. Tustin, Qual. Prog. (Sept. 1990) 61–64.

42 M. J. LuValle, Unpublished private communication, 1992.

Quality through Engineering Design - W. Kuo (Editor)

The Use of Reliability Assessment Techniques to Improve the Quality of Engineering Design

G. A. Bohoris

School of Manufacturing and Mechanical Engineering, University of Birmingham, P.O. Box 363, Birmingham B15 2TT, U.K.

Abstract

This paper summarizes part of the work carried out to date on seeking non-graphical, analytical solutions to the two-sample problem with censored data in the context of quality and reliability applications. For this purpose, non-parametric two-sample tests for censored data are introduced, reviewed, and successfully applied to an engineering problem where design changes are found necessary in order to meet the specified reliability targets.

INTRODUCTION

People from all walks of life have probably suffered one way or another from unreliable products or services. An appliance not working, a car failing to start, a bus breaking down in the middle of a journey are but a few of the very unfortunate and common hassles. Of wider concern, even if one is not directly involved, are plane crashes and the rather frequent, recent train collisions, as well as larger-scale accidents such as the Chernobyl, Bhopal, and Space-Shuttle disasters.

Because of unfortunate events of this kind, a significant amount of management and engineering effort is spent in industry on a more or less daily basis in identifying causes of failure, evaluating replacement policies, introducing design changes, comparing materials, manufacturing methods, vendors, and the like. In many of these cases, decisions are based on comparing life data obtained either from reliability testing or from behavior in service.

This kind of problem is known as the modified two-sample problem. It is mainly concerned with the procedure that has to be followed in order to detect any statistically significant differences between the reliability functions (quality over time) of two populations on the basis of samples drawn at random from them.

This paper summarizes part of the work carried out to date on seeking nongraphical, analytical solutions to the two-sample problem with censored data in the context of either quality or reliability applications. For this purpose, nonparametric two-sample tests for censored data are introduced, reviewed, and successfully applied to an engineering problem where design changes are found necessary in order to meet the specified reliability targets.

NOTATION

Prior to discussing the tests used, it is necessary to define the notation used throughout the paper. The data consists of two random samples, one of size n and the other of size m, denoted as follows:

$$\text{Sample 1:}\quad t_{11}, t_{12}, t_{13}, \ldots, t_{1n}$$

$$\text{Sample 2:}\quad t_{21}, t_{22}, t_{23}, \ldots, t_{2m}$$

All the observed lifetimes t_{ij} are a result of either a failure or a censoring. In distinguishing between the two events, observations due to the former are noted f_{ij} and those due to the latter c_{ij}. Within each sample (i = 1, 2) there are nf_i failures and nc_i censorings. Therefore,

$$n = nf_1 + nc_1 \tag{1}$$

$$m = nf_2 + nc_2 \tag{2}$$

Combining the observations of the two samples results in a sample of size N = n + m and in k-ordered distinct failure times written as

$$t_1 < t_2 < t_3 < \ldots < t_k$$

In general, due to possible failure ties

$$k \le nf_1 + nf_2$$

In addition to the above, let d_{1j}, d_{2j} denote the number of failures occurring at time t_j (j = 1, . . . , k) in samples 1 and 2, respectively. If there are no failure ties, d_{1j} and d_{2j} are either equal to 0 or 1. When failure ties are present, d_{1j} or d_{2j} is greater than 1, taking the value of the number of failures occurring at time t_j.

$$d_j = d_{1j} + d_{2j}$$
$$d_0 = d_{10} = d_{20} = 0, \text{ i.e., no failures at time zero.}$$
$$nf_i = \sum_{j=1}^{k} d_{ij}$$

Let e_{1j}, e_{2j} be the number of right-censored observations in samples 1 and 2, respectively, that fall in the interval (t_j, t_{j+1}) for j = 0, 1, . . . , k.

$$t_0 = 0 \text{ and } t_{k+1} = +\infty$$
$$e_j = e_{1j} + e_{2j}$$
$$nc_i = \sum_{j=0}^{k} e_{ij}$$

Let n_{1j}, n_{1j} denote the number of items at risk at time t_j (j = 0, 1, . . . , k), in samples 1 and 2, respectively. Risk sets are constituted from items with lifetimes greater than or equal to t_j.

$$n_j = n_{1j} + n_{2j} \Rightarrow n_0 = n + m = N$$
$$n_{ij} = \sum_{l=j}^{k} (d_{il} + e_{il})$$
$$n_j = \sum_{i=1}^{2} \sum_{l=j}^{k} (d_{il} + e_{il})$$

A REVIEW OF THE MODELS USED

Linear rank order statistics

Several generalizations of the Wilcoxon statistic [1] for the solution of the two-sample problem with censored data have been proposed—among them the Gehan test, the Peto and Peto test, and the Prentice test, all of which along with the Logrank test are discussed more fully in Ref. [2].

In this paper only the Gehan generalization of the Wilcoxon test and the Logrank test are used. Both tests can be expressed by the general linear-rank statistics for the censored-data formula, which is a weighted summation over the second sample of the observed number of failures d_{2i} minus the conditionally expected number of failures $n_{2i}d_i/n_i$:

$$W = \sum_{i=1}^{k} w_i \left(d_{2i} - \frac{n_{2i} d_i}{n_i} \right) \tag{3}$$

where the weight function w_i takes the value n_i in Gehan's modification of the Wilcoxon test and 1 in the Logrank test.

Varying the value of the weight function has the effect of increasing the sensitivity-to-survival differences occurring at specific points. For example, Gehan's test displays greater sensitivity to differences occurring earlier in the life distributions whereas the Logrank test shows greater sensitivity to differences occurring late.

Under the null hypothesis, the variate in all the rank order tests

$$L = \frac{W}{\sqrt{V}} \tag{4}$$

is asymptotically standard and normally distributed ~ N (0,1). V is the variance of W, and several approaches have been developed to obtain its value. The most frequently used is the conditional permutation variance estimate, because of its suitability when the censoring mechanisms for the two samples are not the same.

$$V = \sum_{i=1}^{k} \left(d_i w_i^2 \frac{n_{1i}}{n_i} \frac{n_{2i}}{n_i} \frac{n_i - d_i}{n_i - 1} \right) \tag{5}$$

where $(n_i - d_i)/(n_i - 1)$ is defined to take the value 0 if $n_k = 1$.

Modified Kolmogorov-Smirnov test

For complete data, the classical Kolmogorov-Smirnov (KS) test statistic is based on the difference

$$D = \max\left[\left|\hat{R}_{1j} - \hat{R}_{2j}\right|\right] \tag{6}$$

for $j = 1, 2, \ldots, k$, where $\hat{R}_{1j}$ and $\hat{R}_{2j}$ are estimates of the reliability functions of the first and second population respectively.

Many authors including Barr and Davidson [3], Koziol and Byar [4], Dufour and Maag [5], Chandra, Singpurwalla, and Stephens [6] and Guilbaud [7] have modified the critical values for the classical Kolmogorov-Smirnov statistic (Smirnov [8], Feller [9], Massey [10, 11], Kim and Jennrich [12], Parsons and Wirsching [13]) to incorporate the presence of censored observations in the two samples. Unfortunately all of them restricted their modifications to singly Type I- or Type II-censored data.[1] Fleming et al. [14] concentrated primarily on modifying the statistic itself rather than trying to establish further sets of tables with more accurate approximations to the distribution of the classical KS statistic. For complete data, they proposed a different formulation of the KS statistic calculated through the product

$$D = \max\left[\frac{\left(\hat{R}_{1j} + \hat{R}_{2j}\right)\left(\hat{H}_{2j} - \hat{H}_{1j}\right)}{2}\right] \tag{7}$$

where $\hat{H}_{ij}$ and $\hat{R}_{ij}$ are estimates of the cumulative hazard function and the reliability function of the first $(i = 1)$ and second $(i = 2)$ population, respectively, calculated by using the cumulative hazard approach suggested by Nelson [15].

---

[1] Data can be censored in a number of ways. Data in which censoring occurs beyond the last failure are said to be **singly censored on the right**. Singly right-censored data arise when units are started on test together and their study is ended before all units fail. If the time at which the study ends is fixed, the resulting data are singly time censored or **Type-I censored** and the test is a time-terminated test. On the other hand, data are singly failure censored or **Type-II censored**, and the test is a failure-terminated test if the study is stopped when a specified number of failures has been recorded.

Fleming et al. [14] proved the increased efficiency of their modified KS statistic by using Monte-Carlo repetitions. The use of Nelson's estimators allowed the statistic to be estimated even when data were multiply censored. Additionally, Fleming et al. [14] provided a generalization of their modification by deriving the procedure to approximate the distribution of their statistic with data that is progressively censored. The formulation of this generalized procedure, as laid down by Fleming et al. [14], includes the steps detailed below for each of the k distinct failure times in the combined sample (j = 1, ... , k), bearing in mind that for any function f_i

$$\sum_{j=0}^{-1} f_j = 0 \tag{8}$$

Step 1. For each sample separately (i = 1, 2), we calculate estimates for the cumulative hazard functions of the failure $\hat{H}_{f_{ij}}$ and censoring distribution $\hat{H}_{c_{ij}}$ as follows:

$$\hat{H}_{f_{ij}} = \hat{H}_{f_{ij-1}} + \sum_{l=0}^{d_{ij-1}} \frac{1}{n_{ij} - l} \tag{9}$$

$$\hat{H}_{c_{ij}} = \hat{H}_{c_{ij-1}} + \sum_{l=0}^{e_{ij-1}-1} \frac{1}{\left(n_{ij-1} - d_{ij-1}\right) - l} \tag{10}$$

Step 2. Using these quantities, we calculate the following functions:

$$n_j = \left[\left(ne^{-\hat{H}_{c_{1j}}}\right)^{-1} + \left(me^{-\hat{H}_{c_{2j}}}\right)^{-1}\right]^{-0.5} \tag{11}$$

$$U_j = U_{j-1} + n_j \left(\sum_{l=0}^{d_{1j}-1} \frac{1}{n_{1j} - l} - \sum_{l=0}^{d_{2j}-l} \frac{1}{n_{2j} - 1} \right), \quad (U_0 = 0) \tag{12}$$

$$D_j = \frac{U_j}{2}\left(e^{-\hat{H}_{f_{1j}}} + e^{-\hat{H}_{f_{2j}}}\right) \tag{13}$$

$$A = \max\left(\left|D_j\right|\right) \tag{14}$$

$$R_2 = 1 - \frac{1}{2}\left(e^{-\hat{H}_{f_{1k}}} + e^{-\hat{H}_{f_{2k}}}\right) \tag{15}$$

Step 3. The quantity A represents the test statistic. Its approximate P-value for the two-tailed Kolmogorov-Smirnov test[2] is

$$P_2 = 2p(A, R_2) = 2\left\{1 - \Phi\left(\frac{A}{\sqrt{R_2 - R_2^2}}\right) + \Phi\left[\frac{A(2R_2 - 1)}{\sqrt{R_2 - R_2^2}}\right] e^{-2A^2}\right\} \tag{16}$$

where Φ(x) is the cumulative-distribution function of the standard normal distribution, evaluated at x.

If, alternatively, a one-tailed situation is considered, the major part of the above procedure remains unchanged with the exception of Eqs. (12, 14, and 16). Within the square brackets in Eq. (12) the direction of the difference is defined by the alternative hypothesis. To preserve the significance of this direction, the use of the absolute value in Eq. (14) is redundant and hence the test statistic now becomes

$$V = \max\left(D_j\right) \tag{17}$$

Finally, instead of using Eq. (16), the P-value for the one-tailed test is calculated as follows:

$$P_1 = p(V, R) = 1 - \Phi\left(\frac{V}{\sqrt{R - R^2}}\right) + \Phi\left[\frac{V(2R - 1)}{\sqrt{R - R^2}}\right] e^{-2V^2} \tag{18}$$

A CASE STUDY

Introduction

Although the following analysis was based on the original information, the data and some of the information relating to it are presented here in a coded form for confidentiality reasons as requested by the project's client. More specifically, this censorship refers to

1. the titles of the organization concerned and of the data sets presented
2. the magnitude of the data, i.e., the recorded times to events, all of which have been scaled down uniformly by a factor, indicated as one lifetime period.

It is quite clear, however, that this coding did not have any significant impact

---

[2] The generalized Kolmogorov-Smirnov statistic, in contrast with all the generalized Wilcoxon tests, is not a straight standard-normal deviate, but its distribution is a rather complicated function of the standard-normal distribution. The effect of this difference will become more apparent when all the two-sample tests are calculated for the same data set, leading to P-values which are fairly close to each other despite having originated from test statistics that may differ significantly.

on the analysis because of the rank invariant attribute of the two-sample tests; the magnitude, and therefore, the interpretation of the results remains unaffected as long as the order of events is not altered, which is the case when a linear transformation is applied to the data.

Transmission gearboxes

Problems have occurred with the gearbox transmitting the drive between the bogie-mounted traction motor and the wheelset of one of Country's XXX National Railways' most intensively used passenger locomotives. The problems were divided between bearing failures and oil-seal failures.

In an attempt to improve the reliability of these gearboxes (henceforth Type I gearbox), a design modification was undertaken to both the bearing and the seals, costing approximately £500 per gearbox. Unfortunately, within 1.5 lifetime periods of the first of the modified gearboxes (hereafter Type II gearbox) being fitted, they started to fail.

A series of investigations was undertaken using Weibull analysis, and it was found that population estimates of the lifetime distributions crossed over after 5 lifetime periods (Fig. 1), with the modified gearboxes having a higher incidence of failure initially. Unfortunately, due to other problems on the wheelsets, disc failures, and treat damage, the average life of the wheelsets was approximately 4.5 lifetime periods.

The graphical comparisons of the lifetime distributions indicated that the modification showed an overall improvement, but if the wheelsets lasted less than 4.5 lifetime periods, then it was not cost effective. Considerable difficulties arose in convincing the engineers that the modification has in fact worsened the reliability problem because it was not possible, using the traditional Weibull approach, to quantify the difference between the two lifetime distributions. Therefore, the data were further studied by using the above-reviewed techniques for the solution of the two-sample problem with censored data.

Survival analysis

Table 1 shows the breakdown of the data made available for the two gearbox types. Cumulative hazard estimates of the reliability function were calculated for each group. Plots of these reliability functions (Fig. 1) were superimposed. From this graph it was apparent that both gearboxes were more or less equally reliable, with the Type I gearbox being slightly more reliable in the first five lifetime periods.

Statistical comparisons of the reliability functions of the two data sets were carried out (see Table 2) by using the two-sample tests for censored data reviewed above. All methods gave very high P-values, and therefore, the null hypothesis—that the two samples came from identical populations—was not rejected at the 5% level of significance.

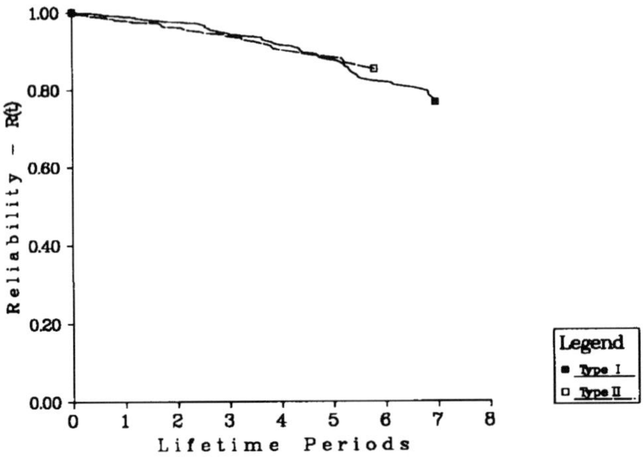

Fig. 1. Gearboxes—original data sets survival curves.

Table 1. Gearboxes—data statistics.

Sample	Failures	Censorings	Sample Size
Type I Gearbox	80	671	751
Type II Gearbox	32	381	413
Combined Sample	112	1,052	1,052

Table 2. Gearboxes—results.

Test	Test Statistic	P-Value
Gehan-Wilcoxon	0.55259	0.5805
Logrank	0.37791	0.7055
Kolmogorov-Smirnov	0.48900	0.4878

Discussion

On the basis of the data sets made available, the following inferences were made:

1. The modified gearboxes are no more reliable than the original ones.
2. All two-sample tests validates our initial conclusions from the graph by showing
   - There was not a significant difference in the reliability of the two gearbox types.

- Any slight difference in the reliability functions was more apparent in the early part of the gearboxes' lifetimes. Gehan's test, which is more sensitive in survival differences occurring early in time, gave a higher test statistic than the Logrank test which is more sensitive to survival differences occurring late in time.
- The modified KS test, being more efficient than the linear rank tests in the 'crossing hazards' situation encountered with the gearboxes, resulted in the smallest P-value.

CONCLUSIONS

A number of two-sample statistical tests capable of dealing with multiply-censored data have been discussed. The tests enabled a formal hypothesis testing exercise to be undertaken with the end result being to reject or fail to reject the null hypothesis depending on the significance probability.

The use of these tests in the case study has shown that the undertaken design changes were not successful in eliminating the observed reliability problems and illuminated the need for further design changes in order to meet the specified reliability targets. It is quite obvious, however, that in every engineering application, reliability-assessment techniques such as the two-samples tests for censored data described in this paper are absolutely necessary tools in evaluating and quantifying the quality of the engineering design.

REFERENCES

1 F. Wilcoxon, Biometrics, 1 (1945) 81–83.

2 G. A. Bohoris and E. M. Aspinwall, Statistical comparison of reliability data sets, Reliability 1989, Conference Proceedings, Brighton, England, 1989, p. 2B/3/1.

3 D. R. Barr and T. G. Davidson, Technometrics, 15, no. 4 (1973) 739–757.

4 J. A. Koziol and D. P. Byar, Technometrics, 17, no. 4 (1975) 507–510.

5 R. Dufour and U. R. Maag, Technometrics, 20, no. 1 (1978) 29–32.

6 M. Chandra, N. D. Singpurwalla and M. A. Stephens, J. Am. Stat. Assoc., 76, no. 375 (1981) 729–731.

7 O. Guilbaud, J. Am. Stat. Assoc. 83, no. 401 (1988) 213–221.

8 N. Smirnov, Ann. Math. Stat., 19 (1948) 279–281.

9 W. Feller, Ann. Math. Stat., 19 (1948) 177–189.

10 F. J. Massey, Jr., Ann. Math. Stat., 22 (1951) 125–128.

11 F. J. Massey, Jr., Ann. Math. Stat., 23 (1952) 435–441.

12 P. J. Kim and R. I. Jennrich, Sel. Tables Math. Stat., Am. Math. Soc., Providence, Rhode Island, I (1970) 79–170.

13 F. G. Parsons and P. H. Wirsching, Microelectron. Reliab., 22, no. 2 (1982) 163–167.

14 T. R. Fleming, J. R. O'Fallon, C. O'Brien and D. P. Harrington, Biometrics, 36 (1980) 607–625.

15 W. Nelson, J. Qual. Technol., 1, no. 1 (1969) 27–52.

A General Reliability Algorithm for Electronic Components

Sudha Ramesh and Paul A. Tobias

IBM Corporation, Rt. 52, East Fishkill, Hopewell Junction, New York 12533 U.S.A.

Abstract

A general, comprehensive, reliability algorithm is used to model components through various stages in their life-cycles. The model content is technology-dependent and uses appropriate fail mechanisms, lifetime distributions, and acceleration models. This mathematical model predicts cumulative component fallout as a function of field power on hours, field operating stresses, system and card test conditions, and effectiveness of these tests. One of the main features of this model is the technical scope it provides for reliability experiments by modeling stress-screen effects. The model also allows for multiple burn-in cycles. Using this model, one can analyze mathematically the effect of burn-in on the failure rate and determine optimal stress-screen conditions to meet the goal of reducing early reliability failures. This paper presents the details of the general reliability model and proceeds to describe the burn-in model.

INTRODUCTION

Comprehensive models for predicting component failure rates and field fallout over time are extremely useful for electronic, computer, and telecommunication development and manufacturing. These models are needed during the design and development stage to judge whether new products will meet reliability requirements and to guide decisions regarding component selection and burn-in strategy. Later on, input from these models helps determine warranty strategy and predict service costs, parts stocking, and service manpower needs. Finally, the output of such models provides targets that are the basis for performance-monitoring plans.

Good models generally take the form of mathematical algorithms that incorporate statistical and engineering information derived from technology-specific studies. The models ideally should give application- and vintage-dependent results and be capable of outputting bathtub-shaped curves for failure rates where experience dictates these will apply.

An example of such a model is the AT&T reliability model [1]. IBM has used its own internal model for nearly a decade, with excellent results across a wide range of technologies and tens of thousands of component part numbers. A simplified version of that algorithm was described in Tobias and Trindade [2], in a section in Chapter 8 entitled "General Reliability Algorithms." The mathematical content of the IBM general reliability algorithm is explained in the following sections.

THE GENERAL RELIABILITY ALGORITHM

A general model evaluates reliability of the product in the field by a mathematical representation of the most significant variables and effects. Component failure is a complex process that is influenced by many variables. The most significant are

- Environmental: temperature, humidity
- Use-related: power on hours, voltage, on/off cycles, percent-rated current
- Technology specific: number of circuits, length of conductor

The basic failure behavior of a component is modeled at the failure-mechanism level by base-failure distributions defined at a set of reference conditions. The most commonly used base-failure distributions are Weibull, lognormal, and exponential. If the actual conditions differ from the reference conditions, the base distribution must be modified to account for the differences. A set of use-related and environmental values are used as inputs to the model. This general model divides the component population into three distinct subpopulations:

- defective escapes that are not fully functional but pass imperfect tests
- reliability defective that are just barely functional
- "normal parts"

Basic reliability concepts

The theoretical population models used to describe device lifetimes are called life distributions. The cumulative distribution function (CDF) denoted by F(t) is called a life distribution.

In the CDF, F(t) is the probability that a component fails by time t. The CDF also represents the population fraction failing by time t. Since in reliability work we deal with failure times t which are nonnegative, the CDF for the population failure times is related to the probability density function via the following relationship.

$$F(t) = \int_0^t f(y)dy \quad 0 \le t < \infty$$

where f(t) is the probability density function (PDF) for the distribution. The reliability function gives the probability that a component survives until time t. The reliability function is given by

$$R(t) = 1 - F(t)$$

The instantaneous failure rate (IFR) at time t, also known as the hazard rate h(t), is defined as

$$h(t) = \frac{f(t)}{1-F(t)} = \frac{f(t)}{R(t)}$$

Frequently it is desirable to relate the CDF of the life of a system to that of its components. A single component with several independent failure modes is analogous to a system with several independent components. There are two special types of systems—series and parallel—for which the system lifetime distribution can be readily derived from the component lifetime distributions, provided that the failures of the different components of the system are independent of each other.

A series system is one in which the failure of any single component causes the system to fail. Let the ith component have the CDF $F_i(t)$. The CDF of the life of a series system comprising n components is given by the "competing risk" model

$$F_s(t) = 1 - \prod_{i=1}^{n} [1 - F_i(t)]$$

A parallel or redundant system is one in which the system fails when all of the components fail. The CDF of the life of a parallel system of n components is given by

$$F_p(t) = \prod_{i=1}^{n} F_i(t)$$

Common lifetime distributions

The basic lifetime distribution can be specified by identifying a distribution family, such as lognormal or Weibull, and giving the values of the parameters, such as T_{50} and σ for the lognormal, which identify a specific distribution within the family. Some commonly used lifetime distributions are the Weibull, exponential, and lognormal.

In the Weibull distribution, m = shape and c = characteristic life.

$$F(t) = 1 - e^{-\left(\frac{t}{c}\right)^m}$$

The characteristic life is the time point at which 63.2% of the units are expected to fail. The Weibull distribution can be theoretically derived as the time of first failure from a set of defect sites and is therefore appropriate for modeling defect-oriented failure modes, as opposed to degradation.

The exponential distribution is a special case of Weibull, with the shape parameter m = 1 and c = characteristic life.

$$F(t) = 1 - e^{-\left(\frac{t}{c}\right)}$$

where c is also the mean time to failure (MTTF) for the exponential distribution. The reciprocal 1/c is the IFR, which is constant for the exponential distribution and is also equal to the average failure rate over any interval. The exponential distribution is used to model random failures that occur at a steady rate as opposed to degradation or wearout failure mechanisms.

The distribution of the failure time is lognormal if the logarithm of the failure time has the normal (Gaussian) distribution.

$$F(t) = \Phi\left[\frac{1n(t) - 1n(t_{50})}{\sigma}\right]$$

where t_{50} = median life and σ = standard deviation of the logarithm of the times to failure. Note: Φ is the CDF of a normal (Gaussian) distribution with mean 0 and standard deviation 1. The lognormal distribution can be derived from a multiplicative degradation process where the rate of degradation at any time is proportional to the previous degradation. Corrosion or diffusion or crack propagation failure mechanisms often can be modeled this way.

THE MATHEMATICAL MODEL

The three components of the model

The general reliability model calculates the total CDF for the component fallout in the field by taking a weighted average of the CDFs for the component subpopulations. The three subpopulations are the defective escapes, the reliability defective, and the normal.

A certain proportion of the population is defective even at time zero. These components are functionally dead but escape detection despite testing and screening. These escapes will be discovered later in the field by the customer. The proportion defective at time zero is denoted by α. The part of α reaching the field is modeled by a discovery distribution or CDF $F_E(t)$. This CDF $F_E(t)$ empirically models the rate at which the customers discover the time zero defects that escape the manufacturing plant.

The second component of the model is the reliability defective subpopulation. There may be several of these reliability process defects, and hence the model will include a term for each of these failure mechanisms. Each of these subpopulations has a manufacturing defect causing them to fail early in life due to a specific failure mechanism. The rest of the components in the population are not susceptible to these fail mechanisms and will never fail due to this mechanism, at least within any time period of practical interest. Let β_1, β_2, . . . , β_j denote the proportions of the population having "defect" type 1, 2, . . . , j. It is assumed that there is no overlap. β_i is also known as the *mortal fraction* or *defect level* for defect i. Let F_{D1}, F_{D2}, . . . , $F_{Dj}(t)$ denote the associated CDFs for the defect subpopulations.

The remaining proportion of the population is called the normal subpopulation. Here γ, the proportion of normal units, is given by

$$\gamma = 1 - \alpha - \sum_{i=1}^{j} \beta_i$$

The CDF F_N for these normal units is derived by applying the competing risk model to the life distributions for all the failure mechanisms that apply to normal (nondefective) components.

$$F_N = 1 - \prod_i \left[1 - F_{n_i}(t)\right]$$

where $F_{Ni}(t)$ is the associated CDF for the normal wearout failure mechanism i.

Combining the Submodels

The total CDF for the entire population is constructed by taking a weighted average of the CDFs for the three subpopulations. Total cumulative failures (defective escapes and reliability) at any point of the field life is fully specified by

$$F(t) = \alpha F_E(t) + \sum_{i=1}^{j} \beta_i F_{d_i}(t) + \gamma \left\{ 1 - \prod_{i=1}^{k} \left[1 - F_{N_i}(t)\right] \right\}$$

where $\alpha + \Sigma\beta_i + y = 1$.

To run an algorithm, the following inputs are needed

- α and $F_E(t)$ (defectives entering test and field defect discovery distribution)
- the β_i and the corresponding $F_{Di}(t)$—mortal fractions and associated CDFs
- environmental conditions for each $F_{Di}(t)$
- γ and the $F_{Ni}(t)$—normal population and associated CDFs
- environmental conditions for each $F_{Ni}(t)$
- the length of the test procedure or point of field lifetime of interest

This model is most effective if acceleration models are known for the parameters of the lifetime distributions of the normal and reliability defective populations. Then the algorithm can be written to give the use CDF as a function of use conditions making the model very useful for products with varied applications.

PHYSICAL ACCELERATION MODELS

Accelerated testing

Very often the parameters for the lifetime distribution are estimated under conditions different from use conditions. This occurs because the products we deal with are so highly reliable that the luxury of testing them under use conditions becomes too expensive. Reliability testing is often done under more severe conditions of temperature, power supply, or other relevant stresses than the normal use conditions. Accelerated life testing reduces the time to failure of units by subjecting them to stresses that would cause early failures. By compressing time one can obtain failure data with practical test times and relatively small test sample sizes. However this calls for additional modeling to relate the failure time distribution under stress conditions to those under actual use conditions. Physical models that provide this bridge are called acceleration models. The common assumptions in the use of acceleration models are as follows:

- The failure mode remains the same under use and stress conditions.
- Increasing stress causes the same events to occur more rapidly.
- All events are speeded up by the same factor. The effect is similar to recording the life of a component and playing it back at a higher speed.

When these conditions hold, we have true or linear acceleration. When every time of failure and every distribution percentile is multiplied by the same constant value to obtain the projected results at another operating stress, we have linear acceleration. When linear acceleration is present,

$$\text{Time to fail at } S_1 = AF \times \text{Time to fail at } S_2$$

where AF is the acceleration constant relating times to fail at the two stresses S_1 and S_2. AF is called the acceleration factor between the two stresses. Let the subscript U denote a typical use set of stresses and the subscript S denote a typical lab set of stresses. Let F(t), f(t) and h(t) denote the CDF, PDF, and failure rate, respectively.

Time to fail: $t_U = AF \times t_S$

Failure probability: $F_U(t) = F_S\left(\frac{t}{AF}\right)$

Density function: $f_U(t) = \left(\frac{1}{AF}\right) f_S\left(\frac{t}{AF}\right)$

Hazard function: $h_U(t) = \left(\frac{1}{AF}\right) h_S\left(\frac{t}{AF}\right)$

These are general equations that hold under the assumption of linear acceleration no matter what the underlying lifetime distribution happens to be. One can easily prove that for exponential, Weibull, and lognormal distributions the shape parameter remains the same under different stresses when linear acceleration is present.

Acceleration models

A physically appropriate statistical model calculates a distribution scale parameter or percentile as a function of the operating stresses. Components and failure modes respond differently to stresses such as temperature, voltage, humidity. Hence there are several acceleration models in vogue in the industry today. These models are generally written in a deterministic form where time to fail is an exact function of the operating stresses and several material and process dependent constants.

There are a dozen or more well-known acceleration functions. Most of them, however, can be considered to be special cases of the general Eyring acceleration function [3]. A simplified version of the Eyring function with temperature and two other stresses is shown below in an acceleration factor form, with respect to specified reference conditions.

- PARAMETERS:
  - ΔH = activation energy
  - B_1, B_2 = stress intensity exponents for stress 1 and stress 2
- REFERENCE CONDITIONS:
  - T_r = reference temperature in degrees Kelvin
  - S_{1r}, S_{2r} = reference values of stress 1 and stress 2
- USE CONDITIONS:
  - T = use temperature in degrees Kelvin
  - S_1, S_2 = use condition values of stress 1 and stress 2

$$AF(T,S_1,S_2)=\exp\left[\frac{\Delta H}{k}\times\left(\frac{1}{T}-\frac{1}{T_r}\right)\right]\times\left(\frac{S_1}{S_{1r}}\right)^{-B_1}\times\left(\frac{S_2}{S_{2r}}\right)^{-B_2}$$

where k (Boltzmann's constant) = (8.617×10^{-5} eV/K).

Typical stresses used in this model may include voltage, current, current density, and relative humidity. The values of the parameters ΔH, B_1, and B_2 may either be theoretically derived or empirically estimated from the data.

The most widely used acceleration function today is the empirically derived Arrhenius reaction rate function. This model projects the rate of reaction (i.e., rate of failure) as a function of use temperature. The equation shown is a relative form with respect to a reference temperature.

$$AF(T)=\exp\left[\frac{\Delta H}{k}\times\left(\frac{1}{T}-\frac{1}{T_r}\right)\right]$$

where ΔH, k, T_r, and T are as defined for the Eyring model.

The modified Coffin-Manson [4] acceleration function projects the acceleration factor as a function of temperature range and frequency of temperature cycling. This model is typically used for failure due to crack propagation in solder or for wire bond fatigue under the stress of repeated temperature cycling as the component is powered and unpowered.

- PARAMETERS:
  - ΔH = activation energy
  - B_1 = stress intensity exponent for temperature range
  - B_2 = stress intensity exponent for cycling frequency

- REFERENCE CONDITIONS:
  - TM_r = maximum reference temperature in degrees Kelvin
  - ΔT_r = reference temperature range
  - CF_r = reference cycling frequency
- USE CONDITIONS:
  - TM = maximum use temperature in degrees Kelvin
  - ΔT = use temperature range
  - CF = use cycling frequency

$$AF(TM,\Delta T,CF) = \exp\left[\frac{\Delta H}{k} \times \left(\frac{1}{TM} - \frac{1}{TM_r}\right)\right] \times \left(\frac{\Delta T}{\Delta T_r}\right)^{-B_1} \times \left(\frac{CF}{CF_r}\right)^{-B_2}$$

Note that the exponent values often used are $B_1 = 2$ and $B_2 = 1/3$.

The power acceleration function is a very commonly used acceleration function [5] for several types of stresses including voltage, current/current density, and relative humidity. In the formula below, the stress is assumed to be voltage.

- PARAMETERS:
  - B = stress intensity exponent
- REFERENCE CONDITIONS:
  - V_r = reference voltage
- USE CONDITIONS:
  - V = use voltage

$$AF(V) = \left(\frac{V}{V_r}\right)^{-B}$$

The exponential acceleration function is another acceleration model used for stresses such as voltage and relative humidity. In the formula below, the stress is assumed to be voltage.

- PARAMETERS:
  - B = stress intensity
- REFERENCE CONDITIONS:
  - V_r = reference voltage
- USE CONDITIONS:
  - V = use voltage

$$AF(V) = \left(\frac{\exp V}{\exp V_r}\right)^{-B} = e^{-B(V-V_r)}$$

The electromigration model uses current density as a key stress parameter. This mechanism produces open or short failures in metal thin film conductors because of the movement of ions toward the anode at high-temperature and high-current densities.

- PARAMETERS:
  - ΔH = activation energy
  - n = current density exponent
- REFERENCE CONDITIONS:
  - T_r = reference temperature in degrees Kelvin
  - J_r = reference value of current density
- USE CONDITIONS:
  - T = use temperature in degrees Kelvin
  - J = use condition values of current density

$$AF(T,J) = \exp\left[\frac{\Delta H}{k} \times \left(\frac{1}{T} - \frac{1}{T_r}\right)\right] \times \left(\frac{J}{J_r}\right)^{-n}$$

Typical values for n are between 1.5 and 2.0. Typical values for ΔH are between 0.5 and 0.7 eV.

Burn-in model

Engineers are often faced with the problem of a bathtub curve which has a steep front end due to the discovery of escapes and reliability defect failures. Burn-in is used to remedy the situation. Burn-in uses intense environmental stress conditions. By stressing or burning in the product before it is shipped, the reliability defects in the front end of the bathtub curve are reduced drastically.

The burn-in model is based on a number of definitions and assumptions:

- Burn-in (BI) consists of a stress screen that may affect every failure mechanism differently.
- Testing may be done as a part of BI and may miss some of the generated failures.
- Some portion of the population may not be aged by the BI.
- Some portion of the failures may "recover."
- The test may or may not catch defective escapes.
- Each failure mode that recovers fails later on according to a mechanism-dependent recovery CDF.

The following notation is used in this model:

E	Proportion of population affected by BI
1 – E	Proportion of population unaffected by BI
R	Proportion of BI failures that recover
1 – R	Proportion of BI failures left for test to catch
C_1	Test efficiency at catching BI failures
C_2	Test efficiency at catching time zero defectives

Assumption: The proportions are not mechanism-dependent.

Burn-in fallout

The general algorithm can be used to do a mathematical analysis of burn-in. Suppose the burn-in is done for T hours at higher stresses than normal use. Then acceleration factors are used to obtain the use equivalent time for every fail mode except the manufacturing defective escapes. For simplicity we assume that none of the α manufacturing escapes are removed or discovered by burn-in. Since the burn-in test efficiency is C_1, burn-in catches C_1 of all the early failures that happen during the burn-in period. The proportion of recovered failures and the proportion that is affected by burn-in also have been taken into account while calculating burn-in fallout.

The amount of fallout removed after burn-in and a normal component test and before going to the field is specified by

$$Fo = C_2\alpha + C_1(1-R)E(F_D + F_N)$$

where

$$F_D = \sum_{i=1}^{j} \beta_i F_{D_i}\left(t_{D_i}\right)$$

and

$$F_N = \gamma\left\{1 - \prod_{i=1}^{k}\left[1 - F_{N_i}\left(t_{N_i}\right)\right]\right\}$$

The t_{Di} and t_{Ni} are the use-equivalent times to the burn-in time based on acceleration from a reference use stress to a burn-in stress.

A certain proportion of the components escape burn-in screening and merge with the manufacturing defect escapes. The merged defect escapes are specified by

$$\alpha' = \alpha + (1-C_1)(1-R)E(F_D + F_N)$$

A certain proportion of the components fail during burn-in but recover before test. The recovered fails for the defective and normal populations are specified respectively by

$$R_{D_i} = RE\beta_i F_{D_i}\left(t_{D_i}\right)$$

$$R_N = RE\gamma\left\{1 - \prod_{i=1}^{k}\left[1 - F_{N_i}\left(t_{N_i}\right)\right]\right\}$$

Burn-in will affect some components, causing them to age but not to fail. The aged population is indicated by

$$A_{D_i} = \beta_i E\left[1 - F_{D_i}\left(t_{D_i}\right)\right]$$

$$A_N = \gamma E \prod_{i}\left[1 - F_{N_i}\left(t_{N_i}\right)\right]$$

The remaining unfailed population, which is not affected by the burn-in, is known as unaged and is indicated by the following formulas.

$$U_{D_i} = \beta_i(1-E)$$
$$U_N = \gamma(1-E)$$

After the initial burn-in, each original CDF [except the discovery CDF $F_E(t)$] gives rise to three new CDFs:

- an unchanged CDF for the unaged population
- an "aged" CDF for the aged and unfailed
- a "recovery" CDF for the recovered failures

In order to calculate the fallout in the field, we have to use the conditional CDF, which is the probability of surviving for t hours given that the component survived for (t_{Di}) hours.

Define the translate function as

$$X_{D_i}\left(t_{D_i}\right) = \frac{F_{D_i}\left(t_{D_i}+t\right) - F_{D_i}\left(t_{D_i}\right)}{1 - F_{D_i}\left(t_{D_i}\right)}$$

and define $X_{Ni}(t_{Ni})$ similarly. The equation for the new CDF after t hours in the field is given by

$$F(t) = \frac{1}{1-Fo}\left(F_{BE} + F_{BD} + F_{BN}\right)$$

where

$$F_{BE} = \alpha' F_E(t)$$

$$F_{BD} = \sum_{i=1}^{j} U_{D_i} F_{D_i}(t) + \sum_{i=1}^{j} A_{D_i} X_{D_i}\left(t_{D_i}\right) + \sum_{i=1}^{j} R_{D_i} F_{RD_i}(t)$$

$$F_{BN} = U_N\left[1 - \prod_{i=1}^{k} F_{N_i}(t)\right] + A_N\left[1 - \prod_{i=1}^{k} X_{N_i}\left(t_{N_i}\right)\right] + R_N\left[1 - \prod_{i=1}^{k} F_{RN_i}(t)\right]$$

The correctional term (1-Fo) appears in the denominator because the population actually shipped has been reduced and the defects are now a higher proportion.

The fallout for the component entering a second or third burn-in is calculated exactly as before with each piece of the total population except the time zero defectives branching into three categories:

- aged
- recovered
- unaffected

The form of the general algorithm remains the same regardless of the number of tests, system run-ins, or burn-ins prior to field use. As has been described in the previous sections, each procedure generates new values for the parameters α, β, γ and conditional distributions for the surviving components. The procedure produces fallout and if prior to field life, generates a new CDF. The new

CDF is input for the next procedure until the final values for α, β_i, γ, F_{Di}, F_{Ni} for the components entering the field are reached.

The burn-in recovery portion of this model has its sources in empirical field data. Field data from memory cards populated with high-volume and high-productivity 64k Drams produced in IBM Burlington, Vermont were analyzed and compared with model predictions [6]. The predictions matched field data once the recovered failures and burn-in escapes were included in the model. In-situ monitoring of experiments of burn-in stresses produced further experimental verification.

CONCLUSIONS

Projecting the field failure rate is relatively complex, especially with the use of burn-in. However, the use of iterative programming makes the task quite simple. All the engineer provides are the environmental and application conditions. The technologist inputs the lifetime distributions and the acceleration models for the product and the vintage. Given these inputs, the iterative program calculates the cumulative fallout. The program and the model have been highly successful and used extensively for a variety of applications within IBM.

REFERENCES

1 D. J. Klinger, Y. Nakada and M. A. Menendez, AT&T Reliability Manual, Van Nostrand Reinhold, New York, 1990.

2 P. A. Tobias and D. T. Trindade, Applied Reliability, Van Nostrand Reinhold, New York, 1986.

3 H. Eyring, S. Glasstones and K. J. Laidler, The Theory of Rate Processes, McGraw-Hill, New York, 1914.

4 A. H. Landzberg and K. C. Norris, IBM J. of Res. Dev., 13, no. 3 (1969).

5 N. R. Mann, R. E. Schafer and N. D. Singpurwalla, Methods for Statistical Analysis of Reliability Life Data, John Wiley & Sons, Inc., New York, 1974.

6 A. J. Wager, D. L. Thompson and A. C. Forcier, Implications of a model for optimum burn-in, in: Reliability Physics 1983, Annual Proceedings, p. 286.

Quality through Engineering Design - W. Kuo (Editor)

Nonintuitive Availability Design and Evaluation for a Generic Flexible Manufacturing System

Way Kuo and Mahammad Rahman

Department of Industrial and Manufacturing Systems Engineering, Iowa State University, Ames, Iowa 50011, U.S.A.

Abstract

This study describes a potential tool to evaluate the performance of flexible manufacturing systems (FMS) on the basis of both conventional design parameters—mean time to failure, mean time to repair, and mean time to replacement—and other factors, including the use of the Markov process. System effectiveness is used as a performance measure, and credits are granted if only some of the systems are functional. To improve performance, the researchers installed a central storage area, where inventory of spares is optimized heuristically to replenish spares for failing equipment. Such a performance measurement technique is likely to be very useful for most real-life FMS environments. The model can be tested for a wide range of applications.

INTRODUCTION

In flexible manufacturing systems (FMS), each system design is unique and depends on the nature of the work parts to be processed through the system. Thus the versatility of FMS concepts make standardization impossible. An FMS has been defined in several contexts through different terminologies. But many overlapping terms identify the characteristics common to most of these systems.

FMS systems are aimed at providing a factory operation that is able to adjust in real time to changes in product design, production goals and scheduling, material handling and routing, and the like. An FMS may consist of a single machine, a small number of machines served by robots, or a very complex set of machines served by automated material handling equipment. In the model presented in this paper, the machine cells are considered primary resources and automatic guided vehicles (AGVs) as secondary resources. This classification is important because AGV is necessary to access any machine cell. Once a job is transferred, AGVs are free to serve the next job in the queue [1–3].

Almost all of the present models for FMSs are treated each in their entirety and not split into subsystems, for example, Vinod and Solberg [4]. Also, current models have difficulties (1) evaluating systems that consist of many subsystems, each performing its specific function, and (2) limiting themselves to a system that is confined to a local region. Studying such performance modeling and justifying these modeling techniques theoretically and practically are important and timely goals.

System effectiveness is a measure to evaluate performance of systems, which can be defined as "a measure of the degree to which an item can be expected to achieve a set of mission requirements and which may be expressed as a function of availability, utilization, production rate, etc." [5]. System effectiveness models are normally expressed in a variety of ways according to their user's requirements and applications. This study describes system effectiveness as a function of system availability and the probability measure of the achievement of a specific mission goal. A nonintuitive availability design and the evaluation for a generic flexible manufacturing system are introduced in the paper. Also described are the modeling philosophies to determine system availability by setting up a central storage area.

PERFORMANCE MEASURE FOR A GENERIC FMS

This study refers to system effectiveness as the performance measure of a generic FMS, which in a compact form may consist of one or more than one subsystem and is intended to perform a single functional purpose. Such subsystems are viewed as local systems in this study. If an environment consists of several local systems that may or may not be of the same kind, then we extend the definition of system effectiveness to local system effectiveness. If an FMS contains several local systems, each serving one kind of customer, then the needs of all kinds of customers should be considered when the performance of the whole FMS is evaluated.

An FMS contains many local systems and shared central storage. Spares for the components of local systems are equipped in central storage. Each local system works independently until it requires spare parts from the central storage. The performance measure for such a system is determined by the local system effectiveness and the adopted sparing policy. System effectiveness for a local system is one special case under our definition of performance measure for an FMS system.

A generic FMS contains M distinct local systems. Each local system contains N identical units as shown in Fig. 1. Fulfilling the requirements that all M local systems function properly is of interest to system engineers because physical links exist among local systems. In a case where a fraction of M local systems work properly while the others are "down" for repairs, partial performance is recognized by the overall environment.

Failures are repaired at the central storage area. The central storage has a limited number of various kinds of spare parts. A failed unit may be routed through the central storage area to factory for repairing if needed. Repaired units are shipped from a central storage area to replenish a down system.

Units at the same site are also referred to as the same equipment type. The central storage area is located at a distance accessible to every local system. The

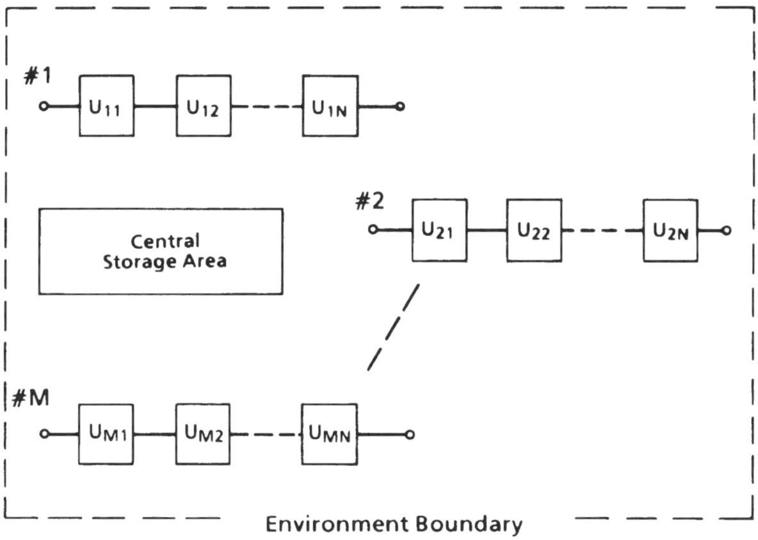

Fig. 1. Diagram of a generic FMS containing M distinct systems.

original storage level of the kth part at the central storage is x_k. The primary concern of this study is to evaluate and fulfill the environmental requirements so that a certain level of overall environmental performance can be met. A proposed weighted availability, along with a stochastic process approach, serves this evaluation purpose.

MARKOV MODELING

Notation

n_i	number of working units in state (i,0)
(i, j)	i represents number of units of a certain type currently in the repair process; j tracks additional failed unit to be replaced
p	probability that a selected spare works
$p_k(i, j)$	the $(i, j)^{th}$ state probability of the kth equipment type
n_{max}	maximum number of the units in the repair process
U_k	the kth equipment type in a given local system, k = 1, 2, . . . , M
x_k	number of allocated spares for the kth equipment type, k = 1, 2, . . . , M
x	number of allocated spares for one equipment type, a shorthand notation for x_k in later sections
$\overline{X}$	$(x_1, x_2, \ldots, x_k, \ldots, x_n)$
λ_i	transitional probability from (i,0) to (i,1) for one type of equipment
μ_i	transitional probability from (i+1,0) to (i,0) for one type of equipment
β_i	transitional probability from (i,1) to (i+1,0) for one equipment type
$\beta ii'$	transitional probability from (i,1) to (i',1) $i \neq i'$ for one type of equipment

Markov modeling for one type of equipment

An FMS may or may not consist of many different types of equipment. However, one can easily classify an FMS into subsystems or local systems. Each local system consists of one kind of equipment functioning independently. It is realistic to assume that the failures of various types of equipment occur independently; hence, the Markov approach can be used to analyze each equipment type separately. This results in a separate application of the mathematical model for each type of equipment. The effect of an equipment type on the operation of any FMS is characterized by the states of the model. Solving for steady-state probabilities gives the percentage time that the local system spends in each state. These probabilities are used to drive the steady-state availability of an FMS on the basis of one type.

Events in the interval (t_0,t_9) are outlined below.

t_0 – all equipment types functioning
t_1 – failure of one working unit
t_2 – replacement of failed unit with working spare
t_3 – failure of the second working unit
t_4 – replacement of failed unit with working spare
t_5 – failure of the third working unit
t_6 – replacement of failed unit with working spare after first spare was found defective
t_7 – repaired unit is returned to stock as spares
t_8 – another repaired unit is returned to stock as spares

The overall effect on the equipment type at the site of operation during the time interval (t_0,t_9) is an increase from 0 to 2 units in the repair process.

Two state variables (i, j) characterize the Markov model. The variable i tracks the number of units (of a given type) currently in the repair process; additional units yet to be replaced are reported by the variable j. To illustrate the state variable, the events indicate the following one-to-one correspondence:

t_0 – (0,0)
t_1 – (0,1)
t_2 – (1,0)
t_3 – (1,1)
t_4 – (2,0)
t_5 – (2,1)
t_6 – (4,0)
t_7 – (3,0)
t_8 – (2,0)
t_9 – (2,0)

Implicit in the Markov model is the concept of random failures of FMS equipment items. Statistically, random failures correspond to postulating an exponential distribution for the time between failures (inter-failure distribution).

It is assumed that the working units are operating in the post-infant-mortality, constant-failure-rate region. Some spare parts are defective when they are installed. These spares may be broken in shipment, or they may deteriorate in storage. Notice that a failed spare is discovered at time t_6. Clearly, the spare has failed during the interval (t_0,t_6), but generally no more information is known concerning this spare. This situation is modeled by simply assuming a constant probability that a spare unit will work. Once a working spare replaces a failed unit, this spare is treated as a working unit and thus is assumed to operate in the constant-failure-rate region.

Replacement rates of working units rather than failure rates are used in the Markov model because often units that are replaced haven't necessarily failed. For example, system diagnostics may isolate a problem to three parts of different types. In order to expedite the restoration of service, all three units may be replaced even though only one might have failed. The replacement rate is determined from the failure rate and specification of a constant ratio of the rate of replacement to failure.

Immediate replacement of a failed unit in the plant is not assumed. The concept of replacement (or response) interval—that is, the time taken to reach a failed item and replace it with a working spare—is introduced. The replacement interval involves time to (1) diagnose the failure, (2) travel to the site (from a central storage area to an unattended equipment site), and (3) insert spares until a repair is completed. Replacement intervals are (t_1,t_2), (t_3,t_4), and (t_5,t_6).

Spare repair interval is the time required to replenish the stock of spare units with a repaired or new unit. It involves transportation time to and from the factory and time spent at the factory either for repairs or to obtain a new unit. The spare repair intervals are (t_2,t_7) and (t_4,t_8). A unit in the spare repair interval is described as being in the repair process.

The Markov model is applied separately to each equipment type under the following assumptions:

1. Failure time, replacement time, and spare repair time are distributed exponentially with different parameters.
2. Each spare has constant working probability.
3. The system starts at a working state.
4. Replacements come from spares of identical age as the failing units.
5. M local systems function with statistical independence.
6. N units in a local system function with statistical independence.
7. Repairing two or more spare units simultaneously is impossible.
8. Workpieces in each fixture are independent.

The equations for the steady-state probability are set up and solved from a continuous-time, discrete-state Markov model. Solving the simultaneous equations yields the steady-state probabilities that determine the allocation and system availability of spares.

Let Z(t) be the stochastic process defined by the state variables (i, j) and indexed by time t. The observed value of Z(t) at time t_n is denoted by Z_n. A set of random variables {Z(t)} forms a Markov chain. The transitions between states for a continuous-time Markov chain may take place at any time. These transitions are indexed by $t_0, t_1, \ldots, t_n, \ldots$ [6].

Recall that the system is assumed to be in steady-state equilibrium. Let P(i, j) denote the steady-state probability associated with state (i, j); $i = 0, 1, \ldots, n_{max}$, $j = 0,1$. According to the definition of steady state, the transition rate out of a given state equals the rate into that state. Thus, the following equations are written

$$\lambda_0 P(0,0) = \mu_1 P(1,0) \tag{1}$$

$$(\lambda_i + \mu_i) = p\beta_{i-1}P(i-1,1) + \mu_{i+1}P(i+1,0) \quad \text{for } i = 1, 2, \ldots, x \tag{2}$$

$$(\lambda_i + \mu_i) = \beta_{i-1}P(i-1,1) + \mu_{i+1}P(i+1,0) \quad \text{for } i = x+1, \ldots, n_{\max} - 1 \tag{3}$$

$$\mu_{n_{\max}} P(n_{\max},0) = \beta_{n_{\max}-1} P(n_{\max} - 1, 1) \tag{4}$$

$$p\beta_0 P(0,1) + \sum_{i'=1}^{x} \beta_{0i'} P(0,1) = \lambda_0 P(0,0) \tag{5}$$

$$p\beta_i P(i,1) + \sum_{i'=i+1}^{x} \beta_{ii'} P(i,1) = \lambda_i P(i,0) + \sum_{i'=0}^{i-1} \beta_{i'i} P(i',1) \quad \text{for } i = 1, 2, \ldots, x-1 \tag{6}$$

$$\beta_x P(x,1) = \lambda_x P(x,0) + \sum_{i'=0}^{x-1} \beta_{i'x} P(i',1) \tag{7}$$

$$\beta_i P(i,1) = \lambda_i P(i,0) \quad \text{for } i = x+1, x+2, \ldots, n_{\max} \tag{8}$$

According to the axioms of probability theory, the following equation also holds

$$\sum_{i=0}^{n_{\max}} P(i,0) + \sum_{i=1}^{n_{\max}-1} P(i,1) = 1 \tag{9}$$

By solving Eqs. (1–8) along with the normality condition of Eq. (9), a set of solutions for P(i, j) where i = 0, 1, . . . , n_{max} and j = 0,1 can be obtained.

EVALUATION OF FMS PERFORMANCE MEASURE

The FMS performance requirements include the following:

1. All subsystems are properly functioning.
2. Some of the subsystems are functioning while some are not, given that one equipment type is down.
3. Some of the subsystems are functioning while some are not, given that more than one equipment type is down.

The first performance requirement is a conventional way to treat the effectiveness of the system, that is, a system regarded as a unit in its entirety. Partial fulfillment of the requirements in the system deserves a nonzero performance measure; this is important when each unit/subsystem has an independent identity. Thus, the second requirement accounts for partial credit that may be granted to the overall performance. Because all N pieces of equipment within each M local system are identical, it would be immaterial to assign partial credit to any specific unit; only a weight assigned to the overall effectiveness is important. The third performance requirement also accounts for some partial credit granted to the overall performance. Because more than one equipment type is down, these failing equipment types are spread over several subsystems. The probability that a specific event occurs might affect the weighting factor in the overall system performance.

Effectiveness evaluation

Consider a kth type of FMS equipment, where all of the kth-type units function properly as long as the number of units in the repair process is less than x_k. Let $P_k(0,0)$ be the probability that all of the kth types of equipment are working properly in the environment without any failure in the repairing process. Similarly, $P_k[(i,0) \mid i \leq x_k]$ is the probability that all of the kth types of equipment are functioning properly in the environment, given that i units are in the repairing process, where $i \leq x_k$. Traditionally, the availability of the kth types of equipment [7] is

$$\sum_{i=0}^{x_k} P_k[(i,0)\, i \leq x_k] \tag{9a}$$

Hence, the availability of a given system can be expressed as

$$A_0 = \prod_{k=1}^{M} \sum_{i=0}^{x_k} P_k[(i,0)|i \leq x_k] \tag{10}$$

In case that the kth equipment is in the state $(x_k + 1,0)$ with probability $P_k(x_k + 1,0)$ while all the other equipment types are in the state (i,0) where $i \leq x_i$ for $i \neq k$, then we can assume that (1) one of the M subsystems is in a partially down or down state because of the unavailability of a good spare for the kth type of equipment, and (2) the other M-1 subsystems are functioning properly. The probability associated with events (1) and (2) guarantees that M-1 systems function and can be written as

$$A_1^k = P_k(x_k + 1,0) \prod_{\substack{k'=1 \\ k' \neq k}}^{M} \sum_{i=1}^{x_{k'}} P_{k'}[(i,0)| \leq x_{k'}] \tag{11}$$

If the kth equipment type is in the state $(x_k + 1,0)$ and simultaneously the lth equipment is in the state $(x_l + 1,0)$ while all the other types of equipment are in the state (i,0) where $i \leq x_i$ for $i \neq k$ and $i \neq \ell$, then one can imagine that two of the M subsystems are in the partially down or down state because of the unavailability of good spares for the kth equipment type in one system and for the ℓth equipment type in the other system while the remaining M-2 are functioning properly. Therefore, probability associated with this event is written as

$$A_{11}^{k1} = P_k(x_k + 1,0)P_\ell(x_\ell + 1,0) \prod_{\substack{k'=1 \\ k' \neq k \\ k' \neq \ell}}^{M} \sum_{i=0}^{x_{k'}} P_{k'}[(i,0)|\ i \leq x_{k'}] \tag{12}$$

The probability that at least one subsystem is working properly can similarly be calculated. Unless all M equipment types are in the state $(x_i + 1,0)$, for all $i = 1, 2, \ldots, M$, there exists a positive probability that some of the local systems will function properly, and hence credit should be granted as partial attributes to the overall environmental performance measurement.

In general, if the $k_1{}^{th}$ type is in the state $(x_{k1} + a_1,0)$, the $k_2{}^{th}$ type is in the state $(x_{k2} + a_2,0), \ldots$, the $k_m{}^{th}$ type is in the state $(x_{km} + a_m,0)$, and $a_i < N$, $\forall$ i and $m < M$, then there exists a positive probability that at least one system is functional. The probability associated with the events is

$$A_{a_1 a_2 \ldots a_m}^{k_1 k_2 \ldots k_m} = \prod_{j=1}^{m} P_{k_j}\left(x_{k_j} + a_j\right) \prod_{\substack{k'=1 \\ k' \neq k_j \\ \forall j=1,2,\ldots,m}}^{M} \sum_{i=0}^{x_{k'}} P_{k'}[(i,0)|\ i \leq x_{k'}] \tag{13}$$

Assignment of weights to the effectiveness

Since all N units in each of M local systems are identical, it is realistic to regard each unit and each subsystem as equally important to overall system effectiveness. Therefore, the frequency of failures occurring in a specific system follows the combination law. The relative weight given to the environmental effectiveness under various failures in the repairing process can be readily evaluated. If $i \leq x_k$, the weight of unity is assigned to availability; otherwise a fraction will be assigned.

For the kth equipment type in the state $(x_k + 1,0)$ and all the other types of equipment in the state (i,0) where $i \leq x_i$ for $i \neq k$, the down state can appear in any one of the M local systems with equal probability. Therefore, the weighting factor to be assigned to A_1^k is

$$W_1^k = (N-1)/N \tag{14}$$

For the kth equipment type in the state $(x_k + 1,0)$ and simultaneously, the ℓth equipment in the state $(x_\ell + 1,0)$ (while all the other equipment types are in the state (i,0) where $i \leq x_i$ for $i \neq k$ and $i \neq k$), the weighting factor of

$$W_{11}^{k1} = (N-1)/N \times (N-1)/N = [(N-1)/N]^2 \tag{15}$$

should be assigned to A_{11}^{k1}.

In general, if the k_1th equipment type is in the state $(x_{k_1} + a_1, 0)$, the k_2th equipment type is in the state $(x_{k_2} + a_2, 0)$, . . . ; the k_mth equipment type is in the state $(x_{k_m} + a_m, 0)$; $a_i < N$; $\forall$ i and m < M; and the weighting factor is denoted by

$$W_{a_1 a_2 \ldots a_m}^{k_1 k_2 \ldots k_m} = \prod_{i=1}^{m} (n - a_i)/N \tag{16}$$

FMS performance measurement

To model the performance measurement adequately for all the local systems in an environment, we formulated a weighted availability measure that includes the following:

1. A_0 where all the systems are operating properly, either no failure occurs, or the failures do not exceed the storage level.
2. $W_1^k A_1^k$ for kth types of equipment, k = 1, 2, . . . , M where exactly one local system is in the partially down or down state.
3. $W_2^k A_2^k$ for kth types of equipment, k = 1, 2, . . . , M where exactly two equipment items in a local system are in the partially down or down state.
4. $W_{a_1 a_2 \ldots a_m}^{k_1 k_2 \ldots k_m} A_{a_1 a_2 \ldots a_m}^{k_1 k_2 \ldots k_m}$ where the k_1th equipment type is in the state $(x_{k_1} + a_1, 0)$, the k_2th equipment type is in the state $(x_{k_2} + a_2, 0)$, . . . , and the k_mth equipment type is in the state $(x_{k_m} + a_m, 0)$, as long as $a_i < N \ \forall$ i. A summation of the above attributes forms the performance measurement.

$$\begin{aligned} P = \; & A_0 + \sum_{a_1=1}^{N-1} [(N-a_1)/N] \sum_{k=1}^{M} A_{a_1}^{k} \\ & + \sum_{a_1=1}^{N-1} \sum_{a_2=1}^{N-1} [(N-a_1)/N][(N-a_2)/N] \sum_{k_1=1}^{M} \sum_{k_2=1}^{M} A_{a_1 a_2}^{k_1 k_2} \\ & \ldots\ldots\ldots + \ldots\ldots\ldots + \ldots\ldots \\ & + \sum_{a_1=1}^{N-1} \sum_{a_2=1}^{N-1} \cdots \sum_{a_m=1}^{N-1} [(N-a_1)/N][(N-a_2)/N] \ldots \end{aligned} \tag{17}$$

$$\ldots[(N-a_m)/N]\sum_{k_1=1}^{M}\sum_{k_2=1}^{M}\ldots\sum_{k_m=1}^{M}A_{a_1a_2\ldots a_m}^{k_1k_2\ldots k_m}$$

Equation (17) is a realization of a performance measure for a system consisting of one or more subsystems. The first term on the right-hand side is a regular estimate for effectiveness. The other terms are the effectiveness contributed by the down states.

NUMERICAL ILLUSTRATION

Description of system and terminology

Consider that a company maintains a system in a flexible manufacturing environment that can be broken down into four local systems or subsystems. The first, second, third, and fourth local systems are formed by five units, six units, four units, and six units (equipment items of same type), respectively. Because each subsystem is necessary for production operation, it is reasonable to assume that each subsystem is equally important to the company.

A previous study showed that this company served by local systems would be satisfied with the service if the average down time of the system was less than 6% of the total operating time. It is known from historical data that the first equipment type has a mean time to failure (MTTF) of 300 hours and a mean time to repair (MTTR) of 10 hours; the second equipment type has an MTTF of 650 hours and an MTTR of 125 hours; the third equipment type has an MTTF of 500 hours and an MTTR of 15 hours; and the fourth equipment type has an MTTF of 400 hours and an MTTR of 20 hours. The steady-state availability of such a system is

$$\prod_{i=1}^{M=4}\left(\frac{MTTF_i}{MTTF_i+MTTR_i}\right)=\frac{300}{310}\times\frac{650}{662}\times\frac{500}{515}\times\frac{400}{420}$$
$$=0.878594$$

Therefore, traditionally the probability that all the local systems in the FMS environment are available is

$$\left(\frac{300}{310}\right)^5\left(\frac{650}{662}\right)^6\left(\frac{500}{515}\right)^4\left(\frac{400}{420}\right)^6=0.504249$$

Neither 0.878594 nor 0.504249 satisfies the maximum 6% down-time restriction. Since each subsystem is very expensive and it is unnecessary to achieve the extraordinary high availability at the expense of very high cost, a redundant system configuration is not recommended. A central storage area, in which one spare for each equipment type is stored, can be used for achieving high availability.

Performance evaluation using weighted factors without spares

Traditionally no partial performance accounted for partial operational units. Because this research assigns partial credits if some of its subsystems are operational, the availability of 0.504249 is underestimated. Applying Eq. (17) and substituting values of $x_1 = x_2 = x_3 = x_4 = 0$

$$P = \left[P_1(0,0)P_2(0,0)P_3(0,0)P_4(0,0)\right]$$

In the state of (1,0)

$$\begin{aligned}
&+\left[(4/5)P_1(1,0)P_2(0,0)P_3(0,0)P_4(0,0)\right]\\
&+\left[P_1(0,0)(5/6)P_2(1,0)P_3(0,0)P_4(0,0)\right]\\
&+\left[P_1(0,0)P_2(0,0)(3/4)P_3(1,0)P_4(0,0)\right]\\
&+\left[P_1(0,0)P_2(0,0)P_3(0,0)(5/6)P_4(1,0)\right]
\end{aligned}$$

In the state of (2,0)

$$\begin{aligned}
&+\left[(3/5)P_1(2,0)P_2(0,0)P_3(0,0)P_4(0,0)\right]\\
&+\left[P_1(0,0)(4/6)P_2(2,0)P_3(0,0)P_4(0,0)\right]\\
&+\left[P_1(0,0)P_2(0,0)(2/4)P_3(2,0)P_4(0,0)\right]\\
&+\left[P_1(0,0)P_2(0,0)P_3(0,0)(4/6)P_4(2,0)\right]
\end{aligned}$$

In the state of (1,0) and (1,0)

$$\begin{aligned}
&+\left[(4/5)P_1(1,0)(5/6)P_2(1,0)P_3(0,0)P_4(0,0)\right]\\
&+\left[(4/5)P_1(1,0)P_2(0,0)(3/4)P_3(1,0)P_4(0,0)\right]\\
&+\left[(4/5)P_1(1,0)P_2(0,0)P_3(0,0)(5/6)P_4(1,0)\right]\\
&+\left[P_1(0,0)(5/6)P_2(1,0)(3/4)P_3(1,0)P_4(0,0)\right]\\
&+\left[P_1(0,0)(5/6)P_2(1,0)P_3(0,0)(5/6)P_4(1,0)\right]\\
&+\left[P_1(0,0)P_2(0,0)(3/4)P_3(1,0)(5/6)P_4(1,0)\right]\\
&+\ldots\ldots\ldots\ldots+\ldots\ldots\ldots\ldots
\end{aligned}$$

In the states of (5,0), (6,0), (4,0), and (6,0)

$$+\left[(1/5)P_1(4,0)(1/6)P_2(5,0)(1/4)P_3(3,0)P_4(5,0)\right]$$

$$P = \left(\frac{300}{310}\right)^5\left(\frac{650}{662}\right)^6\left(\frac{500}{515}\right)^4\left(\frac{400}{420}\right)^6$$

Any one equipment type in (1,0) state

$$+\frac{4}{5}\left(\frac{10}{310}\right)\left(\frac{300}{310}\right)^4\left(\frac{650}{662}\right)^6\left(\frac{500}{515}\right)^4\left(\frac{400}{420}\right)^6$$

$$+\frac{5}{6}\left(\frac{12}{662}\right)\left(\frac{650}{662}\right)^5\left(\frac{300}{310}\right)^5\left(\frac{500}{515}\right)^4\left(\frac{600}{420}\right)^6$$

$$+\frac{3}{4}\left(\frac{15}{515}\right)\left(\frac{500}{515}\right)^3\left(\frac{300}{310}\right)^5\left(\frac{650}{662}\right)^6\left(\frac{400}{420}\right)^6$$

$$+\frac{5}{6}\left(\frac{20}{420}\right)\left(\frac{400}{420}\right)^5\left(\frac{300}{310}\right)^5\left(\frac{650}{662}\right)^6\left(\frac{500}{515}\right)^4$$

Any one equipment type in (2,0) state

$$+\frac{3}{5}\left(\frac{10}{310}\right)^2\left(\frac{300}{310}\right)^3\left(\frac{650}{662}\right)^6\left(\frac{500}{515}\right)^4\left(\frac{400}{420}\right)^6$$

$$+\frac{4}{6}\left(\frac{12}{662}\right)^2\left(\frac{650}{662}\right)^4\left(\frac{300}{310}\right)^5\left(\frac{500}{515}\right)^4\left(\frac{400}{420}\right)^6$$

$$+\frac{2}{4}\left(\frac{15}{515}\right)^2\left(\frac{500}{515}\right)^2\left(\frac{300}{310}\right)^5\left(\frac{650}{662}\right)^6\left(\frac{400}{420}\right)^6$$

$$+\frac{4}{6}\left(\frac{20}{420}\right)^2\left(\frac{400}{420}\right)^4\left(\frac{300}{310}\right)^5\left(\frac{650}{662}\right)^6\left(\frac{500}{515}\right)^4$$

Any two equipment types in (1,0) and (1,0) states

$$+\frac{4}{5}\frac{5}{6}\left(\frac{10}{310}\right)\left(\frac{300}{310}\right)^4\left(\frac{12}{662}\right)\left(\frac{650}{662}\right)^5\left(\frac{500}{515}\right)^4\left(\frac{400}{420}\right)^6$$

$$+\frac{4}{5}\frac{3}{4}\left(\frac{10}{310}\right)\left(\frac{300}{310}\right)^4\left(\frac{15}{515}\right)\left(\frac{500}{515}\right)^3\left(\frac{650}{662}\right)^6\left(\frac{400}{420}\right)^6$$

$$+\frac{4}{5}\frac{5}{6}\left(\frac{10}{310}\right)\left(\frac{300}{310}\right)^4\left(\frac{20}{420}\right)\left(\frac{400}{420}\right)^5\left(\frac{650}{650}\right)^6\left(\frac{500}{515}\right)^4$$

$$+\frac{5}{6}\frac{3}{4}\left(\frac{12}{662}\right)\left(\frac{650}{662}\right)^5\left(\frac{15}{515}\right)\left(\frac{500}{515}\right)^3\left(\frac{300}{310}\right)^5\left(\frac{400}{420}\right)^6$$

$$+\frac{5}{6}\frac{5}{6}\left(\frac{12}{662}\right)\left(\frac{650}{662}\right)^5\left(\frac{20}{420}\right)\left(\frac{400}{420}\right)^5\left(\frac{300}{310}\right)^5\left(\frac{500}{515}\right)^4$$

$$+\frac{3}{4}\frac{5}{6}\left(\frac{15}{515}\right)\left(\frac{500}{515}\right)^3\left(\frac{20}{420}\right)\left(\frac{400}{420}\right)^5\left(\frac{300}{310}\right)^5\left(\frac{650}{662}\right)^6$$

$$+\dots\dots\dots+\dots\dots\dots+\dots$$

All four types in (5,0), (6,0), (4,0), (6,0) states

$$\begin{aligned}+\left(\frac{10}{310}\right)^5\left(\frac{12}{662}\right)^6\left(\frac{15}{515}\right)^4\left(\frac{20}{420}\right)^6 \quad &= 0.504249+0.053561+0.001541\\ &\quad +0.002040+0.000035+o(10^{-7})\\ &= 0.561426\end{aligned}$$

Achieving high availability by a central storage area

While Tansel and Kiran [8] discussed optimum storage locations in flexible manufacturing systems, we are investigating the optimal storage area to achieve high system availability. Suppose that a central area is provided that is stocked with x_1 spares of equipment type 1, x_2 spares of equipment type 2, x_3 spares of equipment type 3, and x_4 spares of equipment type 4. Mean time to replacement (MTR) is four hours for equipment type 1, six hours for type 2, eight hours for type 3, and five hours for type 4. It is given by the vendor that the probability of a spare failing is 1%. A summary of the numerical data of the parameters appears in Table 1.

Each equipment type is treated as statistically independent of others. The formulas describing the transitional probabilities are given in Eqs. (1–9).

Table 1. Numerical data of the parameters.

Parameters	Transition Rate			
	Equipment Type 1	Equipment Type 2	Equipment Type 3	Equipment Type 4
λi, i = 0, 1, . . . N_i	1/300	1/650	1/500	1/400
μi, i = 1, 2, . . . n_{max_i}	1/10i	1/12i	1/15i	1/20
βi, i = 0, 1, . . . N_i	1/4	1/6	1/8	1/5
β_{01}	1-P	1-P	1-P	1-P
N_i	5	6	4	6
x_i	1	1	1	1
n_{max_i}	6	7	5	7
M = 4, P = 0.99				

Similarly, the Markov Process for the other three types of equipment are considered, and the formulas resulting from the transitional probabilities give similar equations as above. The state probabilities are computed and listed in Table 2 for all four equipment types. With Table 2, the traditional availability becomes

$$A_0 = \prod_{i=1}^{4} \left[P_i(0,0) + \dot{P}_i(x_i,0)\right]$$

Because it is given that $x_1 = x_2 = x_3 = x_4 = 1$,

$$\begin{aligned} A_0 &= \prod_{i=1}^{4} \left[P_i(0,0) + P_i(1_i,0)\right] \\ &= (0.949980 + 0.031666) \times (0.970012 + 0.017908) \\ &\quad \times (0.949106 + 0.028473) \times (0.930222 + 0.046511) \\ &= 0.981646 \times 0.987920 \times 0.977579 \times 0.976733 \\ &= 0.925986 \end{aligned}$$

which is much better than 0.561426 when the central storage area is not installed. To account for the partial credits from the environmental viewpoint, we must consider the states one by one and assign partial credits for partial availability of the manufacturing subsystem.

Table 2. State probabilities of the numerical illustration.

State	State Probabilities			
	Equipment Type 1	Equipment Type 2	Equipment Type 3	Equipment Type 4
	P1(i, j)	P2(i, j)	P3(i, j)	P4(i, j)
(0,0)	0.949980	0.970012	0.949106	0.930222
(0,1)	1.230×10^{-2}	8.532×10^{-3}	1.420×10^{-2}	1.119×10^{-2}
(1,0)	3.167×10^{-2}	1.791×10^{-2}	2.847×10^{-2}	4.651×10^{-2}
(1,1)	9.094×10^{-4}	6.721×10^{-4}	1.580×10^{-3}	1.135×10^{-3}
(2,0)	4.547×10^{-3}	2.688×10^{-3}	5.927×10^{-3}	9.081×10^{-3}
(2,1)	6.063×10^{-5}	2.482×10^{-5}	9.483×10^{-5}	1.135×10^{-4}
(3,0)	4.547×10^{-4}	1.489×10^{-4}	5.334×10^{-4}	1.362×10^{-3}
(3,1)	6.063×10^{-6}	1.375×10^{-6}	8.534×10^{-6}	1.703×10^{-5}
(4,0)	6.062×10^{-5}	1.099×10^{-5}	6.400×10^{-5}	2.724×10^{-4}
(4,1)	8.083×10^{-7}	1.015×10^{-7}	1.024×10^{-6}	3.405×10^{-6}
(5,0)	1.010×10^{-5}	1.017×10^{-6}	9.600×10^{-6}	6.811×10^{-5}
(5,1)	1.346×10^{-7}	9.386×10^{-9}	–	8.514×10^{-7}
(6,0)	2.020×10^{-6}	1.147×10^{-7}	–	2.044×10^{-5}
(6,1)	–	1.059×10^{-9}	–	2.555×10^{-7}
(7,0)	–	1.482×10^{-8}	–	7.154×10^{-6}

(1) Any one equipment type in failure states

(a) Only one equipment type in the state (x + 1,0) = (2,0)

$$\sum_{k=1}^{4} W_1^k A_1^k = 4/5\, A_1^1 + 5/6\, A_1^2 + 3/4\, A_1^3 + 5/6\, A_1^4$$

$$= 0.0169158$$

(b) Only one equipment type in the state (x + 2,0) = (3,0)

$$\sum_{k=1}^{4} W_2^k A_2^k = 3/5\, A_2^1 + 4/6\, A_2^2 + 2/4\, A_2^3 + 4/6\, A_2^4$$

$$= 0.00146$$

(c) Only one equipment type in the state (x + 3,0) = (4,0)

$$\sum_{k=1}^{4} W_3^k A_3^k \quad = 2/5\, A_3^1 + 3/6\, A_3^2 + 1/4\, A_3^3 + 3/6\, A_3^4$$

$$= 0.000172$$

(d) Only one equipment type in the state (x + 4,0) = (5,0)

$$\sum_{k=1}^{4} W_4^k A_4^k \quad = 1/5\, A_4^1 + 2/6\, A_4^2 + 2/6\, A_4^4$$

$$= 0.000024$$

(e) Only one equipment type in the state (x + 5,0) = (6,0)

$$\sum_{k=1}^{4} W_5^k A_5^k \quad = 1/6\, A_5^2 + 1/6\, A_5^4$$

$$= 0.000003$$

(2) Two or more equipment types in failure states
(a) Any two equipment types in the state (x++ 1,0) = (2,0)

$$\sum_{\substack{1=1 \\ 1\neq k}}^{4} \sum_{k=1}^{4} W_{11}^{k1} A_{11}^{k1} \quad = 4/5\left(5/6\, A_{11}^{12} + 3/4\, A_{11}^{13} + 5/6\, A_{11}^{14}\right)$$

$$+ 5/6\left(3/4\, A_{11}^{23} + 5/6\, A_{11}^{24}\right) + 3/4 \times 5/6\, A_{11}^{34}$$

$$= 0.000107$$

(b) Other terms given in Eq. (17) have numerical values much smaller than 10^{-7}.

Finally, the proposed performance measure is the summation of A_0 and the above mentioned points including their subsections, which are equal to the following:

$$= 0.925986 + 0.018578 + 0.000107 + 0(10^{-7})$$
$$= 0.944672$$

ACKNOWLEDGMENT

This study was supported in part by the Fulbright Foundation.

REFERENCES

1 M. P. Grover, Automation Production System and Computer Aided Manufacturing, Prentice-Hall, Englewood Cliffs, N.J., 1980.
2 A. Kusiak, (ed.), Flexible Manufacturing Systems: Methods and Studies, Elsevier Science Publishers, Amsterdam, 1986.
3 N. R. Greenwood, Implementing Flexible Manufacturing Systems, McMillan Education Ltd., London, 1988.
4 B. Vinod and S. S. Solberg, Omega, The Int. J. of Manage. Sci., 12, no. 3 (1984) 299–308.
5 B. S. Abbas and W. Kuo, IEEE Trans. Sys. Man and Cybern., 20, no. 4 (1990) A-26 to A-34.
6 S. M. Ross, Introduction to Probability Models, Academic Press, Orlando, Fla., 1985.
7 M. A. Mazumada, Nav. Res. Log. Q., 29, no. 3 (1982) 411–418.
8 B. Tansel and A. Kiran, J. of Manuf. Sys., 7, no. 2 (1988) 121–130.

Quality through Engineering Design - W. Kuo (Editor)

Computing Reliability of Noncoherent Systems

Shambhu J. Upadhyaya[a] and Hoang Pham[b]

[a]Department of Electrical and Computer Engineering, State University of New York at Buffalo, Buffalo, NY 14260, U.S.A.

[b]Idaho National Engineering Laboratory, EG & G Idaho, Inc., Idaho Falls, ID 83415, U.S.A.

Abstract

A k-to-ℓ-out-of-n system[1] is a noncoherent system in which no fewer than k and no more than ℓ-out-of-n units are to function for the successful operation of a system. Examples of noncoherent systems are found in communication, multiprocessor, and transportation applications. A noncoherent system is very general and includes the parallel, series, and N-modular redundancy systems as special cases.

The reliability of noncoherent systems has been analyzed by several researchers, but the number of computational steps for the evaluation has not reached the minimum. Earlier techniques based on the generalizations of the inclusion-exclusion principle and the event-space approach for computing reliability are inefficient. After briefly reviewing these and some of the recent and more efficient techniques, an alternative approach based on the elementary event method is presented for computing the reliability of noncoherent systems. A salient feature of this new technique is its adaptability to a simple cellular realization in hardware. This cellular implementation is shown to improve the computational complexity from quadratic to linear-order time. Theoretical results on the optimization issues of noncoherent systems are also presented.

INTRODUCTION

A class of systems called noncoherent systems was first modeled and analyzed by Heidtmann [1]. Very little attention has been focused on noncoherent systems in the reliability literature, however, despite their being general and applicable to modeling a large class of practical systems. A noncoherent system is characterized by three parameters—k, ℓ, and n—and is defined as a k-to-ℓ-out-of-n system wherein no fewer than k and no more than ℓ-out-of-n units are to function for the successful operation of the system. The k-to-ℓ-out-of-n system takes into account the failures caused by the underproduction of units or

---

[1]For ease in reading, the l in the text is the same as the l in the equations.

services and the failures that are caused by overproduction as well. Significantly, a large number of practical systems can be modeled as failed if there is an underproduction or overproduction of units or services.

Consider a multiprocessor system where resources such as memory, I/O units, and buses are shared among the various processors. In a multiprocessor environment, one might wish to employ a large number of processors n to accomplish a specific task. If less than a certain number of processors k are being used in such an installation, the system will not work to its maximum capacity and system efficiency goes down. On the other hand, if the number of processors being used is more than ℓ, the efficiency again suffers because of the traffic congestion caused by the large number of processors in use. Thus, it is desirable to avoid both of these cases, which can be accomplished by modeling the system as failed for these cases. The multiprocessor system is thus a k-to-ℓ

As another example, we consider the space shuttle computer complex. In the space shuttle avionics system, five identical computers are assigned to redundant operations [2]. During the critical mission phases such as boost, reentry, and landing, four of these computers operate in an N-Modular Redundancy (NMR) mode, receiving the same inputs and executing identical tasks. The fifth computer is used to perform noncritical tasks. The shuttle can tolerate up to two failures. If all the five computers are functioning during the critical mission phases, the technique employed to detect a failed computer may not be able to identify the failure, given that a failure has occurred. This will eventually cause the space shuttle to fail. It is therefore a 2-to-4-out-of-5 system. More examples of k-to-ℓ-out-of-n systems can be found in Ref. [1].

The generality of the k-to-l-out-of-n system can be readily demonstrated. If ℓ = n and k = $\left\lfloor \frac{n+1}{2} \right\rfloor$ (where $\lfloor x \rfloor$ is the largest integer smaller than or equal to x) and the units are identical, it is the NMR system. If k = 1 and ℓ = n, it is a parallel system; whereas if k = ℓ = n, it is a series system.

Modeling and analysis of k-to-ℓ-out-of-n systems is essential if optimal configurations are to be designed. The reliability of noncoherent systems has been first analyzed by Heidtmann by a generalization of the inclusion-exclusion method (see Feller [3]). The evaluation of reliability by this approach is computation intensive. Although several techniques have been suggested in the literature to reduce the computational efforts [4–5], the number of computational steps is not yet minimum. In this paper, we present a direct computational algorithm for the evaluation of reliability of a k-to-ℓ-out-of-n system. This algorithm is quadratic in time and further reduces the total number of computational operations. The design of the algorithm is such that very simple schemes can be used if it is to be implemented in hardware. The parallelism in the algorithm has been exploited, and a simple cellular architecture is proposed for implementing the hardware for the algorithm. This implementation evaluates the system reliability in linear-order time.

BACKGROUND

Consider a system composed of n units. We assume that both units and the system are two-state, i.e., either successful or unsuccessful (failed). We also assume that the unit states are statistically independent and the unit reliabilities are not necessarily equal (nonidentical units). Let pi(qi) be the reliability (unreliability) of unit i. Let S(x,n) be the probability that exactly x units out of the n units are successful. Let R(k : ℓ,n) denote the reliability of the k-to-ℓ-out-of-n system.

According to Heidtmann's model [1], the reliability of the k-to-ℓ-out-of-n system is

$$R(k:l,n)=\sum_{x=k}^{l} S(x,n) \tag{1}$$

To evaluate S(x,n), Heidtmann [1] uses a technique based on a generalization of Poincare's formula and Bonferroni's inequalities [3]. Also, each term in these sets has n multiplicands. Clearly, the expansion of Eq. (1) generates a large number of terms, each term having n multiplicands. The computation of the reliability therefore requires a large number of addition and multiplication operations, even for a modest size of n. But, in the expansion of Eq. (1), there are several sets of terms, with terms in the successive sets differing by one failure. Jain and Gopal [4] combine the terms having the same number of multiplicands and differing in one failure using the Boolean identity $A\,B+A\,\overline{B}=A$ With this approach, a fewer number of terms with fewer multiplicands the expression for R(k: ℓ,n) results than the original expression in Eq. (1). The simplified expression corresponding to Eq. (1) is given by Ref. [4]

$$\begin{aligned}R(k{:}l,n) &= S(k,m)+\sum_{i=0}^{n-m-1} p_{m+1+i}S(k-1,m+i)\\ &+\sum_{i=0}^{n-m-1} q_{m+1+i}\,S(k+1+i,m+i)\end{aligned} \tag{2}$$

where m = n – (ℓ – k). It can be verified that the representation for R(k : ℓ,n) shown in Eq. (2) reduces the computation overhead significantly.

Rushdi [5] has proposed a space-economizer, quadratic-time algorithm for computing k-to-ℓ-out-of-n system reliability. In his approach, R(k : ℓ,n) is calculated by using the relation

$$R(k:l,n)=R(k,n)-R(l+1,n) \tag{3}$$

where R(k,n) is the reliability of a k-out-of-n system. A k-out-of-n system is a special case of the k-to-l-out-of-n system that is characterized as successful if at least k out of the n units are functional. From the evaluation of R(k,n), we see that some of the intermediate results produced in computing R(k,n) are also required in the computation of R(l + 1,n). Rushdi's improvements are based on this observation. Rushdi's algorithm for computing the reliability is known to execute in quadratic time in the worst case.

Rushdi has also proposed another algorithm for computing the reliability of noncoherent systems that uses Shannon's expansion of symmetric functions [6]. Although this algorithm performs better than his earlier technique [5] in terms of the time complexity, it has a comparatively poor space complexity. Further it is applicable to only a special case where $(n + k) < 2(\ell + 1)$. Because of its limited applicability, this algorithm is precluded from further discussion.

We have observed that the total number of operational steps in the evaluation of reliability of the k-to-ℓ-out-of-n system can be further reduced by an approach based on the elementary event method [3]. In this approach, we define $(\ell - k + 1)$ events in which k, (k + 1), . . . , ℓ units are successful and the corresponding (n − k), (n − k − 1), . . . , (n − ℓ) units are unsuccessful (failed). The reliability R(k : ℓ,n) is then obtained by adding the probabilities of the occurrence of the above events. Further, this approach lends itself to a very simple architecture for implementing the underlying algorithm.

A NEW APPROACH TO RELIABILITY EVALUATION

This approach adopts the elementary-event method [3] for the evaluation of reliability of the k-to-ℓ-out-of-n system and shows how some of the classical techniques can be employed to increase the efficiency of computation.

The technique

Let e_x be the event in which exactly x units fail and (n − x) units are operational. Then, by the elementary-event method, the reliability of the k-to-ℓ-out-of-n system can be written as

$$R(k:l,n)=\sum_{x=n-l}^{n-k}\Pr\{e_x\} \tag{4}$$

where $\Pr\{e_x\}$ *is* the probability of the occurrence of the event e_x.

Let S_x be the set of all subsets of {1, . . . , n} containing exactly x elements, $0 \le x \le n$, and C be a subset of S_x. By a simple manipulation of terms, the following theorem provides an efficient technique for the computation of the reliability.

Theorem 1: *Let a_i denote q_i/p_i. The reliability of the k-to-ℓ-out-of-n system can be expressed as*

$$R(k:l,n)=P_\pi\sum_{x=n-l}^{n-k}A_x \tag{5}$$

where

$$P_\pi = \prod_{i=1}^{n} p_i, \quad A_0 = 1, \quad A_1 = \sum_{i=1}^{n} a_i$$

and

$$A_x = \sum_{i_1=1}^{n-x+1} \sum_{i_2=i_1+1}^{n-x+2} \cdots \sum_{i_x=i_{x-1}+1}^{n} a_{i_1} a_{i_2} \ldots a_{i_x}, \quad 1 \le x \le (n-k)$$

Proof:
Note that A_x is simply

$$\sum_{C \in S_x} \prod_{i \in C} (q_i / p_i)$$

Thus

$$\begin{aligned} R(k:l,n) &= \sum_{x=n-l}^{n-k} \Pr\{e_x\} = \sum_{x=n-l}^{n-k} \left[\sum_{C \in S_x} \left(\prod_{i \in C} q_i \right) \left(\prod_{k \notin C} p_k \right) \right] \\ &= \sum_{x=n-l}^{n-k} P_\pi \sum_{C \in S_x} \prod_{i \in C} (q_i / p_i) = \sum_{x=n-1}^{n-k} P_\pi A_x \\ &= P_\pi \sum_{x+n-l}^{n-k} A_x \end{aligned}$$

Example 1: Consider a 2-to-3-out-of-4 system.

Equation (4) gives

$$\begin{aligned} R(2{:}3,4) = &\left(p_1 p_2 q_3 q_4 + p_1 p_3 q_2 q_4 + p_1 p_4 q_2 q_3 + p_2 p_3 q_1 q_4 \right. \\ &\left. + p_2 p_4 q_1 q_3 + p_3 p_4 q_1 q_2 \right) \\ &+ \left(p_1 p_2 p_3 q_4 + p_1 p_2 p_4 q_3 + p_1 p_3 p_4 q_2 + p_2 p_3 p_4 q_1 \right) \end{aligned}$$

Using Eq. (5)

$$P_\pi = p_1 p_2 p_3 p_4$$

$$A_1 = \sum_{i_1=1}^{4} a_{i_1} = (a_1 + a_2 + a_3 + a_4)$$

$$A_2 = \sum_{i_1=1}^{3} \sum_{i_2=i_1+1}^{4} a_{i_1} a_{i_2} = (a_1a_2 + a_1a_3 + a_1a_4 + a_2a_3 + a_2a_4 + a_3a_4)$$

Now,

$$\begin{aligned} R(2:3,4) &= P_\pi(A_1 + A_2) \\ &= p_1p_2p_3p_4(a_1 + a_2 + a_3 + a_4 + a_1a_2 \\ &\quad + a_1a_3 + a_1a_4 + a_2a_3 + a_2a_4 + a_3a_4) \end{aligned}$$

establishing the equivalence between Eqs. (4) and (5).

An algorithm

In Eq. (5), the terms in the expansion of A_x can be grouped in such a way that the total number of addition and multiplication operations is reduced drastically when compared to the direct evaluation of Eq. (5). A_x can be rewritten as

$$A_x = \sum_{i_1=1}^{n-x+1} a_{i_1} \sum_{i_2=i_1+1}^{n-x+2} a_{i_2} \cdots \sum_{i_x=i_{x-1}+1}^{n} a_{i_x} \tag{6}$$

Equation (6) clearly indicates that R(k : ℓ,n) can be obtained by a simple recurrence on A_x. We use the following variation of the definition of A_x in order to derive the recurrence relation.

For $0 \le x \le 1$ and $1 \le y \le n - x + 1$ let A_x^y be defined by

$$A_0^y = 1, \quad A_1^y = \sum_{i=y}^{n} a_i$$

and for $2 \le x \le n$

$$A_x^y = \sum_{i_1=y}^{n-x+1} a_{i_1} \sum_{i_2=i_1+1}^{n-x+2} a_{i_2} \quad \cdots \quad \sum_{i_x=i_{x-1}+1}^{n} a_{i_x} \tag{7}$$

Then

$$A_x^y = \sum_{i_1=y}^{n-x+1} a_{i_1} A_{x-1}^{i_1+1} = a_y A_{x-1}^{y+1} + A_x^{y+1}$$

Note that by definition,

$$A_x^t = 0 \; \forall \, t > n - x + 1, \; A_x^1 = A_x.$$

R(k : ℓ,n) can be evaluated by computing A_1 A_2, . . . , A_{n-k} and summing the A_x's from (n – ℓ) up to (n – k) and multiplying the sum by P_π as in Eq. (5). The algorithm for the evaluation of the reliability of a k-to-ℓ-out-of-n system is given in the following:

Algorithm RSYS

Inputs: k, n, p_i, q_i, $1 \leq i \leq n$

compute $P\pi = p_1 . p_2 \cdot \ldots \cdot p_n$;

compute $a_i = q_i/p_i$ /* obtain all a_i's */

$A_0 \leftarrow 1$; /* initialize A_o */

determine ℓ_limit and u_limit; /* these limits are (n – ℓ) and (n – k) respectively */

if ℓ_limit ≠ 0, then

begin

for x ←1 to (n – k) do /* obtain all A_x's */

begin

for y ← (n – x + 1) down to 1 do /* use recursion on A_x */

$A_x^y = a_y A_{x-1}^{y+1} + A_x^{y+1}$

end

end

$R(k:l,n) \leftarrow P_\pi\left(A_{n-l} + A_{n-l+1} + \ldots + A_{n-k}\right)$

end RSYS

Algorithm RSYS requires n temporary storage elements for computation.

Example 2: Consider a (2 : 3,4) system.

$$R(2:3,4) = p_1 p_2 p_3 p_4 (A_1 + A_2)$$

By going through the algorithm RSYS,

$$A_o = 1;$$

$$A_1^4 = a_4;\; A_1^3 = a_3 + A_1^4;\; A_1^2 = a_2 + A_1^3;$$

$$A_1^1 = A_1 = a_1 + A_1^2$$

$$A_2^3 = a_3 A_1^4;\; A_2^2 = a_2 A_1^3 + A_2^3$$

$$A_2^1 = A_2 = a_1 A_1^2 + A_2^2 = a_1(a_2 + a_3 + a_4) + \left[a_2(a_3 + a_4) + a_3 a_4\right]$$

COMPLEXITY CONSIDERATIONS

In the previous section we presented a basic technique for evaluating reliability of noncoherent systems using the elementary event method. Now we obtain expressions for the total number of addition and multiplication operations required by our method. Let M and S represent the total number of multiplication and addition operations, respectively, required for the evaluation of R(k:1,n). We need to compute the terms A_1 $A_2, \ldots,$ $A_{(n-\ell-1)}$ although these terms do not appear in the expression for R(k : ℓ,n) in Eq. (5). This is due to the presence of the recurrence relation in algorithm RSYS. Evaluation of A_o and A_1 in RSYS does not entail any multiplication. However, (n − x + 1) multiplications are required for the evaluation of A_x, $2 \leq x \leq (n - k)$. The evaluation of P_π requires (n − 1) multiplications. Evaluation of a_i for all i requires n divisions. However, if we assume that $(1/p_i)$'s are given, these n division operations can be changed to n multiplication operations. Finally, the evaluation of R(k : ℓ,n) in Eq. (5) requires an additional multiplication. Hence,

$$\begin{aligned} M &= \sum_{x=2}^{n-k} (n - x + 1) + (n-1) + n + 1 \\ &= \tfrac{1}{2}\left[n(n+3) - k(k+1)\right] \end{aligned} \tag{8}$$

From the algorithm RSYS we see that the evaluation of A_x entails (n − x + 1) − 1 addition operations and summing of A_x's from (n − ℓ) to (n − k) requires (ℓ − k) additions . Thus,

$$
\begin{aligned}
S &= \sum_{x=1}^{n-k} (n-x)+(l-k) \\
&= \tfrac{1}{2}\left[n(n-1)-k(k+1)\right]+l \qquad (9)\\
&= M-(n-l)
\end{aligned}
$$

Note that Eqs.(8–9) hold good for all permissible values of k, ℓ, and n except for the case k = ℓ = n. For k = ℓ = n, the system reduces to a series system. For this case, M = (n – 1) and S = 0, which follows directly from Eq. (5).

Generalization

A direct application of our method for the reliability evaluation of noncoherent systems may not give minimal computation for all values of n, ℓ, and k. Consider, for example, the case where ℓ = n and k$\ell$is small compared to *n*. This type of system (where ℓ = n) is known as a k-out-of-n : G system in which the system is characterized as successful if at least k out of n units are successful [7]. From Eqs. (8–9), we can see that the worst case occurs when ℓ = n and k = 1. The number of computational steps can be dramatically decreased for this case by computing the reliability of the dual system, namely, (n – k + 1)-out-of-n : F which is characterized as failed if at least (n – k + 1) units out of n are unsuccessful. The reliability of the k-out-of-n : G system is the complement of the probability of failure of the (n – k + 1)-out-of-n : F system.

The above discussion signifies that our basic scheme can be modified to our advantage depending upon the relative values of k, ℓ, and n. The following two theorems give alternate ways to evaluate system reliabilities such that the number of computational steps is minimized.

Theorem 2: *Let b_i denote p_i/q_i. Then R(k : l,n) can be expressed as*

$$R(k:l,n)=Q_\pi \sum_{x=k}^{1} B_x \qquad (10)$$

where

$$Q_\pi=\prod_{i=1}^{n} q_i$$

and

$$B_x=\sum_{i_1=1}^{n-x+1}\ \sum_{i_2=i_1+1}^{n-x+2} \cdots \sum_{i_x=i_{x-1}+1}^{n} b_{i_1} b_{i_2} \ldots b_{i_x}$$

Proof: Similar to the proof of Theorem 1.

Theorem 3: *Let* $a_i = q_i / p_i$, $b_i = 1/a_i$, *and* $l \neq n$. *Reliability R(k :l,n) can be expressed as*

$$R(k:l,n) = 1 - \left[P_\pi \sum_{x=0}^{n-l-1} A_x + Q_\pi \sum_{y=0}^{k-1} B_y \right] \tag{11}$$

Proof: Using the principle of duality, it can be shown that the probability of failure of the noncoherent system is the sum of the probability of failure of the (n – k + 1)-out-of-n : F system and the probability of success of the (ℓ + 1)-out-of-n: G system. Theorem 3 now follows from Theorem 1 and Theorem 2.

Theorems 1, 2, and 3 give three different methods for evaluating the reliability of the noncoherent systems. Call these methods M1, M2, and M3 respectively. With the objective of minimizing the total number of computational steps, the choice of a particular technique for a given system depends on the relative values of n, ℓ, and k. This is summarized in the following theorem.

Theorem 4: *Let*

$$K_1 = (n+2) - \sqrt{2(n+2)(n-l) = 2}$$

$$K_2 = \sqrt{(n-l)^2 + n}$$

and

$$K_3 = \frac{l(l+4)}{2(n+2)} + \frac{1}{2}$$

When $\ell \neq$ n, the following choice minimizes the total number of computational steps in the reliability evaluation:

M1 if $k \geq K_2$ and $k \geq K_3$

M2 if $K_1 \leq k < K_2$

M3 if $k < K_1$ and $k < K_3$.

Proof: Expressions for the number of multiplication and addition operations required in M1 are given in Eqs. (8–9), respectively. The number of multiplication and addition operations required in M2 and M3 can be derived in a similar manner: M and S for M2 are

$$M = \frac{1}{2}\left[2n(l+1) - l(l-1)\right] \tag{12}$$

and

$$S = M - (n+k) \tag{13}$$

respectively.

M and S for M3 can be shown to be

$$M = \begin{cases} \frac{1}{2}\left[n(n+3) + 2k(n+2) - l(l+3) - k(k+1) - 4\right] & \text{if n} > l \\ \frac{1}{2}\left[2k(n+2) - k(k+1) - 2\right] & \text{otherwise} \end{cases} \tag{14}$$

and

$$s = \begin{cases} \frac{1}{2}\left[n(n+3) + 2k(n+2) - l(l+3) - k(k+1) - 4\right] - 2n, \text{if } n > l \\ \frac{1}{2}\left[2k(n+2) - k(k+1) - 2\right] - n & \text{otherwise} \end{cases} \tag{15}$$

respectively. The theorem now follows by comparing the expressions for M and S for the various methods. When ℓ = n, the noncoherent system becomes a coherent system and existing efficient techniques can be used for the reliability evaluation [7].

We now present a general algorithm for evaluating R(k : ℓ,n). Let RSYS(P) [(RSYS(Q)]represent the algorithm RSYS when M1 (M2) is used in the implementation of the algorithm. Note that the RSYS algorithm given previously is, in fact RSYS(P). Design of RSYS(Q), on the other hand, is straightforward.

Algorithm GENRSYS;

Inputs: k, ℓ, n, pi, qi, $1 \leq i \leq n$;

determine K_1, K_2 and K_3;

if $k \geq K_2$ and $k \geq K_3$, then use RSYS(P) to obtain R(k : ℓ,n);

if $K_1 \leq k < K_2$, then use RSYS(Q) to obtain R(k : ℓ,n);

if $k < K_1$ and $k < K_3$, then use RSYS(P) to obtain the unreliability F1 of the (n − k + 1)-to-n-out-of-n: F system and RSYS(Q) to obtain the reliability G1 of the (ℓ + 1)-to-n-out-of-n: *G* system;

R(k : 1,n) = 1 − [F1+ G1];

end GENRSYS

Comparison

We now show that both our algorithm and Rushdi's algorithm [5] compute in quadratic time. Let an index T be defined as the sum of addition and multiplication operations required for the computation of reliability. Let T_R and T_P denote this index for Rushdi's technique [5] and the proposed technique, respectively. Let T_{Rmax} and T_{Pmax} represent the worst case values of these indices, respectively. In Rushdi's algorithm [5], M and S are given by

$$M = (l+1)(n+k-l) - k^2 \tag{16}$$

and

$$S = 2M + 1 \tag{17}$$

Therefore,

$$T_R = 3\left[(l+1)(n+k-l) - k^2\right] + 1$$

The maximum of T_R occurs when $k = (n+1)/3,\ l = (2n-1)/3$ [5]. Thus,

$$T_{Rmax} \le (n+1)^2 + 1 \tag{18}$$

From Eqs. (8–9, 12–15), it can be shown that the worst case computation for M1and M2 occurs when $\ell = n$ and $k = 1$. However, for M3 the worst case occurs when $k = l = (n/2)$. Thus, the worst case bound on the computation using M1, M2, and M3 is

$$T_{P\max} = n(n+3) - 2 \tag{19}$$

$$T_{P\max} = (n+1)^2 - 2 \tag{20}$$

$$T_{P\max} = \frac{3}{2}n^2 + n - 2 \tag{21}$$

respectively. From Eqs. (18–21), it is clear that both Rushdi's algorithm and the proposed algorithm execute in quadratic time in the worst case. It can easily be verified that the worst-case computational bound of the proposed technique is better than the bound of Rushdi's algorithm. Figures 1–3 show how our approach using Theorem 4 reduces the computational efforts compared to the technique of Rushdi [5] for a set of arbitrarily selected cases.

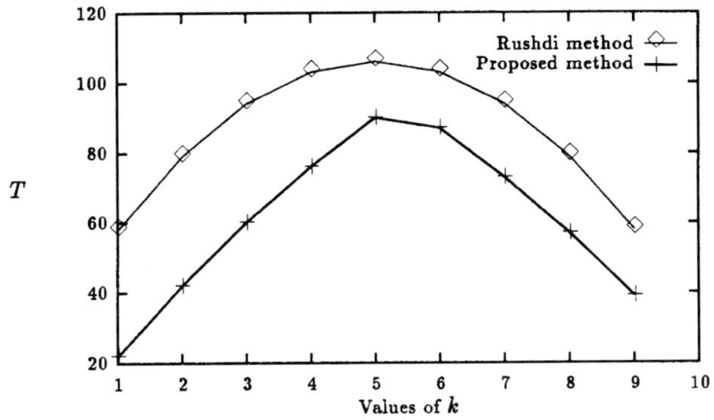

Fig. 1. Computation comparison for $\ell = 9$, n = 10.

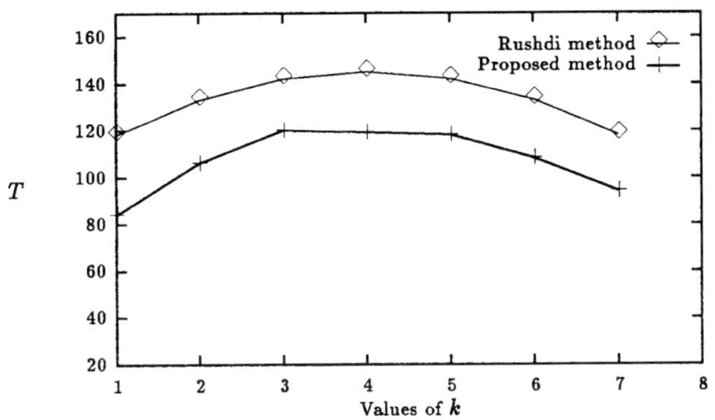

Fig. 2. Compulation comparison for $\ell = 7$, n = 11.

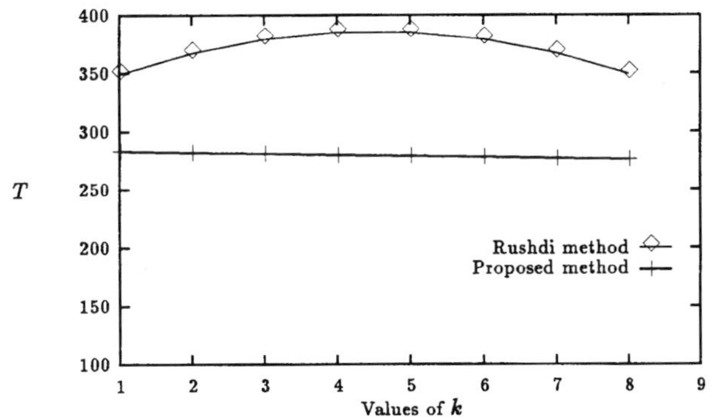

Fig. 3. Compulation comparison for $\ell = 8$, n = 20.

Hardware Implementation

A k-to-ℓ-out-of-n system is very general and readily models most coherent systems in practice, such as k-out-of-n systems, and series, parallel, and the NMR systems. Therefore, it may be worthwhile to build a hardware scheme for the reliability evaluation of such a versatile system. A unique feature of our approach for the evaluation of reliability of the noncoherent system is its adaptability to a simple hardware implementation. It can be seen that the A_x's in the RSYS algorithm presented previously can be computed in parallel by using a cellular hardware structure. Cellular structures are easy to build in very large-scale integration (VLSI), and the design is inexpensive due to the regularity of the structure. A basic cell of the hardware structure for the implementation of RSYS consists of a multiplier ($\prod$) and an adder (Σ) with a predetermined precision. The structure of the basic cell is given in Fig. 4a and is represented symbolically as $\prod\Sigma$ in Fig. 4b. In our representation, if a basic cell inscribes the symbol $\prod$ (Σ) it means that only the multiplier (adder) element needs to be used. If, on the other hand, both the symbols are present, it means that both multiplier and adder are active in that cell. The basic cell has three inputs represented by I_u, I_m and I_ℓ as shown in Fig. 4b.

Figure 5 shows a cellular structure that realizes the A_x's, for n = 5. In this figure, the basic cells are arranged to form a triangle-like structure. All the cells except those which lie on the hypotenuse and the base of the triangle have their I_u and I_ℓ inputs connected to two neighboring cells at the previous level. The third input I_m is fed externally at the same level. The single output is fed to two neighboring cells, one to the right as I_u input and the other diagonally above as I_ℓ input, at the next level. In the cells at the base of the triangle, only the multiplier is active, which is implied by feeding a 0 at the I_ℓ input. On the other hand, in the cells on the hypotenuse of the triangle, a 1 is fed to the I_u input (see Fig. 4a for the internal structure of the cell), implying that the multiplier can be bypassed. This structure needs n – 1 levels, and n(n + 1)/2] – 1 cells to generate the A_x's.

An additional 2n cells are required to generate R(k : ℓ,n). This includes n – 1 cells to compute P_π, n cells to sum the relevant A_x's, and one cell to multiply P_π with this sum. This hardware with the additional switching mechanisms for computing reliability is shown for an example system with n = 5 in Fig. 6. The cells on the right of the triangle structure are chained together and are also connected to a multiplexer to generate the sum of the appropriate A_x's. The row on the base of the triangle generates P_π The generation of the sum of the appropriate A_z's is accomplished by first determining (n – ℓ) and (n – k) and closing the switches $SW_{n-\ell}$, $SW_{n-\ell+1}$, . . . , SW_{n-k} (see Fig. 6) and feeding the binary code of (n – k) to the select inputs of the multiplexer. Note that a 1 is fed to the I_m inputs of those cells in the chain that have the I_ℓ inputs connected to the closed switches. The I_m inputs of the rest of the cells in the chain receive a 0 (corresponding to the open switches at the respective I_ℓ inputs). The I_u input of the cell at the top of the chain is always a 0 except for n = ℓ, in which case it is a 1. Figure 6 shows the appropriate switch connections for a 2-to-3-out-of-5 system. Clearly, the computation of A_x's can be accomplished in less than 2n clock cycles. It is therefore evident that R(k : ℓ,n) can be evaluated in linear-order time.

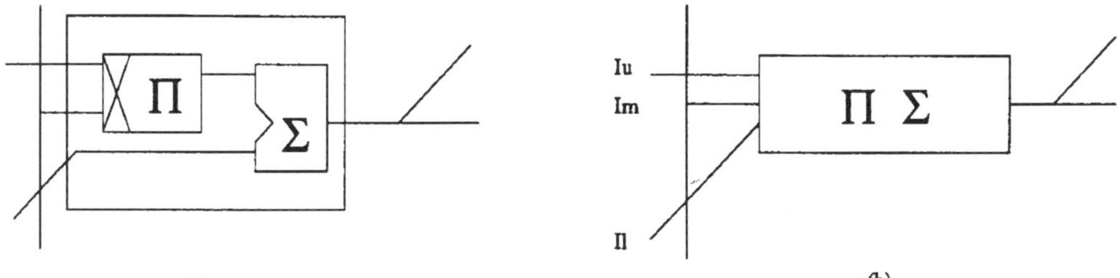

Fig. 4. A basic cell.

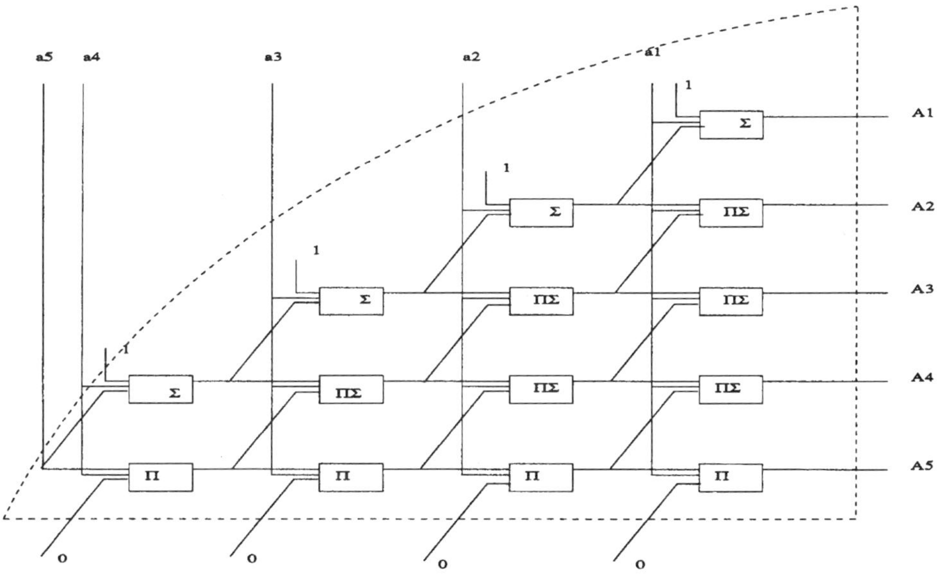

Fig. 5. A cellular structure for generating A_x's.

OPTIMIZATION CONSIDERATIONS

Both Jain and Gopal [4] and Rushdi [5] have observed that in a noncoherent system, there exists an optimum value of n for a given set of values of k, ℓ, and p; and for given values of k, ℓ, and n, there is an optimum value of unit reliability p. In this section we address this optimization issue and provide a theoretical basis. In order to obtain closed-form expressions for the reliability of k-to-ℓ-out-of-n systems, we assume that the units are independent and identically distributed (i.i.d. units) with $p_i = p$, $1 \le i \le n$. Reliability $R(k : \ell, n)$ of the i.i.d. system is given by

$$R(k:l,n)=\sum_{i=k}^{n}\binom{n}{i}p^{i}(1-p)^{n-i}-\sum_{j=l+1}^{n}\binom{n}{j}p^{j}(1-p)^{n-j} \tag{22}$$

Equation (22) follows directly from Eq. (3). Let R(n) represent R(k : ℓ,n) when k, ℓ, and p are fixed; and let R(p) represent R(k : ℓ,n) when k, ℓ, and n are held constant. The following lemma is used in establishing the principal results in the sequel.

Lemma 1: Let

$$f(n)=\frac{\binom{n}{k-1}}{\binom{n}{l}} \tag{23}$$

f(n) is a strictly decreasing function of n for $1 \leq k \leq \ell \leq n$.

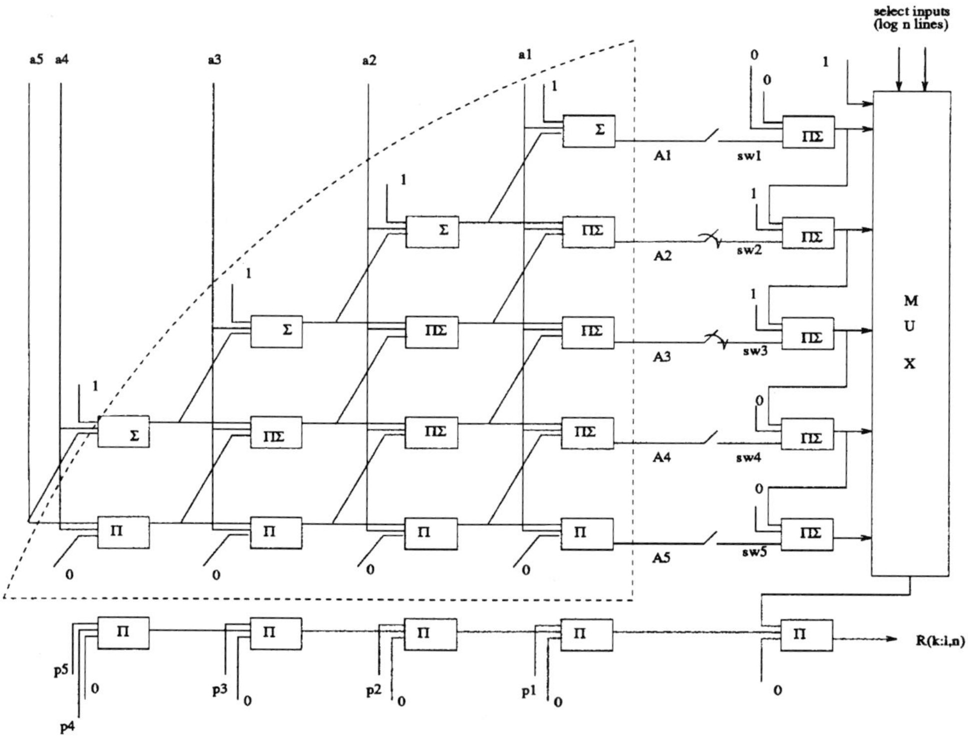

Fig. 6. A hardware scheme to compute R(2 : 3,5).

Proof:

It is sufficient to show that (n – ℓ)! (n – k + 1)! is a strictly decreasing function of n.

$$\frac{(n+1-l)!}{(n+1-k+1)!}-\frac{(n-l)!}{(n-k+1)!}=\left[\frac{(n-l)!}{(n+2-k)!}(k-l-1<0\right]$$

implying that f (n + 1) – f (n) is negative.

Theorem 4: *For fixed k, l, and p, there exists an optimal value n* for n that maximizes the system reliability R(n). The value n* is given by*

$$n^* = \sup\left\{n : f(n-1) > a^{l+1-k}\right\} \tag{24}$$

where

$$a = p/(1-p)$$

and f (n) as defined in Eq. (23).

Proof: Let ΔR(n) = R(n + 1) – R(n). From Eq. (22), it follows that

$$\Delta R(n) = \sum_{j=0}^{l}\binom{n+1}{j}p^j(1-p)^{n+1-j} - \sum_{i=0}^{k-1}\binom{n+1}{i}p^i(1-p)^{n+1-i}$$

$$-\sum_{j=0}^{l}\binom{n}{j}p^j(1-p)^{n-j} + \sum_{i=0}^{k-1}\binom{n}{i}p^i(1-p)^{n-i} \tag{25}$$

Using the identity,

$$\sum_{i=0}^{k}\binom{n+1}{i}p^i(1-p)^{n+1-i} = \sum_{0}^{k}\binom{n}{i}p^i(1-p)^{n-1} - \binom{n}{k}p^{k+1}(1-p)^{n-k}$$

we have,

$$\Delta R(n) = \binom{n}{k-1}p^k(1-p)^{n-k+1} - \binom{n}{l}p^{l+1}(1-p)^{n-l}$$

The function R(n) has a relative maximum at n = n* if the following condition is satisfied:

$$\Delta R(n^*-1) \geq 0 \geq \Delta R(n^*) \tag{26}$$

ΔR(n*) ≤ 0 gives

$$\binom{n^*}{k-1} p^k (1-p)^{n^*-k+1} - \binom{n^*}{l} p^{l+1} (1-p)^{n^*-l} \leq 0$$

That is,

$$(1-p)^{n^*+1} \left[\binom{n^*}{k-1} a^k - \binom{n^*}{l} a^{l+1} \right] \leq 0$$

where a = p/(1 – p). Now from Eq. (23),

$$f(n^*) \leq a^{l+1-k} \tag{27}$$

ΔR(n* – 1) ≥ 0 gives

$$\binom{n^*-1}{k-1} p^k (1-p)^{n^*-k} - \binom{n^*-1}{l} p^{l+1} (1-p)^{n^*-l-1} \geq 0$$

That is,

$$(1-p)^{n^*} \left[\binom{n^*-1}{k-1} a^k - \binom{n^*-1}{l} a^{l+1} \right] \geq 0$$

From Eq. (23),

$$f(n^*-1) \geq a^{l+1-k} \tag{28}$$

Since R(n) is unimodal, the result now follows from Eqs. (26–27).

Corollary 1: *If f(n – 1) =* a^{l+1-k}*, then the optimal value is n* or n* – 1.*

Proof: From Eq. (27), if f(n – 1) = $a^{\ell+1-k}$, then ΔR(n – 1) = 0. That is, R(n) = R(n – 1). Hence there are two values of n which maximize R(n).

The above corollary states that there could be two values of n that maximize the system reliability for the noncoherent systems. In such cases, obviously the lower value should be taken for the optimal configuration of the system

Theorem 5: *For fixed values of k, l, and n, there exists an optimal value p* for p that maximizes R(p). The value p* is given by*

$$p^* = \frac{1}{1+\dfrac{1}{b^{k-l-1}}}$$

where

$$b = \frac{l!(n-l-1)!}{(k-1)!(n-k)!}$$

and

$$m! = 0 \quad for \quad m < 0$$

Proof: Using the incomplete beta function [8]

$$\sum_{i=k+1}^{n} \binom{n}{i} p^i (1-p)^{n-i} = n\binom{n-1}{k} \int_0^p t^k (1-t)^{n-k-1} dt$$

we can rewrite Eq. (22) as

$$R(p) = n\binom{n-1}{k-1} \int_0^p t^{k-1}(1-t)^{n-k}\,dt - n\binom{n-1}{l} \int_0^p t^l (1-t)^{n-l-1} dt$$

$$\frac{\partial}{\partial p} R(p) = \frac{n!}{(k-1)!(n-k)!} p^{k-1}(1-p)^{n-k} \frac{n}{l!(n-l-1)!} p^l (1-p)^{n-l-1} \qquad (29)$$

Setting Eq. (28) to 0, we get

$$\frac{(1-p)^{k-l-1}}{p^{k-l-1}} = \frac{l!(n-l-1)!}{(k-1)!(n-k)!}$$

Let b denote [l!(n − ℓ − 1)]/[(k − 1)!(n − k)!] Then the value of p which maximizes R(p) is given by

$$\left(\frac{1-p^*}{p^*}\right)^{k-l-1} = b$$

The following theorem shows the dependence of p* on n for fixed values of k and $\ell \leq$ n.

Theorem 6: *For a given set of k and l, p* that maximizes R(p) is a strictly decreasing function of n.*

Proof: Follows directly from Lemma 1 and Theorem 3.

DISCUSSION

Noncoherent systems are very realistic in practice. A number of engineering design problems in various disciplines can be modeled and their reliabilities and optimal configurations studied. We consider two such examples—one in power engineering [4] and another in computer engineering [1]. Consider, for example, a power plant where ten generators are available and only six to eight of them are sufficient to feed the network and cope with the demand. There are at least two generators in standby mode at any given time. If all the ten generators are operational, the boiler may not be able to cope and thereby lead to a deterioration of the unit's performance. The system will eventually reach a failed state from the quality of the power supply point of view. This power system problem is essentially a 6-to-8-out-of-10 system. The reliability and availability of such systems can be evaluated by using the method previously discussed .

The second example illustrates the usefulness of our work from the optimal system configuration point of view. Consider a computer installation [1] where a certain number of video display units are to be connected to the computer. The objective in this case is to maximize the reliability/availability of the video display unit subsystem. If the number of display units at a given time is greater than ℓ, for example, then the response time may not be tolerable due to the network congestion. If, on the other hand, the number of units is less than k, then the computer is not being utilized to its maximum capacity. This is a case of underproduction of work. Both of these cases are to be avoided, and hence the video display unit subsystem is a k-to-l-out-of-n system. If the reliability of the video display unit hardware is known, Theorem 2 can be used to obtain the optimal configuration of the display unit subsystem from the point of view of system availability. Figures 7 and 8 give the plot of R(p) and R(n), respectively, for various values of k and ℓ.

Optimization of the noncoherent systems in terms of the mean system profit has been looked at in Pham [9]. Given a profit function fn(i) where i units are functioning, Pham shows that there exists an optimum system size that maximizes the mean system profit.

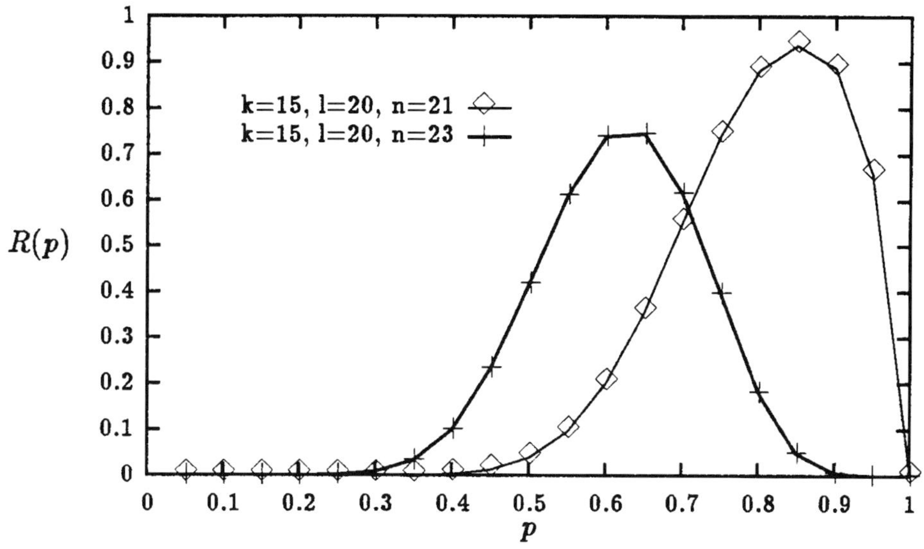

Fig. 7. System reliability for different values of p.

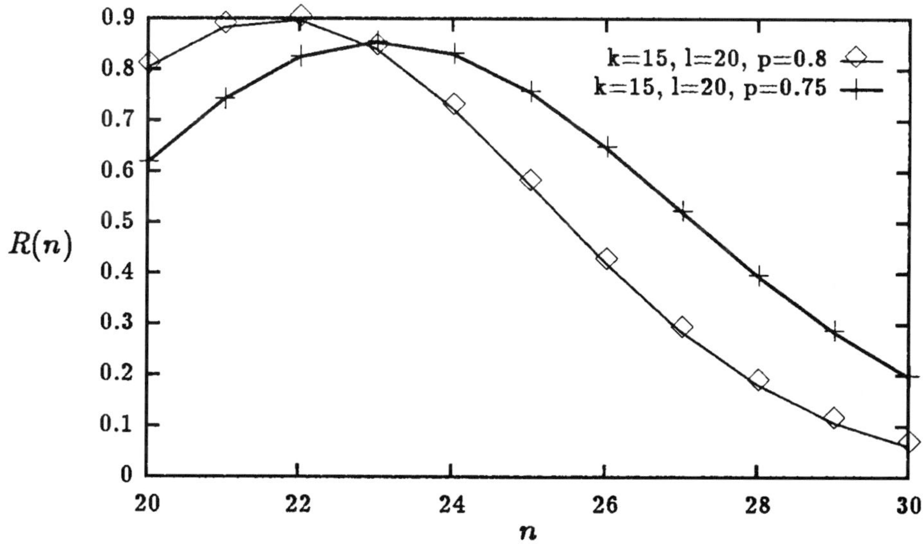

Fig. 8. System reliability for different values of n.

CONCLUSIONS

In this paper, we have presented an efficient technique for the computation of reliability of the k-to-ℓ-out-of-n system. These kinds of systems find application in communication, multiprocessor, and transportation system environments. The k-to-ℓ-out-of-n systems are very general and readily model the coherent systems such as series, parallel, and NMR systems.

The algorithm presented for the evaluation of the reliability of the k-to-ℓ-out-of-n system computes in quadratic time in the worst case. The worst-case bound for computation of our algorithm is much better than the bounds of the existing algorithms and is shown to yield superior results for all permissible values of k, ℓ, and n. However, we must note that the general algorithm GENRSYS presented for the computation of reliability requires the determination of three constants and a few condition checks, which adds slightly to the complexity. On the other hand, it lends itself to easy hardware implementation. We have presented a cellular realization for the implementation of the algorithm in hardware. This scheme is shown to evaluate the reliability in linear-order time. The scheme presented employs a basic cell consisting of a simple multiplier, an adder, and some switches that can be easily implemented in VLSI by using computer-aided design tools. We have also discussed the ways of obtaining optimal configurations for the k-to-ℓ-out-of-n system.

REFERENCES

1 K. D. Heidtmann, A class of noncoherent systems and their reliability analysis, in: 11th Ann. Symp. on Fault Tolerant Computing, Portland, Ore., June 1981, pp. 96–98.
2 P. K. Lala, Fault Tolerant and Fault Testable Hardware Design, Prentice Hall, Inc., Englewood Cliffs, N.J., 1985.
3 W. Feller, An Introduction to Probability Theory and its Applications, John Wiley and Sons, New York, 1968.
4 S. P. Jain and K. Gopal, Reliab. Eng., 12 (1985) 175–179.
5 A. M. Rushdi, Reliab. Eng., 17 (1987) 157–163.
6 A. M. Rushdi and F. M. A. Dehlawi, Microelectron. Reliab., 17 (1987) 157–163.
7 H. Pham and S. J. Upadhyaya, IEEE Trans. Reliab., 37, no. 5 (1988) 521–523.
8 Handbook of Mathematical Formulas, Tables, Graphs, Functions, and, Transforms, M. Fogiel and Staff of the Research and Educations Association, Compilers, Research Education Association, New York, 1980, 515.
9 H. Pham, IEEE Trans. Reliab., 40 (1991) 361–363.

Chapter 5
Tolerances and Discrete Parts Design

Tolerance design and design for assembling parts are discussed. We intend to find ways for hardware to quantify design tolerance and specify design parameters. Excellent case studies are discussed to illustrate both the feasibility and shortfalls of the present design methodology. First, issues in assessing conformity of dimensions of manufactured parts to design tolerance are investigated. Dimensional tolerancing is a language used to communicate product quality specifications. The weaknesses of the language of tolerances is discussed and suggestions for improving research and standards are provided. Readers can find very original work related to tolerances and discrete sampling with coordinate measurements in this chapter.

Quality through Engineering Design - W. Kuo (Editor)
1993 Elsevier Science Publishers B.V.

Issues in Assessing Conformity of Dimensions of Manufactured Parts to Design Tolerances

Dennis A. Swyt[1]

Precision Engineering Division, U.S. Department of Commerce, National Institute of Standards and Technology, Building 220, Room A109, Gaithersburg, MD 20899, U.S.A.

Abstract

After defining dimensions and tolerances of manufactured parts and showing a major trend to tighten tolerances, this paper outlines briefly some of the issues in assessing conformity of dimensions in manufactured parts to design tolerances in terms of five topical areas: (1) mathematical representation of geometric dimensions and tolerances, (2) evaluation of the computational aspects of computerized measuring systems, (3) evaluation of the physical measurement aspect of measuring machines, (4) assignment of uncertainty to measurement results, and (5) estimation of manufacturing deviations relative to tolerances in the presence of measurement uncertainty.

INTRODUCTION

In the manufacture of discrete parts, from automobile engine blocks to quantum-well electronic devices, a principal determinant of quality is conformity of geometrical dimensions of the part as manufactured to those of the part as designed. The degree of this conformity is determined by the relationship of measured deviations of geometrical dimensions to specified tolerances.

Today the historical system of manufacture, which involves relatively low precision, hard gaging of part elements, graphic representations of tolerances, and a willingness to accept high scrap rates, is being supplanted by a new system of manufacture, which involves high-precision manufacture, computer-based coordinate-metrology, mathematical representations of tolerances, and demand for making every part right the first time. This new system has raised major issues for research and standardization across the spectrum of design specification, precision manufacture, and measurement-based quality assessment.

DIMENSIONS OF MANUFACTURED DISCRETE PARTS

Whether aircraft, automobiles, computers or integrated circuits, discrete-part products are able to function and to function properly because the components of these products have specific geometrical shapes, sizes, and related dimensions.

Figure 1 is an engineering drawing of an automobile engine that illustrates some of the principal types of geometrical dimensions of discrete parts: A represents a length or width of the engine block; B represents the diameter of a cylinder bore; C represents the distance between the centers (axes) of the cylinder holes; D represents the depth of a blind hole; E represents the location of the center (axis) of a hole relative to a plane; and F represents the (true) position of a feature relative to the origin of a part coordinate system defined for example by reference holes in the oil-pan plane.

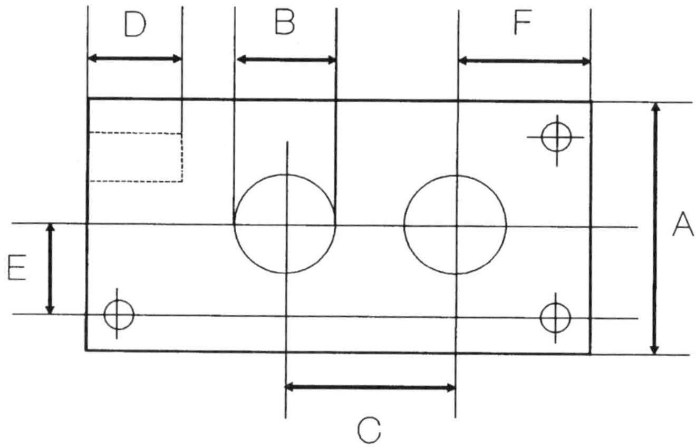

Fig. 1. An engineering drawing representation of the principal geometrical dimensions of a manufactured discrete part.

TOLERANCES ON THE DIMENSIONS OF DISCRETE PARTS

As defined by U.S. ANSI Standard Y14.5 on "Geometric Dimensioning and Tolerancing," a tolerance is a number given in an engineering drawing that specifies the total amount by which a dimension is permitted to vary [1]. Represented here by $\pm T$, a tolerance defines a range of acceptability of the magnitude of the difference between the dimension of a part-as-made (D_M) and the dimension of the part-as-designed (D_D), that is, $|D_M - D_D| \leq T$.

Figure 2 shows tolerance-versus-dimension for selected examples of the wide variety of outputs of the discrete-parts manufacturing industry, spanning a range of dimensions from 1 m to 10 nm and a range of tolerances from 1 mm to 0.5 nm [2].

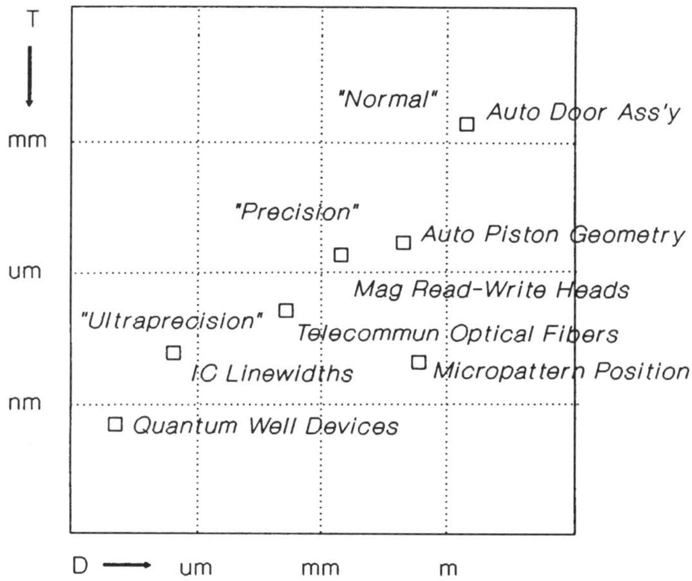

Fig. 2. Log-log plot of tolerance (T) versus dimension (D) for discrete-part products in the normal, precision, and ultraprecision regimes.

At the large-dimension end of the precision-manufacturing spectrum is the state-of-the-art automobile manufacture where a 1-mm tolerance on the fit of a 1-m-sized passenger-car-door assembly is achieved only by the Japanese. Also in state-of-the-art automobile manufacturing is the 7–8-μm tolerance on the metal-matrix composite pistons for a new-generation engine under development by a leading Japanese car company. In the intermediate-dimension range is the 0.25 μm accuracy of a slotting/slicing machine for use in fabricating the new-generation thin-film magnetic read/write heads for computer memory systems. Toward the lower dimensions are the 0.025-μm (25-nm) tolerances on 125-μm-diameter optical fibers for telecommunications systems and the 0.010-μm (10-nm) tolerances representing current demands in electron-beam micropatterning equipment in Japan. Finally at the smallest-dimension end of the precision-manufacturing spectrum shown are 0.0005-μm (0.5-nm) variations on the 10-nm size dimensions of devices such as quantum-well lasers as well as quantum-wire and quantum-dot devices under development.

There is an on-going trend toward dramatically tighter dimensional tolerances in manufactured goods, first reported to the international production community in the early 1980s, which corresponds roughly to a factor-of-three decrease in the size of tolerances every ten years [3]. According to the projection of this trend, between 1980 and 2000 state-of-the-art tolerances will decrease from 7.5 μm to 1 μm in the most-demanding normal machine-tool-based production; from 0.075 μm to 0.01 μm in the most-demanding precision, e.g., diamond turning machine production; and from 0.005 μm (5 nm) to <0.001 μm (<1 nm) in the most-demanding atom-, electron-, or x-ray machining-based production. It is the trend to such tightened tolerances combined with the move in many inspection applications towards use of computerized coordinate measuring machines that has given rise to major current metrological problems in measurement-based quality assessment of precision-manufactured discrete-part products.

PROBLEMS IN ASSESSING CONFORMITY TO TOLERANCES

Shown schematically in Table 1 are the principal aspects of the overall problem of determining with a known degree of confidence whether the dimensions of a manufactured part conform to specified tolerances in an engineering design. Table 1 indicates that with each functional task involved in design, fabrication, point-sample-based measurement, and part evaluation, there is an associated variability that in the best situation is appropriately quantified.

According to this scheme, a bottom-line evaluation of a part consists of the ability to state with confidence C that a part's material deviations D, known with measurement uncertainty U, are within design-specified tolerances T, when the deviations have been determined by means of a point-sample measuring machine that provides coordinate positions with errors X of a number of points distributed at various locations on the surface of the part, when those points are used in conjunction with a software implementation of a fitting algorithm with associated variability Y.

DESIGN: MATHEMATICAL REPRESENTATION OF GEOMETRIC DIMENSIONS AND TOLERANCES

In the part-design phase of Table 1, the basic tasks are to choose an ideal geometric form for the part and to specify its dimensions and tolerances. Current practice and standards on geometric dimensioning and tolerancing, represented by ANSI Y14.5, deal with graphical, engineering-drawing representation of geometry, dimensions, and tolerances. The first major issue associated with the assessment of part-manufacturing deviations relative to tolerances by means of computerized point-sample coordinate measuring systems arises with the tolerances themselves. Absent are the formal representations required for unambiguous interpretation and computation [4]. The required mathematical representation of dimensioning and tolerancing standards is an area of active research underlying new standards [5]. Problems include, for example, achievement of context-free definitions of concepts such as size, which do not involve embedded notions such as diametrically opposed points or fitted cross sections [6], and means for establishing simultaneously formal and common-sense bases for concepts such as the various types of data to which part features are referred [7].

Table 1. Principal tasks and their associated variabilities in comparing point-sample measurements of manufacturing deviations with design tolerances.

Function	Tasks	Variability	Symbol
Design	Choose ideal form		
	Specify dimensions	Design tolerances	T
Fabrication	Choose process		
	shape material	Mfg deviations	D
Data-Taking	Choose number/location of points		
	Probe surface at those points	Coordinate errors	X
Computation	Choose fitting algorithm		
	Compute substitute geometry	Fitting variation	Y
Evaluation	Calculate manufacturing deviations	Uncertainty	U
	Decide if within tolerances	Confidence	C

DATA TAKING: CMM MEASUREMENTS

Given a chosen manufacturing process that is likely to produce minimum deviations from design values and the part fabricated, the next step—that of data taking—is to choose the number and distribution of points about the part surface at which coordinates are to be measured and to actually probe the part. Two issues arise.

The first issue is selecting the appropriate number and location of sample points on the surface of the part being measured. Approaches include, for example, equidistant, equiangular, and stratified. Some standards provide guidelines for sampling [8]. Active research on the design of sampling strategies as a function of part geometry, process-specific part-form errors, and systematic and random probing errors, for example, is being carried out as potential input to U.S. standardization of dimensional measurement inspection techniques [9].

The second issue is taking into account the residual measurement errors associated with each individual probed point. Recent work has lead to standard techniques for characterizing these specific functional aspects of coordinate-measuring machines [10]. These residual errors, which are vectors of varying direction and magnitude specific to each probed point, are complex functions of the coordinate-measuring-machine geometry, location of the probed point within the working volume of the machine, the type of probe employed, and the

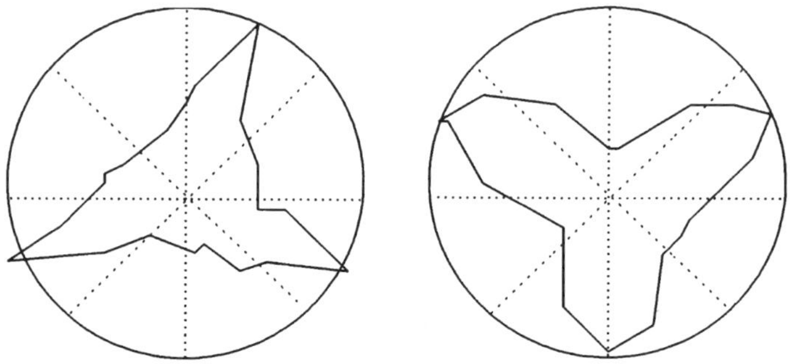

Fig. 3. Lobing in CMM probe response giving rise to direction-dependent error in coordinates of points on the part being measured (full scale of left and right plots: 8 μm and 14 μm, respectively).

direction of the probe's approach to the part surface [11]. On-going research includes comparing the errors of CMM probes that can have systematic, direction-dependent errors, such as the lobing shown in Fig. 3 [12], with an ideal probe having a circular, i.e., nondirection-dependent error pattern of zero radius. As can be inferred by comparing Fig. 1 with Fig. 3, the cumulative effect of such lobed, direction-dependent errors on the apparent dimension of a part being measured depends upon which of the six different types of geometric features is being probed, the number of points probed, and the specific directions of approach to the part relative to the probe's lobing pattern.

COMPUTATION: FIT CRITERIA AND ALGORITHM TESTING

In the computation phase of the scheme shown in Table 1, the basic tasks are (1) to choose an algorithm for achieving a best fit of the set of points measured on the surface of the part and (2) to compute the best-fit substitute geometric feature. Conventional fitting criteria include, for example, least-squares, minimax, maximum-inscribed, and minimum-circumscribed. The issue here is that computed dimensions of the substitute feature, including its size, location, and orientation, depend on the substitute geometric element selected, the best-fit criteria applied, and the software implementation of the fitting algorithm with no established theory of tolerances to rationalize the choice. Figuring out the choices is an active research area with algorithms implemented on the basis of specific criteria, for example, a minimum-deviation zone being evaluated against simulated-measurement points for specific substitute geometries such as cylinders [13]. Also under development are standardized tests for evaluating the ability of various substitute geometry-fitting algorithms to detect particular types of form errors [14].

EVALUATION: MEASUREMENT UNCERTAINTY AND CONFIDENCE

Finally, the evaluation phase of assessing part-manufacturing deviations relative to tolerances by means of computerized point-sample coordinate-measuring systems as shown in Table 1 includes the tasks of actually comparing the substitute geometry with the design geometry to determine whether the manufacturing deviations of a part are within specified design tolerances, given the overall uncertainty in measurement. The conventional approach in dealing with assessing manufacturing deviations is to require that the accuracy of measurements be some specified fraction of tolerances, typically one-fourth (as in the case of the U.S. Department of Defense prescription) or one-tenth (as in the case of the widely applied "gage maker's rule"). However, a major issue here in ascertaining conformance of manufactured parts to design tolerances is how to calculate the confidence factors of the uncertainties in the overall measurement on the basis of coordinate and sampling errors [5].

The main issue is compounded by secondary issues. First is the absence of statistically sound and widely accepted measures of measurement uncertainty. At present there is a move away from the long-established system in the United States of characterizing uncertainty in terms of a random component (one-, two-, or three- standard deviations) and a systematic component (an estimated residual after model-based systematics are removed) [15] to a quadrature-addition, one-sigma representation being pursued by the international metrology organizations [16]. The two approaches are seen to have more or less utility depending on the application [17]. In any case, the question of calculating confidence in measurement uncertainty is still open.

CONCLUSIONS

Achieving quality in the design of discrete parts requires specifying tolerances on manufacturing deviations that can be achieved to an expected degree of confidence. With the press of tightening tolerances and the use of point-sampling coordinate-measuring machines, achievement of a known confidence is a problem. Solving that problem requires resolution of a wide range of issues involving mathematical representation of dimensioning and tolerancing standards, development of point-sampling strategies, analysis of fitting algorithms, testing of software implementations, and development of error propagation and confidence-estimation techniques. While aspects of the overall problem are being addressed within the research and standards communities, much work remains to be done.

REFERENCES

1 ANSI Y14.5m-1982, Dimensioning and Tolerancing: American National Standard Engineering Drawings and Related Documentation Practices, The American Society of Mechanical Engineers, New York.

2 D. A. Swyt, Challenges to NIST in Dimensional Metrology: The Impact of Tightening Tolerances in the U.S. Discrete-Part Manufacturing Industry, NISTIR 91-4757, U.S. Department of Commerce, National Institute of Standards and Technology, Gaithersburg, Md., Jan. 1991.
3 N. Taniguchi, Annals of the CIRP (International Institute of Production Research), 2, no. 2 (1983).
4 S. C. Feng and T. H. Hopp, A Review of Current Geometric Tolerancing Theories and Inspection Data Analysis Algorithms, NISTIR 91-4509, U.S. Department of Commerce, National Institute of Standards and Technology, Gaithersburg, Md., Feb. 1991.
5 V. A. Tipnis, Research Needs and Technological Opportunities in Mechanical Tolerancing, Report of International Workshop, CRTD-15, Center for Research and Technology Development, American Society of Mechanical Engineers, New York, Oct. 1988.
6 V. Srinivasan, A Geometer Grapples with Tolerancing Standards, Research Report RC 16815(74405), IBM Manufacturing Research Department, Yorktown Heights, N.Y., Apr. 1991.
7 T. E. Hopp, The Mathematics of Datums, American Society of Precision Engineering Newsletter, Chapel Hill, N.C., Sept. 1990.
8 British Standard Guide to Assessment of Position, Size and Departure from Nominal Form of Geometric Features, BS 7172-1989, British Standards Institution, Milton-Keynes, U.K., 1989.
9 C. W. Brown, Dimensional inspection techniques specification (DITS), Proc. of Quality Expo Time, Chicago, Apr. 1991.
10 American National Standard, Methods for Performance Evaluation of Coordinate Measuring Machines, ASME B89.1.12M-1990, American Society of Mechanical Engineers, New York, June 1991.
11 D. A. Swyt, New concepts in precision manufacturing, in: Control and Dynamic Systems, Vol. 45, Part 1, Academic Press, New York, 1992.
12 S. Phillips, Properties of touch trigger probes, in: Proc. Measurement Science Conf., Anaheim, Calif., Jan. 1992.
13 W. H. Elmarghy, ASME Manufacturing Review, 3, no. 1 (March 1990) 32–38.
14 M. E. A. Algeo and T. E. Hopp, Form Error Models of the NIST Algorithm Testing System, NISTIR 92-4740, U.S. Department of Commerce, National Institute of Standards and Technology, Gaithersburg, Md., Jan. 1992.
15 C. Eisenhart, J. Res. Nat. Bur. Stand., 67C, no. 2 (Apr.– June 1963).
16 B. N. Taylor and C. E. Kuyatt, Status report on the ISO, IEC, OIML, and BIPM guide to the expression of uncertainty in measurement, in: Proc. Measurement Science Conf., Anaheim, Calif., Jan. 1992.
17 H. H. Ku, Uncertainty and Accuracy in Physical Measurements, NIST Special Publication SE-805, U.S. Department of Commerce, National Institute of Standards and Technology, Gaithersburg, Md., Dec. 1990.

Quality through Engineering Design - W. Kuo (Editor)
1993 Elsevier Science Publishers B.V.

The Language of Tolerances[1]

Theodore H. Hopp

Factory Automation Systems Division, National Institute of Standards and Technology, Bldg. 220, Rm. A127, Gaithersburg, MD 20899-0001, U.S.A.

Abstract

Dimensional tolerancing is a language used to communicate product quality specifications. Effective communication using this language relies on there being a common, shared meaning for tolerances. By this measure, the language of tolerances defined in modern standards does not support good communication. Some reasons for this are the evolution of technology, the lack of research on tolerances prior to the last two decades, and a greater need for rigorous product specifications in a computer environment. I discuss the weaknesses of the language of tolerances and suggest a three-part program of research and standards development to improve it.

INTRODUCTION

This paper is concerned with geometric variability in manufacturing. Designers have always had to deal with the sad fact that parts cannot be made perfectly; or if they could, they would not remain perfect for long during use. So defining the ideal part is only half of a designer's job. The designer must also work out how different an acceptable part can be from the ideal. A part can vary from the ideal in many ways: in its shape, its material strength, surface finish, electrical properties—a virtually unlimited list [1]. Of these, I will be discussing the ways in which variability of shape is handled in manufacturing. I will discuss some of the weaknesses of the language of tolerances and what we might do to make the language more effective.

Over the last century or so, the practice of engineering design has evolved from a variant of aesthetic design into a disciplined, formal method of communication [2]. Today, designers use a highly structured language of engineering drawings to communicate design intent. Various dialects of this language have been defined in national and international standards [3–4]. To date, the language of tolerances has been primarily a graphical symbology; tolerance requirements are expressed as symbols arranged in particular ways on blueprints.

There continues to be considerable energy invested in developing better drawing conventions. Recently there has also been great effort to develop formal computer languages for communicating designs [5]. The first U.S. national standard for digital communication of technical product data was the Initial Graphics Exchange Specification (IGES),[2] developed in 1981 and now in its fifth version [6]. IGES provided a standardized means of exchanging drawing information between computer-aided design systems. A new standard for product data exchange is being developed by the Technical Committee on Industrial Automation of the International Organization for Standardization (ISO). This new standard, called the Standard for the Exchange of Product Model Data or STEP [7], goes far beyond defining a computer representation of blueprints. It is a new language for product specification, as different from drawings as Chinese is different from English.

I have chosen the metaphor of language for more than its evocative power. I would like to analyze the current standards in language-theoretic terms. Language theory tells us that any language has (at least) three constituents: vocabulary, grammar, and semantics. The vocabulary is the set of basic terms or elements of the language. The grammar defines the rules for composition—how the vocabulary terms can be assembled into larger constructs (e.g., sentences). The semantics defines the rules for understanding the meaning of sentences. Applying these definitions to drawings and to STEP, the following comparisons can be made:

	Drawings	**STEP**
Vocabulary	Drawing symbols; call-outs; lines; arrows; and text	Type definitions; entity/relationship models; entity values
Grammar	Drawing conventions as defined in standards such as USA/ANSI Y14.5	The computer language EXPRESS and the mapping to physical-file format
Semantics	As defined by illustrations; as explained in various standards; as interpreted by people in the manufacturing enterprise	As defined by the language semantics of EXPRESS; as explained in STEP documents; as interpreted by application software

---

[2] Throughout this paper, I will use as examples U.S. or ISO standards, since I am most familiar with them. Standards in other countries similar to IGES are SET (AFNOR Z68-300) in France and VDFS (DIN 66 301) in Germany.

We will soon haveiin our standards two fundamentally different languages for specifying tolerances: a language based on drawings (here I include IGES) and a language based on computer models of products. This situation can and will create great confusion unless the two languages share common semantics [8]. It is imperative that tolerances specified in drawings have the same meaning as tolerances specified in computer-data structures.

In this paper, I will discuss how semantics can be developed for drawing tolerances. This material is based on work in the United States on developing standard mathematics for tolerances. I will also discuss several weaknesses of current tolerance semantics. Finally, I will propose how the research and standards communities might cooperate to address these weaknesses.

WHY IMPROVE THE LANGUAGE OF TOLERANCES?

The meaning of a tolerance lies in how it is interpreted. Generally, standards provide explanations of how this should be done for specific tolerance call-outs. By the norms of formal language theory, these descriptions are somewhat casual. The explanations were written to convey the intent of the standards developers, and they strike a balance between the need to be precise and the need to be comprehensible to the average user of the standards.

Increasingly, tolerance information is directly used in computer systems. Computer applications that use tolerance information include computer-aided design, assembly simulation, generative process planning, and coordinate-measuring machine inspection. Software writers are faced with the problem of converting the informal definition of tolerances found in standards into computer codes that carry out very specific actions, with no allowance possible for case-by-case interpretation. The result is that tolerances are interpreted in inconsistent ways by different application systems. This can have serious repercussions, as evidenced by a 1988 advisory published by the U.S. Department of Defense on discrepancies between the requirements of tolerances and the results of coordinate-measuring machine software [9].

While better education can address some of the problems of inconsistent tolerance interpretations, there also are serious technical difficulties in precisely defining the meaning of tolerances. At a 1988 workshop sponsored by the U.S. National Science Foundation [10], the most frequently mentioned need was to provide a sound theoretical basis for mechanical tolerances. As demand increases for tighter tolerances on manufactured goods, the economic impact of misunderstood or misapplied tolerances becomes ever greater [11]. The development of a sound theoretical foundation for tolerances is not only an intellectually demanding challenge; it is rapidly becoming an economic necessity.

Existing tolerance theories

Most theories of tolerance semantics are based on providing a mathematical or logical formulation of tolerance specifications. (For a survey of tolerance theories, see Juster [12] or Feng and Hopp [13].) For instance, a flatness tolerance applied to a nominally planar feature specifies that the actual feature must lie within some pair of parallel planes separated by the tolerance value. The differences between many of the theories have to do with the rules by which bounding geometries (such as the parallel planes) are derived from nominal geometries and tolerance call-outs. Early theories were based on concepts of varying parameters of the nominal geometry; for instance, a position tolerance allowed a variation in the location of the axis of a hole. Researchers found that such theories did not allow consistent geometries or topologies to be maintained. Other researchers struggled to find consistent theories that did not rely on a plethora of special cases. Unfortunately these theories, of which Requicha [14] is perhaps the most widely cited example, did not produce interpretations that corresponded to practical applications of tolerances [15]. A recent proposal by Jayaraman and Srinivasan [16–17] to address this issue is to base tolerance semantics on the notion of *virtual boundaries*. A virtual boundary is a surface that, in essence, corresponds to the worst-case mating part in an assembly. It is very similar to the concept of *virtual condition* found in tolerancing standards. In their approach, Jayaraman and Srinivasan develop methods based on virtual boundaries that closely match practical uses of tolerances and do not require special, case-by-case rules for interpretation of call-outs.

Virtual boundaries appear to hold much promise as a basis for formalizing the semantics of our current tolerancing standards. Work under way in the United States to develop standard tolerance semantics is based, in part, on virtual boundary concepts. However, much is left to be done. The standards development work has made clear that despite considerable advances in tolerance theory there remain significant gaps in our ability to formalize the language of tolerances in a way that matches engineering common sense.

One can identify at least two approaches to developing a theory of tolerances. The first is to treat a tolerance as a constraint on the surface of the part. Under this approach, one may sensibly ask whether a part conforms to a tolerance. The second approach is to associate a measurable quantity—a deviation from ideal—with a tolerance. Under this second approach, one may sensibly deal with the actual value of deviation—actual out-of-flatness, actual position deviation, etc. These two approaches are not of necessity equivalent. We might define conformance and actual value in ways such that both definitions are reasonable, but comparing the actual value to the tolerance does not produce the same decision as the definition of conformance. We have no tolerance theory that provides consistent definitions for both views of product tolerances.

In the remainder of this section, I will discuss two kinds of weaknesses of our current language of tolerances. The weaknesses are at both the semantic level (what the tolerances mean) and at the level of vocabulary and grammar (what can be represented). I first consider some of the semantic weaknesses.

Tolerance semantics

Tolerances grew out of the concepts of geometry. For instance, we freely discuss tolerances on the size and position of features. But what is meant by the size of a manufactured hole that is not cylindrical? Where does the hole end and the surfaces of the adjacent features start? How do we assign a coordinate system to the part in order to verify a position tolerance for the hole? These questions, some of which have been noted before [12,18], are at the heart of the major weaknesses of modern tolerance semantics. I will consider examples of challenges in three areas: boundary, size, and datum reference.

Boundary

The notion of boundary seems fairly simple. If I say that a table top must be parallel to the floor within some tolerance, then I expect the table top to lie within a pair of planes, separated by the tolerance and parallel to the plane of the floor. When we try to formalize this, however, we immediately run into the problem of defining the table top. All tolerance theories, being geometric in nature, require that we abstract the table top as a set of points in space. So the question becomes what are all the points on the part feature called "table top"? Put another way, what separates the top of the table from the sides, and what is meant by a point on the top? Neither question has been answered adequately by existing theories.

The question of identifying features has been studied in terms of models of ideal parts [19]. There has been little attention given to identifying the extent of features on physical parts. The problem can be seen in Fig. 1. There is no clear demarcation of the top of the table from the side. How much of the rounded corner should be allocated to the side and how much to the top? For that matter, should the entire surface of the part be allocated to features controlled by tolerances, or might there be an area of the edge exempt from both the top and side tolerance controls? If the latter, what controls the transition area? One proposal is to claim that a part conforms to its tolerances if there exists some partition of the surface into features such that each feature conforms to its controlling tolerances [20]. This proposal fails to provide a definition for the actual feature; it only resolves the issue of conformance for a part in its entirety. It may be that any definition of actual value of, for example, flatness of the top and side of the table illustrated in Fig. 1, must rely on information external to the part definition.

A second issue of boundary is to define the surface of the part. Clearly we do not mean a surface in the mathematical sense; at sufficiently small scale, there is no surface, but only subatomic particles thinly spread out in a void. In the physical sense, a definition of the surface must involve some kind of smoothing. The difficulty is in providing enough smoothing to provide a basis for defining geometric constraints and actual values, while not smoothing over surface characteristics that we may want to control.

Some guidance for this issue can be found in existing standards. The U.S. standard on surface texture [21] uses a spatial filter that, in principle, is applied before assessing the part. A low-pass filter is used for shape variation, while various band-pass filters are used for roughness, surface texture, and waviness. Each filter has a specific response curve as a function of the scale of the variation. A similar approach is found in the U.S. standard on roundness [22]. This standard recommends that a roundness specification include an instrument frequency response characteristic and a sensor tip radius—values which control the filtering of the surface.

We do not understand how surface-smoothing operations interact with other tolerance semantics. In some cases the issue of smoothing is extremely

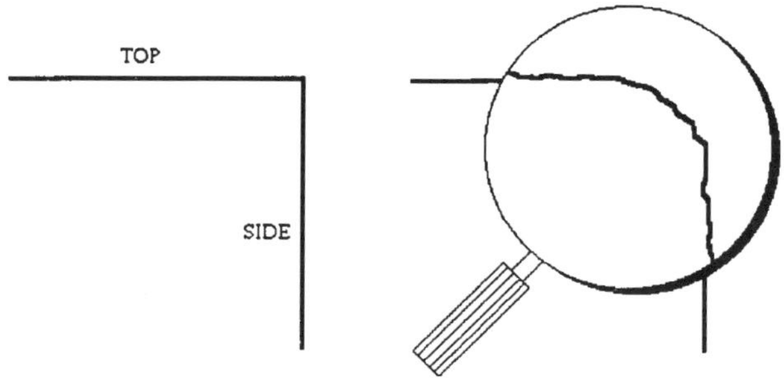

The table as designed The table as produced

Fig. 1. Ideal geometry versus physical feature.

important. For instance, the heat shield tiles on the NASA space shuttle are composed of a porous, glass foam. The tiles must fit together with no gaps that would allow hot gasses to penetrate to the shuttle skin. The form tolerances for the tiles are tighter than the surface-roughness requirements. The tolerances are specified by including a specified measurement procedure for verification. In this case, the semantics of the tolerance is defined as the result of a particular inspection process. Such "process semantics" allow designs to push the limits of manufacturing technology. At the same time, tying a design to a specific manufacturing method can make it harder to introduce better manufacturing tools in the future.

Size

Size is an attribute of certain types of geometry. These include cylinders, spheres, and parallel, opposed planes. Cones and nonopposed parallel planes do *not* have a size. For a physical part, the concept of size, like the concept of feature boundary, is surprisingly complex. The current U.S. standard on dimensioning and tolerancing defines the actual size of a feature simply as the measured size. Unfortunately, there is no standard definition of the measured size of a feature. What is worse, the essence of a measurement is comparing a physical phenomenon to a reference. If we are going to insist on using size as an attribute of actual features, we need a theoretical definition of actual size to use as the reference model. It is useless to define the reference in terms of the measurement.

Consider the pin shown in Fig. 2. It has two sizes, a diameter and a length, each with a tolerance. What are the semantics of the diameter size specification? In the United States where the envelope principle applies, the size specification also limits form deviation. The effect of this is that the physical pin must fit entirely within a cylinder that is of diameter 25.1 mm and at least as long as the pin. But what of the ISO interpretation, and what of the lower limit of size? The standards state that the part must be within the limits of size measured at each cross section. There is no guidance as to how the cross sections should be oriented on the actual part that has no clear axis or how size is measured at a cross section. We would like answers that generalize from the pin of Fig. 2 to all applications of size.

Practical measurement procedures might be used as a guide, but there are significant differences between measurement practice with different instruments. Another approach might be to look at the design objectives represented by each size limit. The upper (maximum material) limit of size is usually intended to control fit and interference. The lower limit is often used to maintain material for strength. Of course, other objectives may apply. There is no way of knowing what was intended by examining the call-out alone.

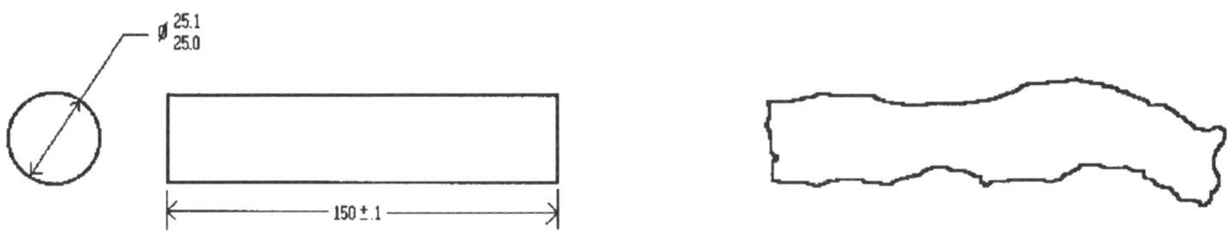

(a) The part specification

(b) The actual part (with errors magnified)

Fig. 2. Typical part with sizes.

Datum reference

Datum referencing[3] is a means of assigning a coordinate system to a physical part. Semantic rules of datum referencing define how to associate datums—ideal geometric elements such as points, lines, and planes—to physical surfaces. Tolerances that require coordinate systems (position, orientation, profile, runout, etc.) are interpreted in coordinate systems established by datum references. The challenge of this scheme is in establishing the rules for assigning datums to datum features. The major difficulty in establishing sensible rules is in defining when a datum feature wobbles or rocks. There are two difficulties, one a question of principle and one a technical problem [23].

The question of principle is illustrated in Fig. 3, which shows three variations of quality of a primary datum feature (the base of the part). If the feature is manufactured as shown in the top sketch, with only a small chamfer at each edge, most people would agree that the part does not wobble on the datum. As the surface form degrades, however, a definite wobble eventually appears, as shown in the bottom sketch. There is a transition state, shown in the middle sketch, which is highly ambiguous. There are absolutely no guidelines—in standards, the technical literature, or common engineering practice—for resolving this ambiguity.

[3] In the "standards English" of tolerancing standards, a *datum feature* is a physical part surface and a *datum* (plural: *datums*) is a geometric element assigned to a datum feature.

Putting aside the question of what wobble is, we are still faced with the technical problem of developing a mathematical definition of "all rocked positions" that allows only those datums that make engineering sense. In other words, suppose we were agreed on when the progression illustrated in Fig. 3 changed from a stable to a rocking datum. How would we describe this in a general way that would apply when the actual datum feature was domed, when it was shaped like the top of a bread loaf, when it was concave, etc.? All formulations proposed to date, when tested against a large number of problems, have either excluded datum assignments that one would want to use in practical applications or have allowed datum assignments that are patently nonsensical.

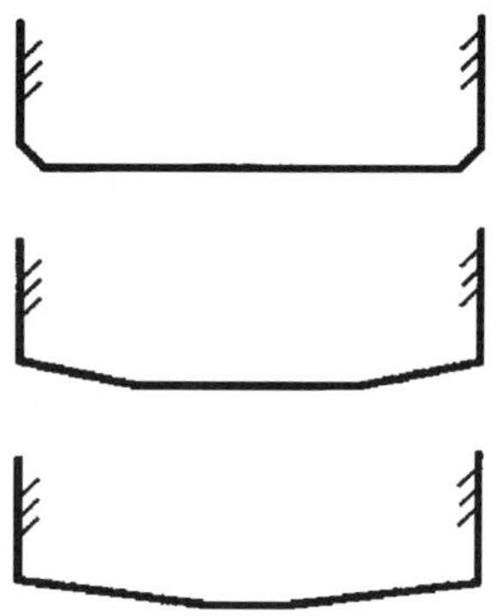

Fig. 3. The difficulty of defining a rocking datum.

I believe that the question of when a datum rocks is inherently unanswerable. (We will never reach consensus on just when the transition occurs in Fig. 3.) We should concentrate on developing a technical definition of limits of rock that resolves the problems known for the existing proposals. The technical definition of rocked positions will then provide a *de facto* definition of when a datum rocks.

The vocabulary of tolerances

The previous section focused on weaknesses in the semantics of our tolerancing language, that is, the challenges facing us in defining what tolerances mean when they appear on a drawing. There are, however, also weaknesses in the vocabulary and grammar of tolerances—what kind of tolerancing information can be placed on the drawing in the first place. The main difficulty is that while our tolerancing language captures the results of the design process, that is, the end decisions, designers often need to communicate something about how those decisions were reached. We will consider three examples of this.

Statistical tolerancing

Suppose I want to build a wall 10-m high out of blocks each 10-cm thick. If the tolerance for the height of the wall is 1 cm, then each block must be manufactured with a size tolerance of 0.1 mm for thickness. This is an example of worst-case tolerancing, a common practice in industry. If I were to relax my requirement for full interchangeability of blocks—that is, if I allow myself to pick and choose blocks from a large selection—then I can relax the tolerance requirement for each block. In fact, I may never have to swap blocks, because variations in block thickness are usually going to average out, not accumulate in a worst-case way. The amount by which I can relax the tolerance on each block depends on what risk I am willing to accept for having to try different block combinations during assembly and what I expect by way of distribution of block thicknesses. This technique of using statistical analysis instead of worst-case analysis is called statistical tolerancing. It is gaining in popularity as designers recognize the savings in manufacturing cost that it provides.

There is a difficulty, however. Use of statistical tolerancing requires the designer to use information about the capability of the manufacturing process—the expected distribution of shape variation. Unfortunately, there is no formal way for the designer to record, as part of the tolerance, the process-capability information that makes the tolerance a faithful representation of design intent. Leaving out this information can have as much impact as not placing tolerances on the drawing in the first place [24]. A typical problem occurs when a manufacturing process is improved, such as occurs when a new machine tool is installed or better process-control techniques are introduced. The process distributions used for setting tolerances are no longer valid and parts no longer assemble as expected. The problem becomes acute when processes for one component are improved without a corresponding change in the processes used for mating parts.

One suggestion being considered by the standards community is to provide a notation whereby the designer can document the assumed process characteristics on the drawing. While this would help communication, it has the danger of *requiring* certain process variability. Manufacturing would not be able to use a better (or worse) process than what was assumed by the designer. A more difficult approach, but one with perhaps fewer economic side effects, is to develop new tolerance notation for describing assembly tolerances. This would generalize the work on standard notation for limits and fits [25]. Either approach will require expanding the vocabulary of tolerances.

Tolerance analysis

Tolerance analysis is the process of studying how the shape variations allowed by the part tolerances affect part function. Tolerance analysis distinguishes between the *design function* of a product and its *model variables* [26–27]. The design function is the property of the product that the designer wishes to study. The model variables are the parameters of the product design that define its shape. Shape tolerances are a way, albeit indirect, for the designer to control the design function. Tolerance analysis involves simulating changes to the shape of the part, as allowed by the tolerances, and studying the effect of those changes on the design function.

Tolerance analysis is used by product designers to develop part designs: to allocate tolerances, to develop more robust shapes, and in general help to optimize the design. Tolerance analysis is also used by production engineers to help define the manufacturing process: to allocate a tolerance budget to each process, to study the effects of process variability on production yield, and to minimize manufacturing costs.

The language of tolerances is incomplete because it does not give designers a way of communicating the relationship between tolerances and the design function. That relationship must be communicated by some means other than the formal product specification. The alternative is having to infer the design function on the basis of design data and "common sense" about what the designer meant. We will need to find a better way of communicating tolerance analysis information as tolerance analysis tools become more widely accepted and used in the production life cycle.

Economic models

The previous two examples, process characteristics for statistical tolerancing and design function for tolerance analysis, can be seen as special cases of a more general shortcoming of the language of tolerances. The design process can be viewed as an effort to maximize (or at least to achieve a threshold for) the value of a product while satisfying a number of constraints—functional requirements, limitations of material properties, manufacturing capability, etc. Many factors considered during the development of a design are of value to later processes. Similarly, much of the information used by the designer is culled from processes occurring later in the life cycle. Much of the interest in concurrent engineering, product data sharing, and enterprise integration is based on the recognition that design decision data must be shared throughout a manufacturing enterprise.

Many design decisions are based on economic factors. For instance, Taguchi frames the entire task of product and process design as an economic problem of minimizing the loss to society due to purchase and use of the product [28–29]. Tolerance design is one of three key components of Taguchi's method, the others being experimental design and parameter design. Tolerance design influences the loss to society because of the combined effects of manufacturing costs and deviations by the product from ideal performance. As with statistical tolerancing and tolerance analysis, tolerance design relies on information—in this case, models of process distributions and loss functions—that should be communicated between design and other life-cycle functions.

The Taguchi method (also called off-line quality control) is only one technique of tolerance design. Another aspect of tolerance design is the intent of the designer in specifying a particular tolerance value. For instance, a tolerance may represent a functional limit, or it may include the anticipated effects of measurement uncertainty in product acceptance [30].

Any systematic tolerance-design practice, however, relies on information that must be part of the design data for effective communication of product definition. The language of tolerances must be expanded if we are to communicate economic models, tolerance goals, and other design decision data effectively.

IMPROVING THE LANGUAGE OF TOLERANCES

I have discussed several weaknesses of our language of tolerances. This is far from a complete list. There are many other areas of active research related to tolerances. Ongoing research has led to many advances in tolerance theory and in the practical application of tolerances. I believe that much more progress can be made much more effectively if we develop better mechanisms for relating progress in one area to problems in another.

We can step back from the issues involving one or another standard and think of ourselves producing a single product—a language of tolerances. The language has many components: tolerance-related standards, technology, and practice. These components have traditionally evolved in a serial manner. We identify a needed change in standard tolerancing practice and write the change into the tolerancing standard. This change impacts other areas: metrology standards, computer software, training, etc. Eventually, after several years, the new practice permeates our entire system and the cycle repeats.

The practice of concurrent engineering within a manufacturing setting is a method of greatly shortening the cycle time of product development [31]. I believe we should apply concurrent engineering concepts to the development of a better tolerancing language. Concurrent engineering is primarily a management approach. It involves identifying the work needed to develop a product, who will do the work, and how the workers can be brought together to work as a team throughout the entire product life cycle. In this section I will discuss how these concepts might apply to the development of an improved tolerancing language. I will discuss a program of work to improve the language and then discuss the roles of various players in this program.

A program for improving tolerances

In this section, I describe a three-part program: develop a framework for the use of tolerances; develop formal semantics for each use; and develop a coordinated set of standards that incorporate the formal semantics. Taken together, these efforts will be an effective way to improve our tolerancing language.

Develop a framework for tolerances in manufacturing

The first area of work is to develop an overall framework for the application of tolerances in manufacturing. This framework should identify the manufacturing life-cycle functions, the role of tolerances in each function, and the standards that codify the language of tolerances as used in each application area.

One can identify four broad areas of manufacturing in which tolerances are used: design, manufacturing, inspection, and product data sharing. In design, tolerances are used to study and specify the range of parts that will satisfy design objectives. Typical tolerance data include limits and fit, dimensions and tolerances, and process capacity models. In manufacturing, tolerances are used to develop error budgets for processes, characterize and measure process capacities, and control quality. Typical tolerance data include machine tool performance measures, manufacturing tolerances for individual parts, and process control limits. In inspection, tolerances are used to select inspection equipment, develop inspection programs, and assure product quality. Typical tolerance data include product tolerances, measurement uncertainties, and product acceptance limits. Shared product representations are used throughout the manufacturing enterprise to communicate product definitions and to correlate manufacturing data with product data. Tolerances are used for cost estimation, design retrieval, variant process planning, and group technology. Tolerances are typically expressed either in graphical form on drawings or electronically in product data bases.

Developing a framework for tolerance use in manufacturing will contribute to efforts to develop standard frameworks for the integration of manufacturing enterprises. Such efforts are underway in a number of countries, as well as within the international standards community [32–34].

Develop formal semantics

The second effort is to develop formal semantics for tolerances in each life-cycle function. While tolerance semantics may differ among applications, the meanings should be compatible across the manufacturing enterprise. The tolerance framework will provide a blueprint for achieving this harmonization.

I have already discussed some of the issues in defining tolerance semantics. I have focused on semantics related to design. Tolerances may have quite different semantics for other applications. Consider, for instance, the selection of data-analysis algorithms for coordinate-measurement systems. One goal is to select an algorithm that minimizes uncertainty of the reported measurement results. This involves studying the relationship between the design semantics of the tolerances, the uncertainty of coordinate values acquired by the sensor, and the sensitivity of the data analysis algorithm to such errors. The semantics for actual value of tolerances for coordinate measurement may, when formally defined, be significantly different from actual value when measurement uncertainty is not an issue. Nevertheless, there should be a well-defined relationship between the two semantics.

Develop coordinated standards

The end goal of the program I am suggesting is the development of a coordinated set of standards that codify the tolerance languages used at each stage of the manufacturing life cycle. This will place the language of tolerances in a form in which it can be transferred to commercial manufacturing applications.

Many tolerance-related standards are already in use and others are under development. In design, standards exist for limits and fits and for dimensioning and tolerancing. In manufacturing, languages for numerical control, such as APT [35], define specialized representations for tolerances. Machine tool performance standards define tolerance capabilities of manufacturing equipment [36]. Similar roles are held in inspection applications, with inspection languages such as DMIS [37] and performance standards for coordinate-measurement systems [38]. Shared product representation standards include drawing standards, IGES, and STEP.

Roles for implementing the program

The program I suggest involves increased interaction between four communities: manufacturers, standards developers, researchers, and technology vendors. Each sector has a role to play.

Manufacturers are the end users of standards and supporting technology. Standards are developed in response to perceived needs or opportunities in manufacturing. Standards are developed either by or with strong participation from industry. The role of manufacturing is to recognize and act on the need for an improved language of tolerances, to support standards efforts, and to demand improved technology from vendors.

Standards development is sometimes pulled along by changing technology, and it sometimes pushes new uses of technology. In the case of tolerances, we have both situations. Drawing standards are fairly mature and change only slowly in response to introduction of new manufacturing technology (e.g., CAD/CAM) or new business practices (e.g., international trade). STEP, on the other hand, presages the introduction of new technology and is, in fact, viewed as a new technology itself that will enable new engineering practices [31,39]. It is easy to understand how standards efforts that push technology can identify substantial research issues. STEP, for example, has spawned research in data modeling, object-oriented data bases, and data-driven manufacturing. It is less obvious that evolution of mature standards can also be the source of a strong research agenda. This has, in fact, happened in the United States as a result of attempts to formalize the semantics of engineering drawings, to develop measurement methods based on these semantics, and to develop performance standards for software that inspects parts specified according to drawing standards. The close working relationship that has begun between some of the tolerance-related standards committees and several researchers from industry, government, and academia should be expanded. Standards committees should become much more proactive in recruiting the involvement of researchers, identifying barriers to improving standards, and using the results that researchers obtain.

In a similar vein, the research community is always interested in problems that are intellectually challenging, will attract support, produce useful results, and help develop better curricula. Researchers should seek support to become involved in standards efforts as one of the most effective ways to identify and gain support for substantive research topics. The products of research may, in some cases, be delivered directly to standards committees or, more often, be made available to developers of new manufacturing technology. That this has happened in several areas of tolerance standards shows that such involvement by researchers is both feasible and productive. Unfortunately, standards participation, at least in the United States, has been traditionally viewed as strictly an industry activity. We need to find ways to increase the career value to researchers of participating in standards work.

Technology vendors are the primary consumers of research results. Vendors, having long ago recognized the relationship of standards to manufacturers, are strong participants in standards efforts. They participate as full partners in understanding and analyzing industry needs and in developing appropriate standards. Technology vendors are in the best position to support participation by researchers. Vendors can transfer research results into practical commercial applications. It is my hope that future generations of CAD systems, machine tools, inspection software, shared product data bases, and other manufacturing technology will be based on a stronger, more unified language of tolerances.

CONCLUSIONS

To date, there has been very little work on developing a common technical basis for tolerance semantics among the many tolerance-related standards we use. As a result, we spend an inordinate amount of effort, time, and money resolving disputes and disagreements about what tolerances mean. Tolerances established by designers are understood differently by manufacturing engineers. Suppliers believe they have shipped high-quality goods, only to find that the customer believes the parts do not comply with specifications. Inspection engineers develop plans to measure part attributes that are irrelevant to the needs of quality control or quality assurance.

Many of the problems we have in using tolerances effectively are, at root, caused by the weaknesses of our tolerancing language. These weaknesses, some of which I have discussed in this paper, need further research. Improvements will require contributions from all phases of manufacturing. I have suggested an integrated program of developing a framework, defining tolerance semantics within that framework, and coordinating development of standards to codify the definitions. Many of the components of the program are already in place, but we must recognize their proper role in relationship to one another. This program will increase the effectiveness of our research efforts and will yield great benefits to manufacturing throughout the world.

REFERENCES

1 C. Alexander, Notes on the Synthesis of Form, Harvard University Press, Cambridge, Mass., 1964.

2 H. Voelcker, Modeling the Design Process, in: Design and Analysis of Integrated Manufacturing Systems, W. Dale Compton (ed), NAE Press, Washington, D.C., 1988.

3 ANSI/ASME Y14.5, Dimensioning and Tolerancing, American Society of Mechanical Engineers, New York, 1982.

4 ISO 8015, Technical Drawings —Fundamental Tolerancing Principle, Int. Org. for Std., 1985. Available from American National Standards Institute, New York.

5 C. Eastman, A. Bond and S. Chase, Res. Eng. Des., 2 (1991) 65–80.

6 ANSI/ASME Y14.26, Digital Representation for Communication of Product Definition Data, NCGA, Merifield, Va., 1987.
7 ISO CD 10303-1, Product Data Representation and Exchange—Part 1: Overview and Fundamental Principles, Int. Org. for Std., 1992. Available from Secretariat, ISO/TC 184/SC 4, National Institute of Standards & Technology, Gaithersburg, Md.
8 S. C. Feng, Comparison of ISO 10303 Part 47 Draft with ANSI and ISO Tolerancing Standards for Harmonization and Completion of Part 47, NISTIR 4774, National Institute of Standards & Technology, Gaithersburg, Md., 1992.
9 R. Walker, GIDEP Alert No. X1-A-88-01, Government-Industry Data Exchange Program, August 22, 1988.
10 V. Tipnis, ed., Research Needs and Technological Opportunities in Mechanical Tolerancing, Results of an International Workshop, October 1988, Orlando, Fla., Report CRTD-15, American Society of Mechanical Engineers, New York, 1990.
11 D. Swyt, Challenges to NIST in Dimensional Metrology: The Impact of Tightening Tolerances in the U.S. Discrete-Part Manufacturing Industry, NISTIR 4757, National Institute of Standards & Technology, Gaithersburg, Md., 1992.
12 N. Juster, Comput. Aided Des., 4, no. 1 (1991) 3–17.
13 S. C. Feng and T. H. Hopp, A Review of Current Geometric Tolerancing Theories and Inspection Data Analysis Algorithms, NISTIR 4509, National Institute of Standards & Technology, Gaithersburg, Md., 1991.
14 A. Requicha, Representation of Tolerances in Solid Modeling: Issues and Alternative Approaches, in: Solid Modelling by Computers: From Theory to Applications, J. Boyse and M. Pickett (eds), Plenum, New York, 1984, pp. 3–22.
15 F. Etesami, Computers in Engineering, 2 (1987) American Society of Mechanical Engineers, 327.
16 R. Jayaraman and V. Srinivasan, IBM J. Res. and Dev., 33, no. 2 (1989) 90–104.
17 V. Srinivasan and R. Jayaraman, IBM J. of Res. and Dev., 33, no. 2 (1989) 105–124.
18 V. Srinivasan, A Geometer Grapples with Tolerancing, in: Proc. of the CIRP Int'l Working Seminar on Computer-Aided Tolerancing, I. Ham and E. A. Lehtihet (eds.), Penn. State University, May 16–17, 1991.
19 M. Henderson, Extraction of Feature Information from Three Dimensional CAD Data, Doctoral diss., Purdue University, 1984.
20 V. Srinivasan, personal communication.
21 ANSI/ASME B46.1, Surface Texture, American Society of Mechanical Engineers, New York, 1973.
22 ANSI/ASME B89.3.1, Measurement of Out-Of-Roundness, American Society of Mechanical Engineers, New York, 1988.
23 T. Hopp, ASPE Newsletter, (Sept. 1990) 4–9.

24 K. Ma, personal communication.
25 ANSI/ASME B4.1, Preferred Limits and Fits for Cylindrical Parts, American Society of Mechanical Engineers, New York, 1987.
26 K. Chase and A. Parkinson, Res. Eng. Design, 3 (1991) 23–27.
27 J. Turner and A. Gangoiti, Concurrent Eng., 1, no. 2 (1991) 11–23.
28 G. Taguchi, The ASI Journal, 1, no. 1 (1988) 5–29.
29 N. Singpurwalla, A Bayesian Perspective on "Taguchi's Approach" to Quality Engineering and Tolerance Design, TR91-2, George Washington University, Washington D.C., 1991.
30 R. Easterling, M. Johnson, T. Bement and C. Nachtsheim, Absolute and Other Tolerances, Report SAND88-0724, Sandia National Laboratory, Albuquerque, N.M., 1989.
31 D. Carter and B. Baker, CE: Concurrent Engineering, The Product Development Environment for the 1990's, Addison-Wesley, New York, 1992.
32 MCC, Technical Issues of Enterprise Integration, Report CAD-349-91, Microelectronics and Computer Technology Corp, Austin, Tex., 1991.
33 AMICE (eds.), Open Systems Architecture for CIM, Springer-Verlag, New York, 1989.
34 H. Jorysz and F. Vernadat, Int. J. Comput. Integrated Manuf., 3, no. 3–4 (1990) 144–156.
35 ANSI X3.37, Programming Language APT, American National Standards Institute, New York, 1977.
36 ANSI/ASME B5.54, Methods for Performance Evaluation of Computer Numerically Controlled Machining Centers, American Society of Mechanical Engineers, New York, 1992.
37 ANSI/CAM-I 101, Dimensional Measuring Interface Standard, Version 2.1, Computer-Aided Manufacturing, International, Arlington, Tex., 1990.
38 ANSI/ASME B89.1.12, Methods for Performance Evaluation of Coordinate Measuring Machines, American Society of Mechanical Engineers, New York, 1990.
39 G. Carver and H. Bloom, Concurrent Engineering Through Product Data Standards, in: Manufacturing and Automation Systems: Techniques and Technologies, Part 1 of 5—Three Pillars of Manufacturing Technology, pp. 31–109, Academic Press, San Diego, Calif., 1992.

Quality through Engineering Design - W. Kuo (Editor)

Optimal Sampling for Coordinate Measurement: Its Definition and Algorithm

Tony C. Woo and Ren Liang

Industrial and Operations Engineering, University of Michigan,
1205 Beal Avenue, Ann Arbor, MI 48109–2117, U.S.A.

Abstract

Accuracy in discrete sampling with coordinate-measuring machines is first defined as the discrepancy of a finite-point set, the lower bound for which was established by a Fields medalist. Then from number theory, a particular sequence of numbers is used to compute the sampling locations, resulting in a nearly quadratic reduction in the sample size over the uniform or random distributions at the same level of accuracy. Finally, experimental results in the measurement of machined surfaces modeled as random processes are compiled, validating the number-theoretic predictions.

INTRODUCTION

Quality has many connotations and implications, such as customer satisfaction and management style. In the engineering context, however, quality can be associated with precision in manufacturing. In the life cycle of a product, precision is expressed as tolerances in dimensioned drawings during the design stage. Viewing the ideal shape (with nominal dimensions) as the input to a manufacturing channel [1], we see that its output (the realized shape) is corrupted by noise in the channel. Since there are encoding and decoding schemes in signal communications, tolerance specification serves as a decoding scheme; inspection detects errors but does not correct them. The difficulty in error correction is deeper than one might think. The signal in the manufacturing channel is complex not only in its dimensionality (since the products are usually three-dimensional rather that one-dimensional bit streams) but also in the semantics that are embedded in the signal as the functionality of the product.

Inspection, as an error-detection scheme, should be error free in itself. Confidence in the inspection process is, however, questionable. In 1988, for example, R. Walker of Westinghouse issued an alert through a medium called GIDEP (government-industry data exchange program) [2] that the vendors of coordinate-measuring machines (CMMs) using different algorithms in their machines, had a disagreement of as high as 60% in their results! (An immediate conclusion one can draw is that good parts may have been rejected and bad parts may have been in service.) This phenomenon was subsequently termed the *divergence of methods* [3].

There are actually two places where methods are invoked in surface measurement: for data gathering (or *sampling*) and for data interpretation (or *fitting*). This paper reviews these methods and presents an important finding: the choice of a metric is subjective and is dependent on the implicit function of the design, but within the same metric there exists an absolute (hence objective) bound in accuracy leading to an optimal sampling strategy. Next this paper clarifies optimality as accuracy and time in coordinate measurement. On the basis of the work by K. F. Roth, a Fields medalist, accuracy in approximation is expressed by a mathematical notion called the *discrepancy* of a finite set of sample points, for which a lower bound exists [4]. For a finite set of N sample points, the time for sampling and for fitting can also be quantified in terms of N. Then, a sampling strategy based on the Hammersley sequence [5] is compared to uniform sampling. It shows a remarkable improvement with nearly a quadratic reduction in the number of samples and hence the time spent, yet it maintains the same level of accuracy. (Loosely speaking, a quadratic reduction of 250,000 points yields 500.) Finally, implementation of the algorithm and simulation ofiinspection are conducted on machined surfaces [6]. The results confirm the theoretical prediction. This paper thus contributes to the convergence of sampling methods for surface measurement.

SOURCE OF DIVERGENCE

Because the notions of distance and metric are central to the source of divergence, it is useful to review them first. Then, divergence in fitting algorithms is given a context.

Distance and metric

The basic idea in fitting is to elevate dimensionality. Sample points are not only numerous but also zero-dimensional topologically. By fitting a curve through points or a surface through curves and hence elevating the dimensionality, two objectives are achieved: (1) the size of data storage is reduced—for later algorithmic processing, and (2) a group behavior emerges—for interpretation and decision making.

While there is abundant literature on approximation such as Powell [7], distance is the basic element in computation. Consider the familiar Euclidean *distance* between two points (x_1, y_1, z_1) and (x_2, y_2, z_2). For each of the coordinates, the difference, e.g., $(x_1 - x_2)$, is first captured. Then, the differences are squared, summed, and a square root taken over the sum; hence the root-mean-square (or RMS) difference:

$$d_2 = \left[(x_1 - x_2)^2 + (y_1 - y_2)^2 + (z_1 - z_2)^2\right]^{1/2} \tag{1}$$

The *metric* applied to Eq. (1) is the so-called L_2 metric with the following general form under the Lp metric [8]:

$$d_p = \left[|x_1 - x_2|^p + |y_1 - y_2|^p + |z_1 - z_2|^p\right]^{1/p} \tag{2}$$

Absolute value is taken so as to accommodate the odd and even nature of the exponent p with respect to the sign of the differences. When $p \geq 2$, one finds a variety of distances, each having its own merit. In particular, when $p = \infty$, Eq. (2) becomes a mechanism for selecting the largest of the three components as the supremum, or the maximal element in the set:

$$d_\infty = \sup\left[|x_1 - x_2|, |y_1 - y_2|, |z_1 - z_2|\right] \tag{3}$$

Transition from Eq. (2) to Eq. (3) can be understood by assuming that one of the terms, for example, $|y_1 - y_2|$, is the largest of the three. Dividing the entire right-hand side of Eq. (2) by this term yields

$$d_p = \left\{\left[\left(\frac{|x_1 - x_2|}{|y_1 - y_2|}\right)^p + \left(\frac{|y_1 - y_2|}{|y_1 - y_2|}\right)^p + \left(\frac{|z_1 - z_2|}{|y_1 - y_2|}\right)^p\right]|y_1 - y_2|^p\right\}^{1/p} \tag{4}$$

The first and the third terms in Eq. (4) vanish when p becomes very large and hence effectively render the supremum as in Eq. (3). Figure 1 illustrates the various d_p.

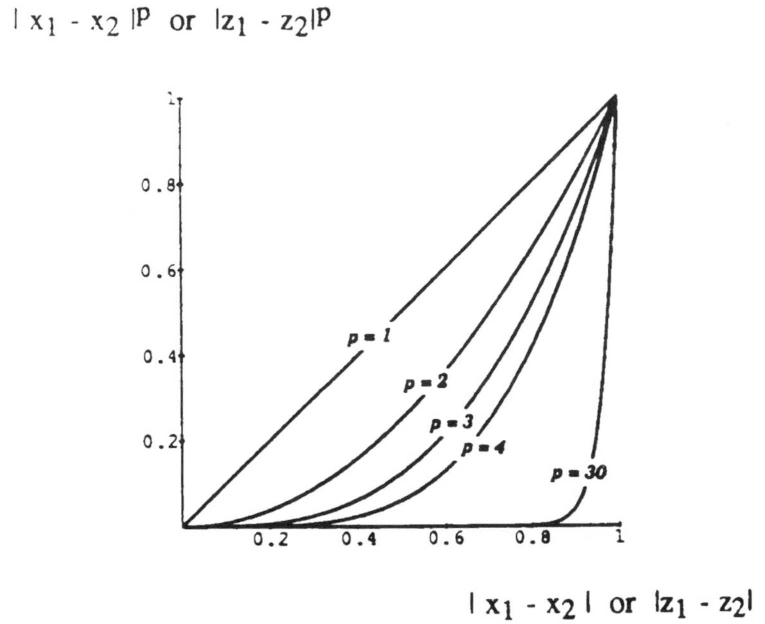

Fig. 1. Relative importance of the individual items in Eq. (4) for different p.

Fitting algorithms

Having seen that distance can adopt infinitely many measures, it should not be surprising that there are a variety of algorithms for elevating the dimensionality of point data by fitting them into curves and surfaces [9]. The word fitting implies a preconceived notion of some ideal geometry. Indeed, there must be some polynomial of a fixed degree into which the data fits; otherwise, fitting becomes arbitrary.

The kinds of ideal geometries in computer-aided design (CAD) are linear, quadratic, cubic, and higher degree. Examples of linear geometry are lines and planes, and examples of quadratic geometry are cylinders, cones, and spheres. When the geometry is cubic or higher, there are few familiar shapes with the possible exception of a torus (which can be used to blend surfaces as fillets). In CAD, these high-degree surfaces are called free-form surfaces. By virtue of the local control the designer has on the shape, there are various interpolators [10,11] such as Hermite, Bernstein, and other polynomials, giving rise to Coons, Bezier, B-spline, and nonuniform B-spline curves, and surfaces that are familiar to the CAD community.

Conceptually, there are three basic criteria to fit data to these ideal geometries: INNER, OUTER and THROUGH. (By the distinction of INNER and OUTER, there is an implied direction that indicates the material side of the workpiece under inspection.) Data can be fitted so that they all lie on the inside (the material side) of the ideal geometry, or conversely, on the outside. While one can take the two extremes by taking both the INNER and OUTER, giving rise to a tolerance zone, fitting an ideal geometry THROUGH the data so that some lie on the inside and others on the outside is a third alternative that gives a sense of an "average." (It may be conjectured that the criterion for fitting data and the choice of a metric depend on the functionality of the component as implied by the intent of the design [12].) To ensure that the ideal geometry satisfies any of the three criteria, one must invoke a distance function, which can be linear as in Eq. (3), or nonlinear as in Eqs. (1) or (2). Thus, fitting algorithms is seen as a linear or nonlinear optimization problem in general [13], whereby some form of a (linear or nonlinear) distance is minimized [14] as constrained by an ideal (linear or nonlinear) geometry. When the ideal geometry is simple, as in the case of lines, circles, planes, and cylinders, analytic solutions are possible [15].

After the data has been fitted, it is compared to the ideal geometry. Here, again, the notion of distance must be considered. For speed, one may choose the largest component, e.g., in Eq. (3) as an approximation to distance. For accuracy, the RMS distance, shown in Eq. (1), is adopted. Recall that the objective of fitting point data is to elevate the dimensionality. Now that the zero-dimensional points have been raised to one-dimensional curves or two-dimensional surfaces, the distance between the fitted and the ideal curves and surfaces once again invokes that between the points. To find the distance between two curves, for example, one computes the distance between a pair of closest points on the two curves and treats it as the distance between the curves. Adopting the paradigm of the Lp metric, one can take the square of the closest points—the cube, . . ., and so on—as distances. Without compounding the divergence in the methods for assessing the distance between the elevated entities, the irony of elevation may be noted: the dimensionality of point data is elevated from zero to one (or two), yet the distance used is for zero-dimensional entities.

SAMPLING ALGORITHM

Discrete sampling is inherently an approximation process. If the number of samples is infinite, one may imagine that the error in approximation approaches zero. But when the sample size is finite, the error must be nonzero. This observation induces two practical questions. For any sampling sequence of size N, how small can the error be? And, in what way should the N samples be located? Answering these questions will address the accuracy and the time for coordinate measurement.

Sampling error

Once a metric is adopted, the variety of methods is reduced. But this still leaves open the question: Suppose one adopts the L_2 metric —what is the best method for sampling? In the last section, the measure of goodness has been heuristic (or subjective), taking the least of the squares or minimizing the maximum. Now we report on an objective criterion for assessing sampling errors and hence accuracy.

Such a criterion does not contradict the works of Gauss, Newton, and Tschebychev [7,9,14]; it was discovered at least a century later by K. F. Roth of the Imperial College, who received the equivalent of a Nobel in mathematics (the Fields metal) in 1958. Roth's contribution was to establish a lower bound in the *discrepancy* D between a finite set of N points and any function in d dimensions [4].

$$D \cong O\left[N^{-1}\left(\log N^{\frac{d-1}{2}}\right)\right] \tag{5}$$

The significance of Eq. (5) is that the lower bound is expressed only in terms of N and d; it is devoid of any constraint on the function that the samples approximate. Most important to coordinate measurement is that one can determine if a finite set of samples of size N, giving rise to a discrepancy D, meets Roth's bound and hence gives the least amount of error possible. For calibrating the sampling sequences to be discussed later, Roth's bounds in low dimensions are

$$d = 2, \quad D \cong O\left(\sqrt{\log N}/N\right) \tag{6}$$

$$d = 1, \quad D \cong O(1/N) \tag{7}$$

As important as discrepancy is to surface measurement, the concept may be unfamiliar. To outline the basic idea, it is convenient to begin in two dimensions with a unit square [0,1] x [0,1] in which a region of interest bounded by some function resides. The area of the region can be approximated with points

$\mathbf{p}_1, \mathbf{p}_2, \ldots, \mathbf{p}_N$ by the fraction:

$$area \cong \frac{1}{N}\sum_{i=1}^{N}\chi(\mathbf{p}_i) \tag{8}$$

where c ($\mathbf{p}_i$) is the characteristic function indicating whether a point $\mathbf{p}_i \in [0,1]^2$ is in or not in the region:

χ ($\mathbf{p}_i$) = 1 if $\mathbf{p}_i$ is in the region
0 otherwise

Generalizing Eq. (8) to d dimensions, whereby the unit square becomes a unit hypercube $I^d = [0,1]^d$ bounding a function f($\mathbf{t}$), $\mathbf{t}$ Σ I^d, one has

$$\int_{I^d} f(\mathbf{t})d\mathbf{t} \cong \frac{1}{N}\sum_{i=1}^{N}\chi(\mathbf{p}_i) \tag{9}$$

Calling the difference between the two sides of Eq. (9) as the residual,

$$R(N,\mathbf{p}_i) = \int_{I^d} f(\mathbf{t})d\mathbf{t} - \frac{1}{N}\sum_{i=1}^{N}\chi(\mathbf{p}_i) \tag{10}$$

the discrepancy in the L_2 norm, or *L_2 discrepancy* [16], is defined as

$$D_2 = \left\{\int_{I^d}\left[R(N,\mathbf{p}_i)\right]^2 d\mathbf{t}\right\}^{1/2} \tag{11}$$

Note the square root, the mean [by taking the integral in the sense of Eq. (9)], and the squaring of R. Likewise, in the L_∞ norm, the *L_∞ discrepancy* is defined by

$$D_\infty = \sup_{I^d}\left|R(N,\mathbf{p}_i)\right| \tag{12}$$

While there is a similarity between d_2 in Eq. (1) and D_2 in Eq. (11) and between d_∞ in Eq. (3) and D_∞ in Eq. (12), f($\mathbf{t}$) in Eq. (10) is a real-valued function that must be Riemann-integrable. A variation of the residual in Eq. (10) is to take the difference between the volume of a hypercube and its approximation [16]:

$$R^*(N,\mathbf{p}_i) = \prod_{j=1}^{d}\mathbf{t}_j - \frac{1}{N}\sum_{i=1}^{N}\chi(\mathbf{p}_i) \tag{13}$$

giving rise to D_2^* and D_∞^*, respectively:

$$D_2^* = \left\{ \int_{I^d} \left[R^*(N, \mathbf{p}_i) \right]^2 d\mathbf{t} \right\}^{1/2} \tag{14}$$

$$D_\infty^* = \sup_{I^d} \left| R^*(\mathbf{t}, N, \mathbf{p}_i) \right| \tag{15}$$

And Roth [14] established the lower bound:

$$D_\infty^* \geq D_2 > kN^{-1} (\log N)^{\frac{d-1}{2}} \tag{16}$$

where the constant k is expressed in terms of the dimension d only. The existence of such a bound triggered a flurry of activities in the construction of sequences of numbers, one of which is the subject of the next section.

Sample size and locations

It may be intuitive that the larger the sample size, the more accurate the measurement is. Such an intuition is confirmed by Eq. (7) in that the discrepancy is inversely proportional to the sample size N in one dimension. But as the dimensionality of the sample points increases, the error is also directly proportional to some power of the logarithm of N, as in Eq. (6). The outstanding question is: Are there sequences of numbers that approach or even reach Roth's bound, as in Eq. (5)? From the work of researchers in number theory, the answer is yes.

Before going into sequences, a word on the dimensionality d of the samples is in order. For a surface under inspection to be visible to an inspection machine, the surface can only be assumed to be single-valued. When the machine is a point-sampling machine, we may assume that a command-coordinate pair (x_i, y_i) is issued and a value z_i is read. When the machine is a line-sampling machine, we may assume that a command-coordinate (x_i) is issued and a pair (y_i, z_i) is read. The dimensionality of a sample, though always three after the data is read in, differs by the type of machinery used. In this work, sampling with the CMM is discussed; hence d = 2. To choose freely among the pairs (x_i, y_i), (x_i, z_i), and (y_i, z_i), one identifies the command coordinate (s_i, t_i) with the term sample.

The sequence of numbers that Roth used to exhibit the tightness of his bound [17] is the *Hammersley sequence* [5]. Its coordinates (s_i, t_i) in two dimensions are defined by

$$\begin{aligned} s_i &= i/N \\ t_i &= \sum_{j=0}^{k} b_j 2^{-j-1} \end{aligned} \tag{17}$$

where

i	$\in$	[0, N − 1]
N	=	total number of points
k	=	$\lceil \log i \rceil$ = ceiling of (log i)
b_j	=	binary digits for representing i

As the binary representation of integers is used, the logarithm will be taken to the base 2.

Suppose N = 10 Hammersley points are used for sampling. The procedure for computing the Hammersley coordinates are as follows:

1. Determine k: Since i $\in$ [0,9], $\lceil \log i \rceil$ = 4.
2. Determine b_j: The four-bit representation ($b_3\, b_2\, b_1\, b_0$) for i is (0 0 0 0), (0 0 0 1), . . . (1 0 0 1).
3. Compute s_i: According to Eq. (17), the s_i coordinates for the N = 10 points are 0/10, 1/10, 2/10, . . . 9/10.
4. Compute t_i: The binary representation for the t_i coordinate is obtained by taking the mirror image of the binary representation for i, about the decimal point. For instance, for i = 1 = (0 0 0 1) or s_1 = 1/10, the mirror image is (1 0 0 0) or t_1 = 8/16. For i = 7 = (0 1 1 1) or s_1 = 7/10, the mirror image is (1 1 1 0) or t_7 = 14/16.

The N = 10 Hammersley points are illustrated in Fig. 2. The 10 × 10 grid is to be interpreted as the locations for the 100 uniform samples.

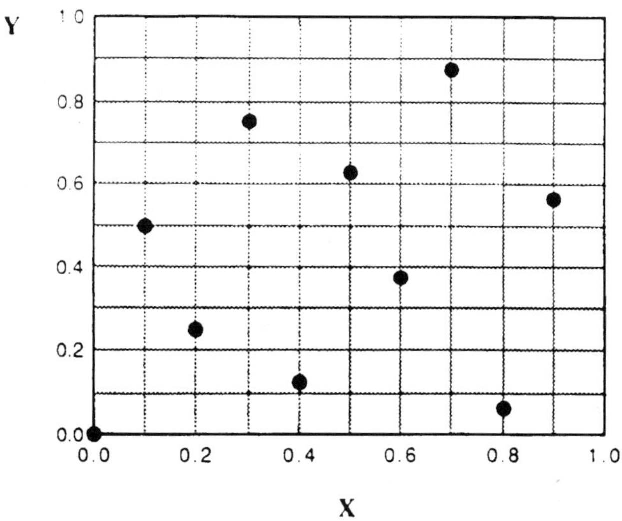

Fig. 2. The distribution of ten Hammersley points.

The discrepancy of the Hammersley points in d dimensions is of the order

$$D_H \cong O\left[N^{-1}(\log N)^{d-1}\right] \tag{18}$$

which is off by a factor of one-half power as compared to the absolute lower bound established by Roth in Eq. (5). While slightly suboptimal, the Hammersley sequence shows a nearly quadratic reduction in the number of points needed by uniform sampling for the same level of accuracy as measured by discrepancy.

More precisely, it is known from Monte Carlo methods that as the number of samples N approaches infinity, the error becomes indistinguishable from that of a uniform sequence [16]. In particular, the error for a uniform sequence is $O(N^{-1/d})$. To compare the number of uniform points N_U against the Hammersley points N_H, suppose that they yield the same discrepancy in two dimensions:

$$\frac{1}{\sqrt{N_U}} = \frac{1}{N_H}\log N_H \tag{19}$$

where N_U denotes the number of uniform points and N_H the number of Hammersley points. It becomes immediately obvious that there are quadratically almost as many N_U as there are N_H:

$$N_U = \left(\frac{N_H}{\log N_H}\right)^2 \tag{20}$$

A plot of the relation between N_U and N_H is given in Fig. 3.

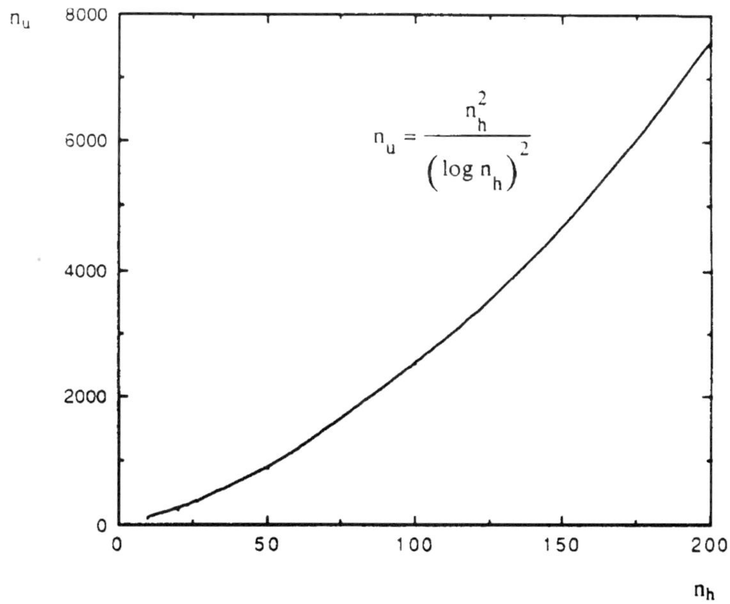

Fig. 3. Hammersley points (n_h) versus uniform points (n_u) under the same accuracy.

IMPLEMENTATION

The confluence of mathematics and metrology still needs many bridges. In his article, R. Hochen [18] reports other needs in the characterization of errors in the kinematics and dynamics of the CMM, lobing errors in the probe of the CMM, and the uncertainty in the manufacturing processes that create the workpiece. Because the characterization of machine by machine, probe by probe, and workpiece by workpiece can be laborious, the classical mathematization of machined surfaces as random processes [19] once again seems attractive.

Not only do random processes offer first-order (mean), second-order (variance), and higher-order statistics [20] that capture ensemble behaviors, evidenced in a beautiful catalog of machined surfaces [21], but also they provide a crucial link to Roth's bound and Hammersley's sequence [22–24]. In this connection the measurement for the roughness of a machined surface, modeled as a random process, permits the employment of discrepancy, and hence the Hammersley and other sequences [25].

A number of experiments have been conducted to assess the accuracy of the Hammersley points for surface measurement. The number of samples ranges from 16 to 10,201 for two reasons. First, the uniform distribution of $N_U = 11 \times 11$, 26×26, 51×51, and $101 \times 101 = 10{,}201$ points is used as the benchmark. Secondly, the Hammersley points are chosen at $N_H = 16, 32, 64$, and 128 to approximate the reduction in Eq. (20).

Machined surfaces are simulated by both modified Wiener and isotropic Gaussian models. For each of the two-surface models, the uniform and the Hammersley points are used as samples so that their (respective) accuracies can be assessed. To give meaning to discrepancy, we measure the average roughness of a surface R_a defined as the average height h at a point (s,t) on a surface approximated by N discrete measurements:

$$R_a = \int_0^1\int_0^1 |h(s,t)| ds\, dt \cong \frac{1}{N}\sum_{i=1}^{N} |h(\mathbf{p}_{si}, \mathbf{p}_{ti})| \tag{21}$$

where $\mathbf{p}_{si}$ is the s-coordinate of the ith point in a sequence of N points. [Absolute value is taken on h(s,t) in Eq. (21) so as to ensure that the asperities do not cancel out.] The similarity between Eqs. (21) and (9) may be noted here. Calling the difference between the two sides of Eq. (21) as the error,

$$\varepsilon(N, \mathbf{p}_i) = \int_0^1\int_0^1 |h(s,t)| ds\, dt - \frac{1}{N}\sum_{i=1}^{N} |h(\mathbf{p}_{is}, \mathbf{p}_{it})| \tag{22}$$

By invoking the L_2 metric, the *root-mean-square* error in surface measurement ε_{RMS}.

$$\varepsilon_{RMS} = \left\{ \int_0^1 \int_0^1 \left[\varepsilon(N, \mathbf{p}_i) \right]^2 f(s,t) ds\, dt \right\}^{1/2} \tag{23}$$

where f(s,t) is the probability density function for the Wiener or the Gaussian surfaces. Again, note the similarity between Eqs. (23) and Eq. (11).

Figures 4a and 4b show 32 Hammersley points superimposed on a Wiener surface and on a Gaussian surface. It may be seen that the Wiener surface is smooth and periodic, simulating single-point rotary cutting with a rotary tool, while the Gaussian surface is more textured and independent, simulating such processes as electro-discharge machining or sand-blasting.

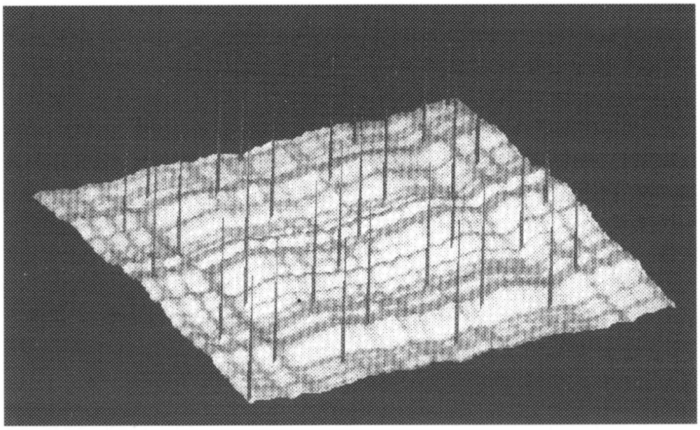

Fig 4a. Thirty-two Hammersley points on a modified Wiener surface.

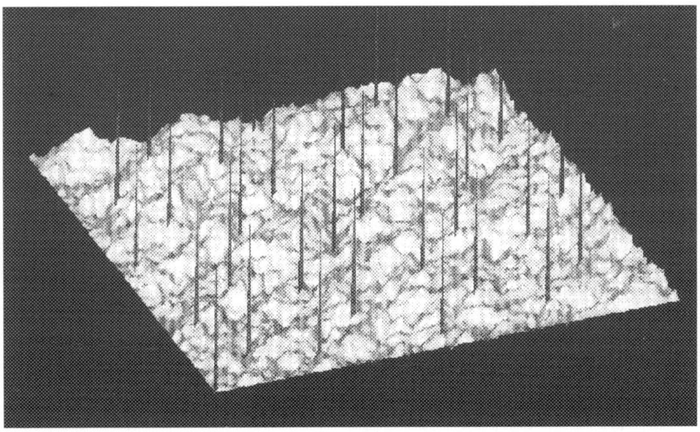

Fig. 4b. Thirty-two Hammersley points on an isotropic Gaussian surface.

One hundred runs are made for each of the Wiener and the Gaussian surface models, and the RMS error in using the Hammersley points versus the uniform points is calculated on the basis of Eq. (23). Table 1 shows the experimental results for RMS errors on Wiener surfaces with correspondences in the reduced number of sample points.

Table 1. RMS error on Wiener surfaces.

Uniform		Hammersley	
Number	RMS error	Number	RMS error
101×101	0.001	128	0.003
51×51	0.003	64	0.006
26×26	0.005	32	0.010
11×11	0.014	16	0.019

Table 1 shows that the Hammersley samples are as accurate as uniform samples with dramatically less points. Results on the one hundred Gaussian surfaces are given in Table 2.

Table 2. RMS error on Gaussian surfaces.

Uniform		Hammersley	
Number	RMS error	Number	RMS error
101×101	0.007	128	0.013
51×51	0.018	64	0.022
26×26	0.040	32	0.028
11×11	0.101	16	0.038

CONCLUSIONS

This paper introduces the application of the Hammersley sequence to surface measurement and hence brings some convergence to sampling methods. The Hammersley sequence is applicable because it exhibits a discrepancy approaching the number-theoretic limit asserted by Roth. And discrepancy is related to the RMS error in measuring average surface roughness.

For point sampling, users of the CMM may wish to examine the reported strategy over random or uniform sampling. For line sampling, practitioners in computer vision may take notice that a scan of 500 x 500 points can now be dramatically compacted [26]. A similar saving in data storage is possible for plane sampling, such as in computer tomography.

ACKNOWLEDGMENT

This work was supported in part by the Fulbright Foundation, grants from the National Science Foundation DMC-8903029 and DMC-9113791, and a grant from the University Research Fund, Scientific Research Laboratory, Ford Motor Company. The authors are grateful to the managing editor, Marcia Martens Pierson, for her skillful editing and to Ruby Sowards for her typing.

REFERENCES

1 U. Roy, C. R. Liu and T. C. Woo, Computer-Aided Design, 23, no. 7 (1991) 466–483.
2 R. Walker, Government Industry Data Exchange Program Alert, X1-A-88-01, Aug. 22, 1988.
3 R. Schreiber, Manuf. Eng., 104 (1990) 10.
4 K. F. Roth, Mathematika, 1 (1954) 73–79.
5 J. M. Hammersley, Annals of New York Academy of Sciences, 86 (1960) 844.–874
6 G. M. Zhang and S. G. Kapoor, ASME Trans., J. Engineering for Industry, 113 (1991) 145–153.
7 M. J. D. Powell, Approximation Theory and Methods, Cambridge University Press, Cambridge, U.K., 1981.
8 Y. A. Schreider, What is Distance, University of Chicago Press, Chicago, 1974.
9 P. Lancaster and K. Salkauskas, Curve and Surface Fitting, Academic Press, New York, 1986.
10 M. Mortensen, Geometric Modeling, John Wiley and Sons, Inc., New York, 1985.
11 G. Farin, Curves and Surfaces for Computer Aided Geometric Design, Academic Press, New York, 1988.
12 S. Y. Chou, T. C. Woo and S. M. Pollock, On Assessing Circularity, Technical Report 92-16, Department of Industrial and Operations Engineering, University of Michigan, Ann Arbor, 1992.
13 K. G. Murty, Linear Complementarity, Linear and Nonlinear Programming, Heldermann Verlag, Berlin, 1988.
14 W. H. Press, B. P. Flannery, S. A. Teukolsky and W. T. Vetterling, Numerical Recipes: The Art of Scientific Computing, Cambridge University Press, Cambridge, U.K., 1989.
15 V. B. Le and D. T. Lee, IEEE Trans. Pattern Analysis and Machine Intelligence, 13, no. 3, (1991), 217–223.
16 L. Kuipers and H. Niederreiter, Uniform Distribution of Sequences, John Wiley and Sons, Inc., New York, 1974.
17 K. F. Roth, Acta Arithmetica, 37 (1980) 67–75.
18 G. Caskey, Y. Hari, R. Hocken, D. Palanvelu, J. Reja, R. Wilson, K. Chen and J. Yang, Sampling techniques for coordinate measuring machines, Proc. NSF Design and Manufacturing Systems Conference, Atlanta, Ga. 1992, pp. 779–786.

19 J. Peklenik, Annals of the CIRP, 12 (1963) 173–176.
20 S. Ross, Stochastic Processes, John Wiley and Sons, Inc., New York, 1983.
21 K. J. Stout, E. J. Davis and P. J. Sullivan, Atlas of Machined Surfaces, Chapman and Hall, London, 1990.
22 B. A., Cipra, AAAS Science (1991) 165.
23 H. Wozniakowski, Bull. AMS, 24, no. 1 (1991), 185–194.
24 R. Liang and T. C. Woo, Accuracy and Time in Measuring Surface Roughness, Part 1: Mathematical Foundations, Technical Report 92-17, Department of Industrial and Operations Engineering, University of Michigan, Ann Arbor, 1992.
25 T. C. Woo and R. Liang, Accuracy and Time in Measuring Surface Roughness, Part 2: The Uniform, Hammersley and Zaremba Sequences, Technical Report 92-18, Department of Industrial and Operations Engineering, University of Michigan, Ann Arbor, 1992.
26 T. C. Woo, G. Lang and R. Liang, Square-root Sampling of Image Arrays, Technical report 92-19, Department of Industrial and Operations Engineering, University of Michigan, Ann Arbor, 1992.

Chapter 6
Finite-Element Method for Design

The finite-element method is a widely applied tool used in solving engineering problems. In this chapter, the robust-design concept is formulated in the context of the finite-element method to ensure response on target and to render the design insensitive to variation. Because of its versatility and adaptability, the finite-element method is also illustrated in analyzing performance of multipad disc brakes and tilting-pad journal bearings. All these applications permit a wider extension of CAD for robust design.

Finite-Element Implementation of Robust Design

H. L. Oh and M. B. Motwani

CPC Group, General Motors Corporation, CPC Headquarters,
General Motors Technical Center, Warren, MI 48090, U.S.A.

Abstract

Quality is currently viewed not as a problem to be solved but as a competitive opportunity to seize. In this view, we focus on the design target and try to reduce variation from it. The strategy is to choose design parameters that not only ensure response on target but also render the design insensitive to variation. This is the concept of robust design. In this paper, we formulate the robust-design concept in the context of the finite-element method, which permits a wider application of computer-aided robust design. A real-life problem illustrates the implementation.

INTRODUCTION

Traditionally, quality is viewed as a problem to be solved. Consider for example the latching of a van seat to two pins on the floor shown in Fig. 1. For the seat to latch, the distance between the two pins must equal the distance between the two latches. This means that with the front latch latched onto the front pin, the rear pin must fall inside the jaw of the rear latch between points (2) and (4) in Fig. 2. If the pin falls outside the jaw, e.g., point (1) or (5), the seat will not latch. The seat is then deemed to have poor quality. Since variability is a fact of life in mass production, the probability does exist that the rear pin may fall outside the jaw of the rear latch. Efforts are therefore made to reduce this probability. In other words, the traditional approach to quality focuses on the tails of the distribution, that is, the probability of failure, and tries to minimize them. The aim is to eliminate customer complaints.

Recently, a new view of quality has emerged. Quality is no longer viewed as a problem to be solved but as a competitive opportunity to seize. In this new view, the focus in the seat-latch design now shifts from point (1) or (5) to a spot marked (3). This spot is the "sweet spot" of the latch, much like the "sweet spot" on a tennis racket or on a golf club. When the pin hits this spot, the seat will latch smoothly and the customer will experience a sweet sensation. Therefore, effort should be spent toward maximizing the probability of hitting the "sweet spot." In other words, the new view of quality focuses on the target, the center of the distribution, and tries to reduce variation from it to maximize the probability of meeting customer expectation. The aim is to elicit customer compliments. Note that in maximizing the probability of hitting the target, the probability of failure is automatically reduced. The quality, as we traditionally know it, therefore becomes a secondary by-product in the new view.

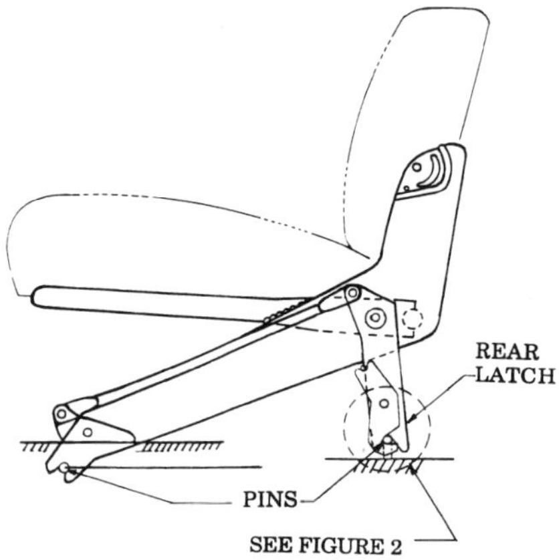

Fig. 1. Latching of a van seat.

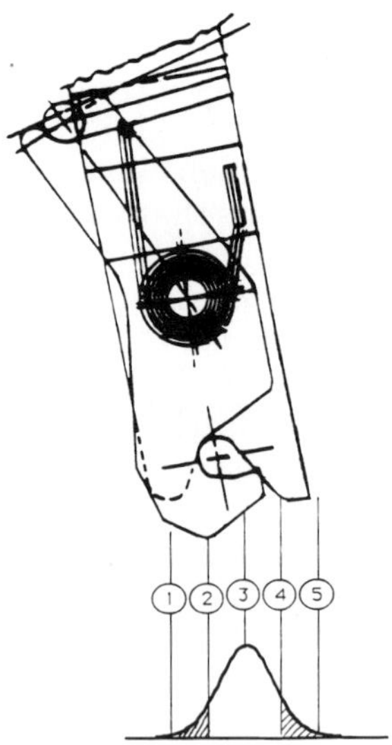

Fig. 2. Possible rear pin positions relative to the latch.

This new view dictates different emphases in design. One emphasis is the definition of the target. The key is to faithfully deploy and translate the voice of the customer into engineering terms. Another emphasis is the reduction of variation from the target. The wisdom is not to control the sources of variation affecting the design as we traditionally do but to make the design insensitive to the sources through a strategic choice of the nominal values of the design variables. A design using such a strategy is called a robust design. It usually results in a quality design at reduced cost.

The new view also requires different methodologies in design. In structural design, methods such as the stochastic finite-element method [1] and the probabilistic structural-analysis method [2] are not applicable because they were developed to assess the uncertainty in design from the traditional view of quality. In Ref. [3], the new view of quality, specifically the concept of robust design, is formulated in the context of optimization. This enables us to do robust design with the traditional optimization techniques. To effect a wider application of the new view, however, the formulation in Ref. [3] must be extended so as to integrate robust design into general-purpose codes such as the finite-element method (FEM). FEM has been used previously to achieve robust designs; in those cases, however, FEM was used merely as a simulation tool to carry out a planned numerical experiment. The concept of robust design was never integrated into the FEM. Consequently, the power of computer-aided robust design was not fully realized. The purpose of this paper is to show how to do robust design in the context of FEM.

FORMULATION

In this section, we derive the mathematics for FEM implementation of robust design. Let g denote the value of the design response of interest and g^* its target. Note the starred quantity denotes the nominal value of the quantity. Let $\mathbf{x} = [x_1, x_2, \ldots, x_k]^T$ denote the k factors affecting g. Here, a bold lower case symbol denotes a vector; a bold upper case symbol a matrix. Of these k factors, m of them are the sources of variation. We call these m factors the noise factors $\mathbf{n}$. Because of the variability in $\mathbf{n}$, g will exhibit a random deviation from g^*. We expand g about $\mathbf{x}^*$ in Taylor series neglecting higher-order terms:

$$g(\mathbf{x}) = g^* + [\nabla_{\mathbf{n}} g(\mathbf{x}^*)]^T (\mathbf{n} - \mathbf{n}^*) \tag{1}$$

The symbol $\nabla_{\mathbf{n}} g(\mathbf{x}^*) = [\partial g/\partial n_1, \partial g/\partial n_2, \ldots, \partial g/\partial n_m]^T$ denotes an m-component column vector of the first-partial derivatives of g evaluated at $\mathbf{x}^*$. The second term on the right-hand side of Eq. (1) represents the contribution to the random deviation of g from g^* by the noise factors. The mean and variance of g are

$$mean\ (g) = E(g) = g^* + [\nabla_{\mathbf{n}} g(\mathbf{x}^*)]^T [E(\mathbf{n}) - \mathbf{n}^*] \tag{2}$$

$$\text{var}\ (g) = E\{[g - E(g)]^2\} = [\nabla_{\mathbf{n}} g(\mathbf{x}^*)]^T \mathbf{S}\, [\nabla_{\mathbf{n}} g(\mathbf{x}^*)] \tag{3}$$

In the above equations, the symbol E denotes the expectation or an averaging over **n**. **S** denotes the variance-covariance of **n:**

$$\mathbf{S} = E\{[\mathbf{n} - E(\mathbf{n})][\mathbf{n} - E(\mathbf{n})]^T\} \quad (4)$$

Traditionally, we chose **x*** such that response is on target, g(**x***) = g*. This is done on the basis of deterministic calculations. Once **x*** is chosen, $\nabla_{\mathbf{n}}$ g(**x***) is fixed. This fixes the variance of g in Eq. (3). If the variance is unacceptable, we try to reduce it by tightening **S.** This is an expensive approach. By contrast in the robust-design approach, we try to find **x*** that not only ensures response on target but also makes $\nabla_{\mathbf{n}}$ g(**x***) small. This reduces variation in g at no cost since there is no attempt to tighten **S.** This is the less expensive approach.

In FEM, the design response is most likely expressed in terms of basic FEM entities, e.g., displacement **u.** In turn, **u** is typically dependent implicitly on the noise factors. Hence, the vector $\nabla_{\mathbf{n}}$g can be obtained using the chain rule:

$$\nabla_{\mathbf{n}} g = \nabla_{\mathbf{n}}\mathbf{u}\nabla_{\mathbf{u}}g \quad (5)$$

Since g is expressed explicitly in terms of **u,** the vector $\nabla_{\mathbf{u}}$g can always be obtained. Also, from the finite-element equation,

$$\mathbf{f} = \mathbf{K}\,\mathbf{u}$$

$$\partial\mathbf{f}/\partial n_i = (\partial\mathbf{K}/\partial n_i)\,\mathbf{u} + \mathbf{K}\,(\partial\mathbf{u}/\partial n_i) \qquad i = 1, \ldots, m$$

$$\partial\mathbf{u}/\partial n_i = \mathbf{C}\,[\partial\mathbf{f}/\partial n_i - (\partial\mathbf{K}/\partial n_i)\,\mathbf{u}] \qquad i = 1, \ldots, m \quad (6)$$

where $\mathbf{C} = \mathbf{K}^{-1}$ is the inverse of the stiffness matrix. Hence, the matrix $\nabla_{\mathbf{n}}\mathbf{u}$ can be constructed from Eq. (6). With $\nabla_{\mathbf{n}}\mathbf{u}$ and $\nabla_{\mathbf{u}}$g known, the mean and variance of g in the context of FEM may now be obtained:

$$mean\ (g) = g^* + [(\nabla_{\mathbf{u}}g)^T\,(\nabla_{\mathbf{n}}\mathbf{u})^T]\,[E(\mathbf{n}) - \mathbf{n}^*] \quad (7)$$

$$\text{var}\ (g) = [(\nabla_{\mathbf{u}}g)^T\,(\nabla_{\mathbf{n}}\mathbf{u})^T]\,\mathbf{S}\,[\nabla_{\mathbf{n}}\mathbf{u}\,\nabla_{\mathbf{u}}g] \quad (8)$$

Equations. (7) and (8) form the basis for robust design in FEM. The aim is to choose **x*** that ensures that mean (g) = g* and minimizes var (g).

IMPLEMENTATION STEPS

The first step is to translate the voice of the customer into engineering terms. That is, to formulate the function g and its target g*. To illustrate, consider the voice, "The car door must close flush." Figure 3 shows the side view of a car door. The lower half of the door may be considered rigid compared to the upper half. The upper half is shown in two profiles, one with a radius of curvature R, the other with a larger radius R'. The first profile corresponds to the door as opened.

On closing, the door will be deflected outward by the seal force and will take on a profile with a larger radius. This deflected profile must match the body curvature indicated by the second profile. The voice can therefore be translated as follows.

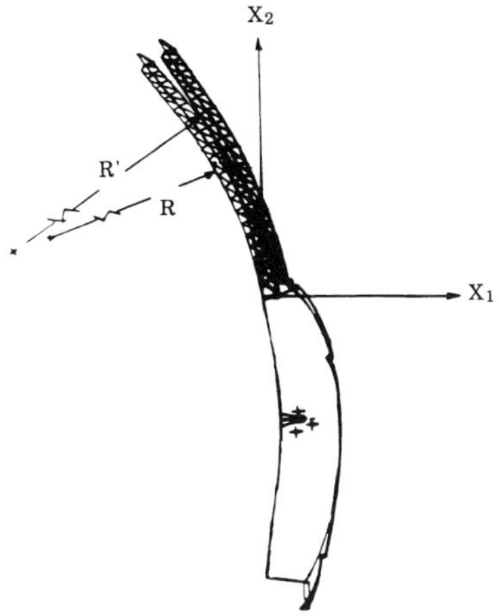

Fig. 3. Profile of a car door.

Set up the coordinate system shown in Fig. 3. Let **u** denote the displacement vector. Then the different profiles may each be approximated by a parabola as

opened: $$x_1 + (x_2)^2/R = 0$$

closed: $$(x_1 + u_1) + (x_2 + u_2)^2/R = 0 \quad (9)$$

target: $$x_1' + (x_2)^2/R' = 0 \quad (10)$$

Subtracting Eq. (10) from Eq. (9) and neglecting the higher-order term, we have the function g describing

flushness: $$g = x_1 - x_1' = u_1 + x_2^2\,(1/R - 1/R') + 2u_2x_2/R \quad (11)$$

The function g is the engineering translation of the voice of the customer. "The door must close flush" means g must equal zero. Hence, the target g* is zero.

The second step is to identify the factors x affecting the function g. Of particular importance is the identification of the noise factors **n** in **x** that are the sources of variation. For the door example above, the obvious factors are the two radii R and R'. R' is decided by body styling. It is probably fixed at this design stage and is no longer a factor. R will exhibit variation from door to door because of error in the fabrication. It is therefore a noise factor. Other factors not obvious from Eq. (11) are those implicit in the displacement **u.** These are the stiffness of the weather-strip that is compressed between the door and the body and the inboard or outboard positions of the door hinges and the latch. The stiffness of the weather-strip is a factor, but not a noise factor. This is because the batch-to-batch variation in dimensions and material properties of the weather-strip are small. The inboard or outboard positions of the hinges and the latch determine the amount of compression in the weather-strip. They are definitely noise factors because they exhibit variation during assembly. Once the noise factors are identified, their variance-covariance should be determined because they are used in subsequent steps.

The third step is to determine which noise factors are the big contributors. An FEM procedure is performed to construct the matrix $\nabla_{\mathbf{n}}\mathbf{u}$. For the car-door example, the positioning of the two hinges and the door latch can be modeled as applied loads. Hence, $\partial(\mathbf{K})/\partial n_i = 0$; and for $n_i = f_l$

$$\partial f_k/\partial n_i = \partial f_k/\partial f_l = \delta_{kl}$$

where l is the nodal number assigned to the ith noise factor. Following Eq. (6), an element of the matrix $\nabla_{\mathbf{n}}\mathbf{u}$ is given by

$$\partial u_j/\partial n_i = c_{jk}\ \delta_{kl} = c_{jl}$$

With $\nabla_{\mathbf{n}}\mathbf{u}$ now known and $\nabla_{\mathbf{u}}g$ able to be computed from Eq. (11), $\nabla_{\mathbf{n}}g$ can be computed from Eq. (5) and var (g) from Eq. (8). Furthermore, if the noise factors are statistically independent, the covariance of **n** is zero. The variance of g simplifies to

$$\text{var}\ (g) = \sum\ [(\partial g/\partial n_i)\ (\sigma_i)]^2 \qquad i = 1, \ldots, m \tag{12}$$

where σ_i is the standard deviation of the ith noise factor. Each term of the series in Eq. (12) represents the contribution by each noise factor to the total variance of g. By comparing their relative magnitude, we can determine which noise factors are the big contributors.

The fourth step is to optimize the design. A new set of nominal values $\mathbf{x}^*$ is sought to reduce the sensitivity to the big contributors and thus further reduce the variance of g. At the same time, tolerances of the small contributors can be opened up to reduce the design cost. Steps three and four are repeated until a design with an acceptable level of variation and cost is achieved.

For existing designs, where opportunity for optimization is not available, the first three steps should still be performed to analyze the design sensitivity to variability. It helps to define key noise factors and promote effective control.

The first two steps comprise the problem-definition phase. They are the most important steps because the success of meeting the customer expectation hinges on the proper translation of the customer's voice into engineering terms. More effort should be spent on this phase. The last two steps comprise the problem-solution phase. They can be made routine by interfacing optimization codes with the existing FEM code.

AN EXAMPLE

In this example, we will investigate the variation in flushness of a sliding door of a van relative to its neighboring panels. As the design is already fixed at this stage, we will not be able to optimize the design to achieve robustness. Instead, we will analyze the sensitivity of flushness to variability and identify key factors responsible for it. This will lead to effective process control.

Figure 4a shows the schematic of a van. Shown shaded is the sliding door. Two brackets with rollers are attached to the sliding door at locations marked (a) and (d). These brackets support the sliding door. The rollers run along tracks built in the body and guide the door along its trajectory during opening or closing. Two pins are attached to the sliding door at locations marked (b) and (c). These pins engage corresponding pockets in the body pillar to secure the door when closed. The latch, marked (e) in the figure, is the third point in the door that is secured to the body. The positioning of the upper and lower brackets, the upper and lower pins, and the latch determine the position of the sliding door relative to the body opening when the door is closed.

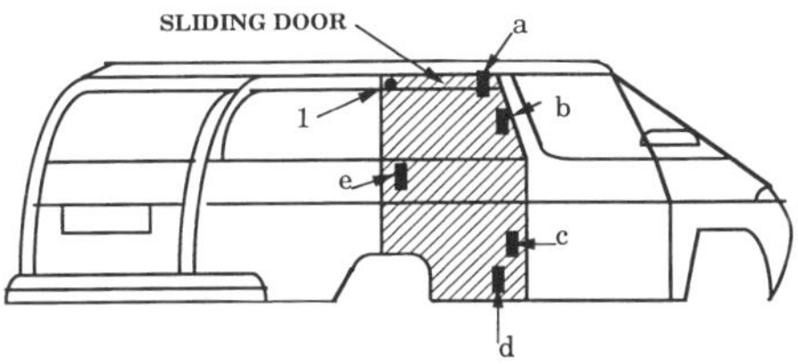

Fig. 4a. Side view of van.

Door positioning has important consequences on the function of the door. Since it determines the gap between the door and body, door positioning affects the amount of compression in the weather-strip between the door and the body. Too little compression results in water leak and wind-noise transmission. Too large compression results in high door-closing effort. Door positioning also affects the door flushness relative to its neighboring panels. A flush door looks good. It also eliminates aerodynamic noise. Experience indicates that a flushness that is off by 1 mm is sufficient to induce turbulence and generate perceptible wind noise. Flushness at point (1) is particularly important because the point is close to passengers' ears. Figure 4b shows flushness at this point for forty sliding-door assemblies. It exhibits a wide scatter in flushness: from 2.5 mm inboard (negative) to 2.5 mm outboard (positive). We shall use flushness as a surrogate to door positioning because assemblers use it as a visual indicator to position the door on the assembly line.

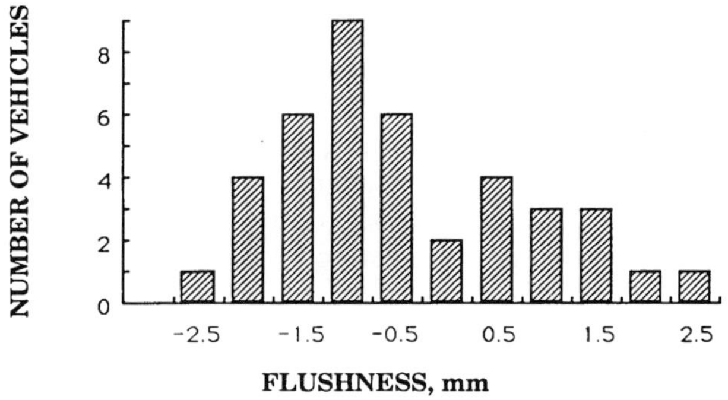

Fig. 4b. Flushness of door at point (1).

The variation in door flushness during assembly comes from the deviation from the nominal positions of the two brackets, the two pins, and the latch. These five deviations constitute the noise factors, which we denote by **n**. To assess their influence, we built an FEM model of the door as shown in Fig. 5. Locations of the brackets, the pins, and the latch are as shown. Also shown as a solid curve around the door is the location of the weather-strip. We modeled the weather-strip as springs on rigid foundation. With the door on top of the springs, we enforce a uniform displacement of nodes at the brackets, the pins, and the latch to push the door against the springs until the door assumes its nominal flush position. Displacements are then monitored at the outer edge at points (1–9). Next, we displace nodes at each of the five locations, one location at a time, an additional 1 mm and observe the change in displacement of nine monitor points. The results are shown in Table 1. A row on this table represents a row vector$[\nabla_{\mathbf{n}} g]^T$ in Eqs. (2–3). It is the sensitivity of flushness to **n**.

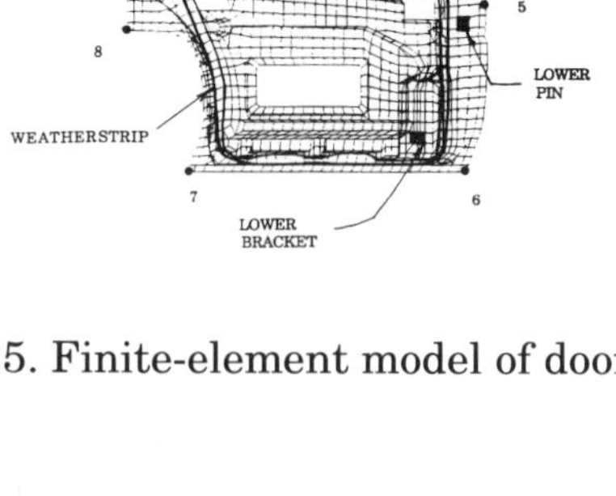

Fig. 5. Finite-element model of door.

Table 1. Change in deflection (mm) at door periphery due to 1 mm change in inboard displacement.

	At Upper Bracket	At Upper Pin	At Lower Pin	At Lower Bracket	At Latch
Point 1	-1.081	1.327	-1.891	1.597	-0.686
Point 2	-1.306	0.883	-0.913	0.654	-0.213
Point 3	-0.670	-0.718	0.635	-0.381	0.099
Point 4	0.202	-0.778	-0.845	0.445	-0.015
Point 5	0.008	-0.039	-0.982	0.051	-0.037
Point 6	-0.079	0.316	-0.451	-0.911	0.116
Point 7	0.332	-0.670	2.367	-2.275	-0.675
Point 8	-0.233	0.301	-0.568	0.548	-1.006
Point 9	-0.600	0.746	-1.239	1.105	-0.869

The next step is to obtain the variance of **n**. During fabrication, doors are periodically checked against a master for flushness. Figure 6 shows the 13 locations where a door is checked. Shown in parenthesis is a typical set of data for one door. These 13 numbers, which we denote by $\Delta\mathbf{z}$, represent the deviations of the points from their nominal cross-car positions. We can see from this set of data that if an assembler were to fit this particular door to a perfect body opening, he would have to rotate this door about a line connecting point (2) to point (8) and translate the entire door a bit outboard. In so doing, the brackets, the pins, and the latch will deviate from their nominal positions by an amount **n.** We can estimate **n** by modeling the door-fitting process as a least square fitting of the door to the opening as follows.

A point in the door that deviates Δz amount can be brought back to its (x,y) nominal position by a rigid-body motion of the door of T_z translation, R_y rotation about the y-axis, and R_x rotation about the x-axis:

$$\Delta z = -[1, -x, y]\ [T_z, R_y, R_x]^T$$

Thirteen points that deviate $\Delta\mathbf{z}$ amount can be brought back to their $(\mathbf{x},\mathbf{y})$ nominal positions by a rigid-body motion in the least square sense:

$$\Delta\mathbf{z} = -\mathbf{A}\ [T_z, R_y, R_x]^T$$

$$[T_z, R_y, R_x]^T = -[\mathbf{A}^T\mathbf{A}]^{-1}\ \mathbf{A}^T\ \Delta\mathbf{z}$$

Here **A** is a 13 × 3 matrix, each row of which has the form $[1, -x_i, y_i]$; (x_i, y_i) is the nominal position of the ith check point; i = 1, . . . , 13. For example, the nominal position of the 5th check point in mm is (4329,1743). The 5th row in the **A** matrix is [1,–4329,1743].

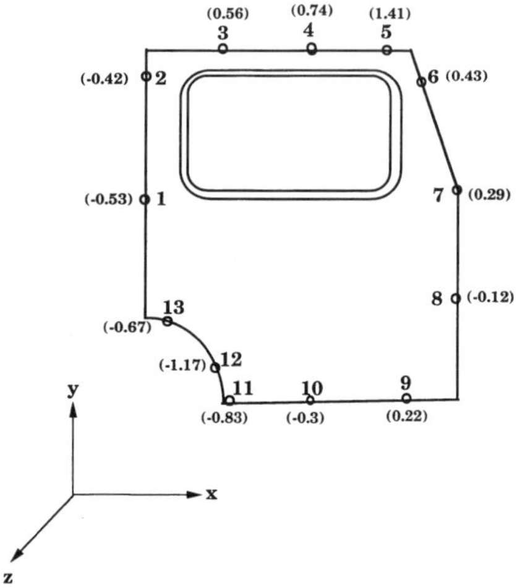

Fig. 6. Flushness around the door.

As the door goes through the rigid-body motion described by $[T_z, R_y, R_x]^T$, the five noise factors will deviate from their nominal positions by

$$\mathbf{n} = \mathbf{B}\ [T_z, R_y, R_x]^T = -\mathbf{B}\ [\mathbf{A}^T\mathbf{A}]^{-1}\ \mathbf{A}^T\ \Delta\mathbf{z} \qquad (13)$$

Here **B** is a 5×3 matrix, each row of which has the form $[1, -x_i, y_i]$; (x_i, y_i) is the nominal position of n_i, $i = 1, \ldots, 5$. For example, the nominal position of n_1, the upper bracket, is (3629,1757). The 1st row of the **B** matrix is [1,–3629,1757].

Since **A** and **B** are known, Eq. (13) can be solved for **n** given $\Delta\mathbf{z}$. For example, the solution **n** to the given set of data $\Delta\mathbf{z}$ in Fig. 6 is shown in the first row of Table 2. Fifteen additional sets of **n** were solved for 15 additional doors. These are shown in Table 2. Also shown are the average and standard deviation of **n**.

The contribution to variance in flushness at check points, Eq. (1 – 9) by the five noise factors can now be computed by using Eq. (12). For example, the contribution to check point (1) by the deviation in the nominal position of the upper bracket n_1 is obtained by squaring the product of the sensitivity to n_1 (first number of the first row in Table 1); and the standard deviation in n_1 (first number of the last row in Table 2). This gives $(-1.081 \times 0.180)^2 = 0.038$, the first number of the first row in Table 3. Table 3 summarizes the computations. The last column in Table 3 shows the summation of contributions by the five noise factors. Note that the variance in flushness for check point (1) is 0.394 mm^2. This gives a three sigma value of 1.88 mm. This explains 75% of the observed value shown in Fig. 4b. The remaining 25% most likely comes from the variation in the body opening, which this analysis did not address.

Table 4 recaps Table 3 on a percentage basis. In this form, the table is useful in identifying key control factors. For example, Table 3 shows large variations at points (1) and (7). Table 4 pinpoints the lower pin and the lower bracket as

Table 2. Outboard deviation (mm).

Sample	At Upper Bracket	At Upper Pin	At Lower Pin	At Lower Bracket	At Latch
Door–1	-0.548	-0.450	0.008	0.211	0.379
Door–2	-0.506	-0.396	0.012	0.182	0.230
Door–3	-0.430	-0.222	0.164	0.276	-0.150
Door–4	-0.328	-0.435	-0.344	-0.245	0.358
Door–5	-0.588	-0.514	-0.024	0.208	0.526
Door–6	-0.576	-0.517	-0.078	0.134	0.452
Door–7	-0.377	-0.443	-0.257	-0.124	0.427
Door–8	-0.361	-0.417	-0.200	-0.057	0.484
Door–9	0.077	0.050	-0.014	-0.037	-0.003
Door–10	-0.558	-0.437	0.072	0.292	0.423
Door–11	-0.385	-0.277	0.061	0.195	0.157
Door–12	-0.113	-0.117	0.191	0.361	0.790
Door–13	-0.279	-0.337	-0.133	0.005	0.539
Door–14	-0.425	-0.452	-0.072	0.147	0.774
Door–15	-0.333	-0.024	0.574	0.756	0.158
Door–16	-0.635	-0.533	-0.104	0.082	0.196
Average	-0.398	-0.345	-0.009	0.149	0.359
Std Dev	0.180	0.173	0.204	0.222	0.246

Table 3. Contribution to variance in flushness.

At	By Upper Bracket	By Upper Pin	By Lower Pin	By Lower Bracket	By Latch	Total Variance
Point 1	0.038	0.053	0.149	0.126	0.028	0.394
Point 2	0.055	0.023	0.035	0.021	0.003	0.137
Point 3	0.015	0.015	0.017	0.007	0.001	0.055
Point 4	0.001	0.018	0.030	0.010	0.000	0.059
Point 5	0.000	0.000	0.040	0.000	0.000	0.040
Point 6	0.000	0.003	0.008	0.041	0.001	0.054
Point 7	0.004	0.013	0.233	0.256	0.028	0.534
Point 8	0.002	0.003	0.013	0.015	0.061	0.094
Point 9	0.012	0.017	0.064	0.060	0.046	0.198

the big contributors. They are therefore the key control variables. Table 4 also helps identify the less important factors whose tolerances may then be opened up to reduce manufacturing cost. Finally, Tables 3 and 4 serve as a starting point for robust design of a new generation of doors.

Table 4. Percent contribution to variance.

At	By Upper Bracket	By Upper Pin	By Lower Pin	By Lower Bracket	By Latch
Point 1	9.58	13.43	37.75	32.03	7.21
Point 2	40.19	17.09	25.29	15.43	2.00
Point 3	26.59	28.40	30.75	13.17	1.08
Point 4	2.23	30.82	50.33	16.60	0.02
Point 5	0.01	0.11	99.36	0.32	0.20
Point 6	0.38	5.60	15.80	76.70	1.52
Point 7	0.67	2.53	43.67	47.98	5.15
Point 8	1.87	2.90	14.30	15.84	65.09
Point 9	5.87	8.44	32.22	30.48	22.99

REFERENCES

1 W. K. Liu, T. Belytschko and S. Mani, Int. J. Numer Meth. Eng., 23 (1986) 1383–1845.
2 T. A. Cruse, O. H. Burnside, Y. T. Wu, E. Z. Polch and J. B. Dias, Comput. and Struct., 29, no. 5 (1988) 891–901.
3 H. L. Oh, Variation tolerant design, in: Proc. ASME Winter Annual Meeting, Boston, Mass., PED-Vol. 27, 1987, pp. 137–145.

Quality through Engineering Design - W. Kuo (Editor)

Performance Analysis of Multipad Disc Brakes by Using the Finite-Element Method

Dampur Srinivasulu

Development Engineer, Research and Development, BEML, KGF, India

Abstract

The actuating disc of a multipad disc brake receives most of the heat developed during braking. Thermal expansion causes the actuating disc to deflect and move away from the friction disc, which results in reduction of the contact area and leads to inefficient braking performance. By providing a suitable rake angle on the friction disc, adequate contact area can be maintained during deflection of the actuating disc at higher temperatures.

The objective of this work is to find an optimum rake angle for the friction disc. This calls for complete thermal analysis of multipad disc brakes, which will be carried out by using finite-element analysis. The finite-element method is a unified procedure for the numerical solution of physical problems in high-technology engineering. It has gained acceptance as a powerful analysis and design aid in diverse areas because of its versatility and adaptability.

INTRODUCTION

A multipad disc brake consists of two actuating discs separated by steel balls located in conical recesses provided in the inner surface of the actuating discs. Two rotating friction discs having a large lining area are placed on either side of the actuating discs. These friction discs are driven by a splined shaft that passes through the housing. A schematic diagram of the multipad disc brakes used in a tractor is shown in Fig. 1.

The brake is actuated by means of a mechanical linkage connected to the actuating disc. The principle of operation is shown in Fig. 2. When the mechanical linkage is pulled, the actuating discs rotate in opposite directions to one another, and the rotation of the actuating disc is limited by a lug in the housing. This causes the balls to move up along the ramp in the conical recesses. The plates are displaced axially outward, which results in the friction disc coming in contact with the actuating disc and housing. When the brake is released, the retracting spring will restore the friction discs to their original position. This type of brake is equally effective in forward and reverse. The brake unit is totally enclosed in a casing to provide protection against mud and water.

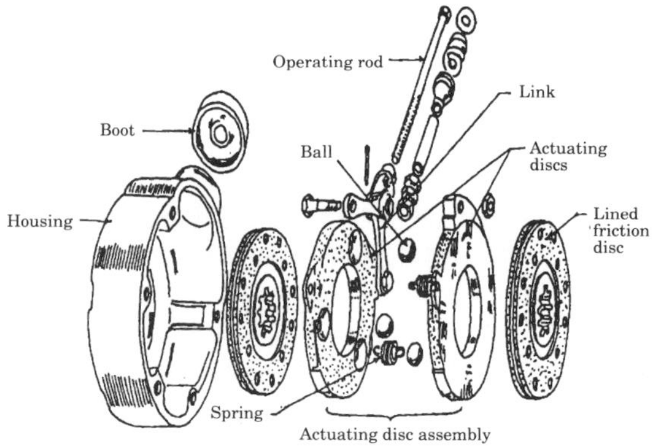

Fig. 1. Multiple-plate disc brake.

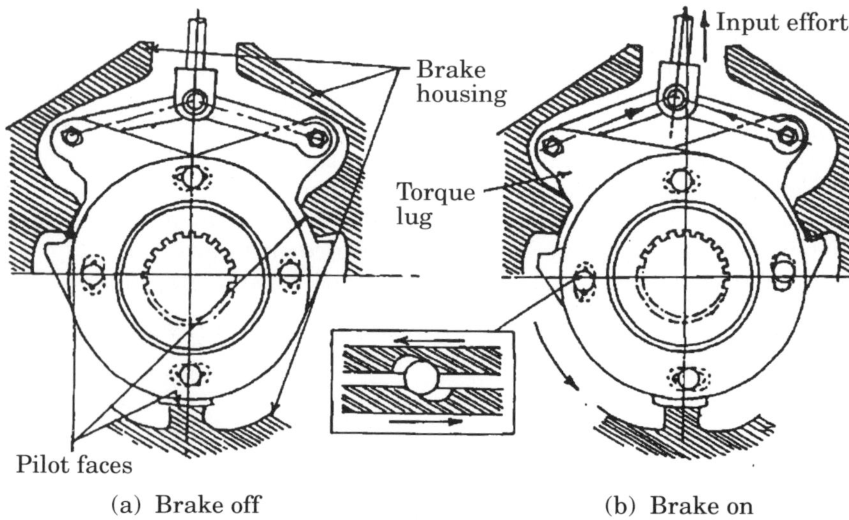

Fig. 2. Schematic diagram showing the principal operation of multiple-plate disc brake.

Continuous application of brakes causes deflection of the lining and disc because of variation in thermal expansion, leading to loss of contact area. Also, the coefficient of friction between the lining and the disc decreases as the temperature increases. Repeated application of the brakes results in less braking efficiency. This phenomenon is known as brake fade.

The actuating disc of a multipad disc brake receives most of the heat developed during braking. To prevent brake fade, the actuating disc must rapidly dissipate heat by free convection. The rate of heat dissipation depends on the amount of air flow and the surface area over which heat is dissipated. In a multipad disc brake, the rate of heat dissipation is very poor because tractors are slow and hence the amount of air flow is not appreciable; the brake is completely enclosed in the casing, and heat generation increases because there is a loss of contact area caused by the thermal deflection of the actuating disc.

In the present work, thermal deflections of actuating discs are determined theoretically by simulating the conditions of a drag test on the brakes. Then a drag test is used to evaluate braking performance. In this test, the vehicle is allowed to descend on a 10% gradient over a distance of 1 km at a speed of 24 kmph. Brakes are applied to maintain the speed of 24 kmph. Temperatures are predicted at every 10-sec time step of a 150-sec drag test. After each time step, the loss of contact area is taken into consideration. It is assumed that when the axial displacement exceeds 0.05 mm, no heat input is given to the deflected area, which will not be in contact during the remaining time interval. Reduction in contact area results in increased heat input. This increase in heat input is taken into account for the next time step.

ANALYTICAL FORMULATION

To arrive at the variations in temperatures and the corresponding thermal deflections with time, we carry out a complete thermal analysis for the actuating disc of the multipad disc brake. A computer program is developed in FORTRAN on a VAX system. The flow chart of the program is shown in Fig. 3.

The governing differential equation for the heat conduction process is

$$\frac{\partial}{\partial x}\left(K_x \times \frac{\partial T}{\partial x}\right)+\frac{\partial}{\partial y}\left(K_y \times \frac{\partial T}{\partial y}\right)+\frac{\partial}{\partial z}\left(K_z \times \frac{\partial T}{\partial z}\right)+Q=c\times\frac{\partial T}{\partial t} \tag{1}$$

where K_x, K_y, K_z = conductivities in x, y, z directions.
T = temperature, C
t = time, sec
Q = heat generated, kcal/mm^3- C

This equation can be solved and can be expressed in matrix form as

$$[C]\frac{\partial T}{\partial t}+[K]\{T\}+\{F\}=0 \tag{2}$$

where $[C]$ = capacitance matrix

$$=\int_V [N]^T[N]dV$$

$[K]$ = stiffness matrix

$$=\int_V [B]^T[D][B]dV+\int_S h[N]^T[N]dS$$

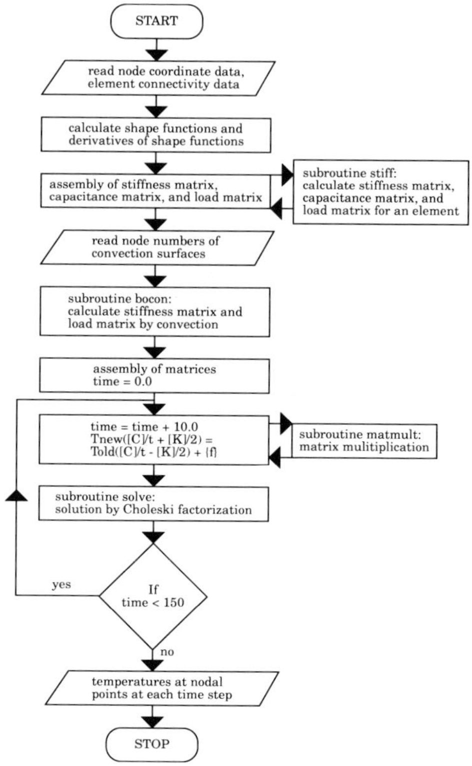

Fig. 3. Flow chart.

$$\{F\} = \text{load matrix}$$
$$= \int_V Q[N]^T dV + \int_S Q[N]^T dS - \int_S hT_a[N]^T dS$$

where $[N]$ = matrix of shape functions
$[D]$ = matrix of conductivities
$[B]$ = derivatives of shape functions
T_a = surrounding temperature in the housing, C

The shape function for the eight-noded brick element is

$$N_B = \frac{1}{8}(1+E_0)(1+Z_0)(1+S_0) \quad (3)$$

where E = EE_B; $Z_0 = ZZ_B$; $S_0 = SS_B$; and where E_B, Z_B, S_B are local coordinates for the unit cube. The stiffness matrix for an element is

$$[K]=\int_V [B]^T[D][B]dV=\int_{-1}^{1}\int_{-1}^{1}\int_{-1}^{1}[B]^T[D][B]dEdZdS\det[J]$$

where det[J] = determinant for Jacobian matrix. The capacitance matrix for an element is

$$[C]=\int_V [N]^T[N]dV=\int_{-1}^{1}\int_{-1}^{1}\int_{-1}^{1}[N]^T[N]dEdZdS\det[J]$$

The load matrix for an element caused by heat generation is

$$\begin{aligned}\{F_1\}&=\int_V Q[N]^T dV+\int_S Q[N]^T dS\\&=\int_{-1}^{1}\int_{-1}^{1}\int_{-1}^{1}Q[N]^T dEdZdS\det[J]+\int_{-1}^{1}\int_{-1}^{1}Q[N]^T dEdZ\det[J]\end{aligned}$$

The load matrix for an element caused by convection is

$$\{F_2\}=\int_S hT_a[N]TdS=\int_{-1}^{1}\int_{-1}^{1}hT_a[N]^T dEdZ\det[J]$$

Load matrix of $\{F\}=\{F_1\}+\{F_2\}$

These matrices are assembled into a single matrix by a subroutine of size n × n, where n is the total number of nodes. The general equation for the finite-element method is

$$[C]\frac{\partial T}{\partial t}+[K]\{T\}+\{F\}=0$$

By using the Crank-Nicholson finite-difference formulas, the equation can be written as

$$\frac{[C]}{\Delta t}+\frac{[K]}{2.0}\{T_{new}\}=\frac{[C]}{\Delta t}-\frac{[K]}{2.0}\{T_{old}\}-\{F\} \tag{4}$$

This equation is used to calculate the temperatures in time steps. The method adopted to solve the final linear equations is Choleski factorization.

INPUT DATA

The details of the multipad disc brake used for analysis include:

Brake size = 165 × 89 mm
Vehicle speed = 24 kmph
Number of actuating discs = 2
Number of friction discs = 2
Number of friction surfaces = 4
Total energy input = 50 HP-min
Energy absorbed per actuating-disc surface, with 95% efficiency
= 211.06*0.95/4 = 50.13 kcal.

Heat generated per unit volume per unit time

$$= 2.215*10^{-4} \text{ kcal/mm}^3\text{-sec}$$

The rate of heat dissipation is low in multipad disc brakes because the brake is completely enclosed in the casing. Thus, the temperature inside the casing is greater than the atmospheric temperature. A probable approximation is taken between the surrounding temperature T_a in the casing and time, as shown in Fig. 4.

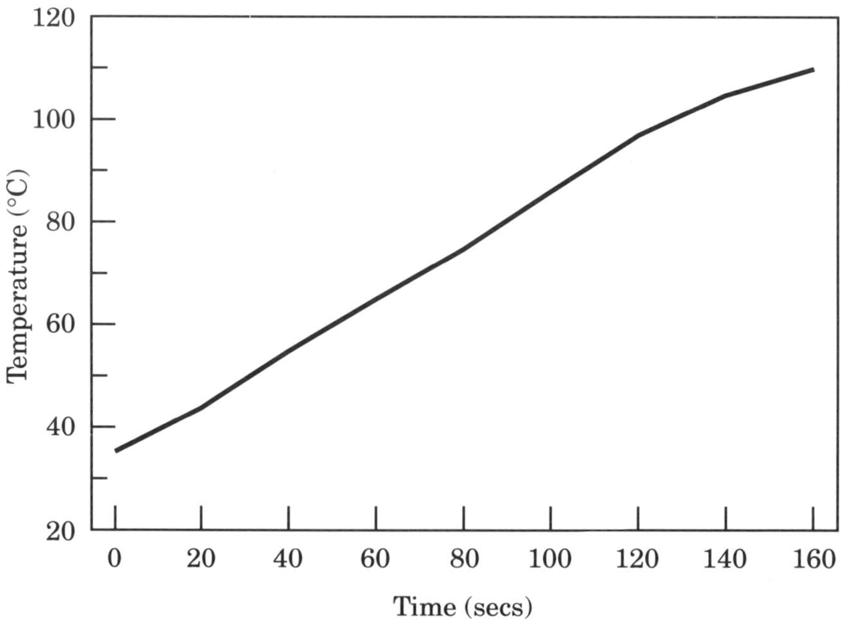

Fig. 4. Variable surrounding temperature.

As the temperature rises, the resulting deflection of the disc leads to loss of contact area. This apparently results in turn in an increase of cooling area, which causes very significant changes in the values of the heat-transfer coefficient h. The variable heat-transfer coefficient is calculated by Newton's differential equation. The relation between time and the heat-transfer coefficient is given in Fig. 5.

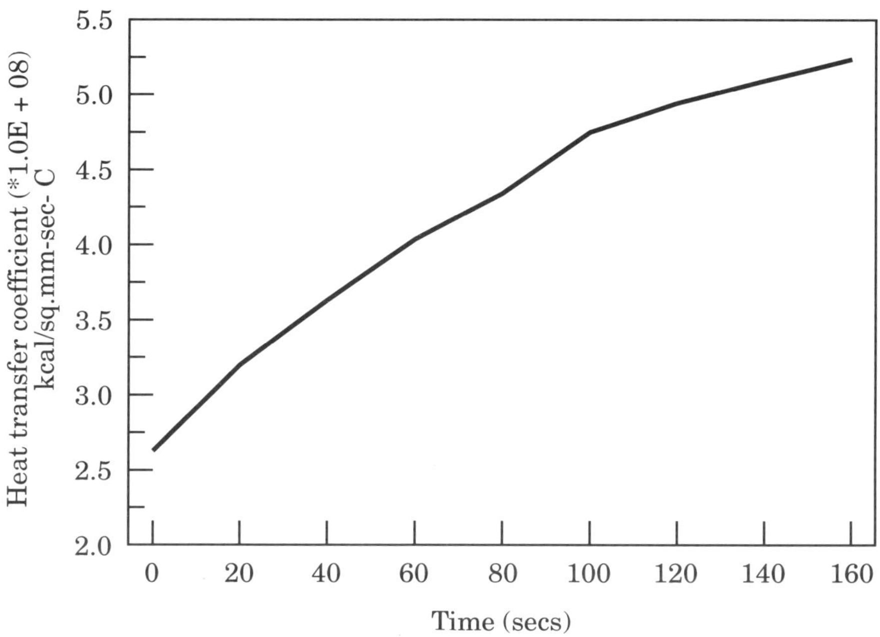

Fig. 5. Variable heat-transfer coefficient.

EXPERIMENTAL PROCEDURE

An experiment, using a microprocessor-controlled dynamometer, is carried out to determine the temperatures in the actuating disc of a multipad disc brake. The experimental setup is briefly described: Two chromel-alumel thermocouples are installed in the actuating disc, 3 mm from the contacting surface. The brake is assembled and mounted on the adaptor of the dynamometer. The brake is adjusted to ensure a 0.05-mm clearance between the friction disc and the actuating disc.

The conditions of a drag test are simulated. When the flywheel of the dynamometer reaches 408 rpm, which corresponds to vehicle speed of 24 kmph, the brake is applied to achieve a drag torque of 357 N-m. The torque is applied constantly to maintain the speed. The temperatures are measured every 10 secs of the 150-sec drag test. Figure 6 shows temperature distribution across the thickness of the actuating disc.

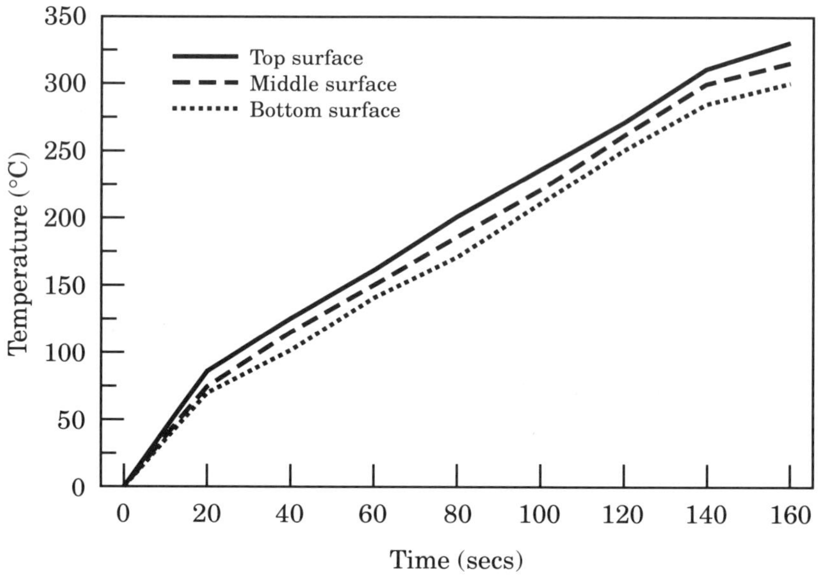

Fig. 6. Temperature distribution in the actuating disc during 150 secs of drag test.

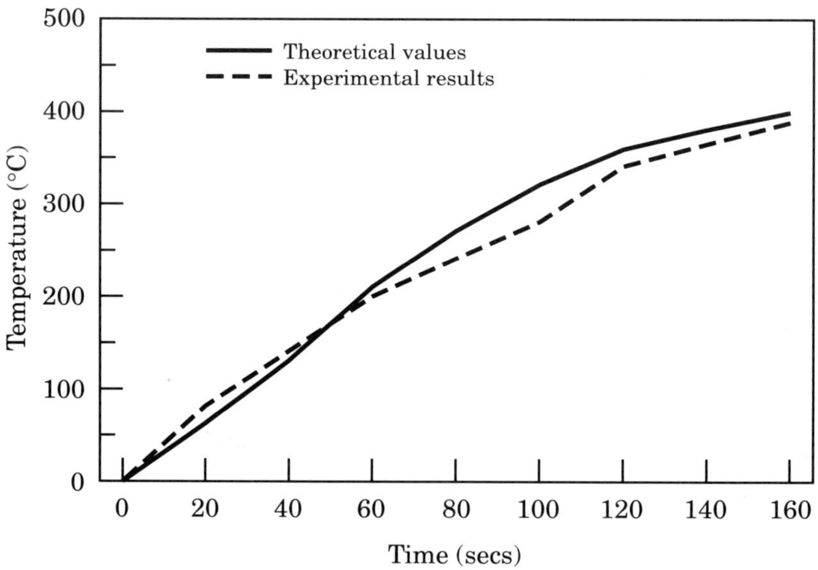

Fig. 7. Comparison of theoretical and experimental results.

Temperatures are obtained from the experimental analysis by using a microprocessor-controlled dynamometer. Figure 7 compares the variation of temperature with time at a particular location, both theoretically and experimentally. From the figure, we see that the theoretically determined temperature values are close to the experimentally obtained values.

CONCLUSIONS

Theoretical and experimental thermal analysis of multipad disc brakes indicates that the heat dissipation can be effectively improved by optimizing the design parameters of the friction discs. The analysis of temperature distribution and the corresponding thermal deflections of an actuating disc calls for an optimum rake angle on the friction disc. This rake angle ensures adequate contact area for heat dissipation at higher temperatures, resulting in improved brake efficiency.

Quality through Engineering Design - W. Kuo (Editor)

Design and Analysis of Tilting-Pad Journal Bearings Using the Finite-Element Method

S. R. Jayaram[a] and B. S. Prabhu[b]

[a]Department of Mechanical Engineering, Malnad College of Engineering, Hassan, 573 201, Karnataka, India.

[b]Department of Applied Mechanics, Indian Institute of Technology, Madras, 600 036, India.

Abstract

This paper presents an approximate solution technique for the design and analysis of tilting-pad journal bearings, considering the thermal effects on viscosity and the effect of pad deformation on film thickness. The method is based on the numerical solution of the Reynolds' equation by the finite-element method, using a one-dimensional temperature profile. The pressure, elastic, and energy equations are solved simultaneously at each step of iteration; and the Vogel-Cameron equation is used to determine local viscosity. Performance parameters are evaluated according to conventional methods.

Performance data for the tilting-pad journal bearing of a horizontal rotor have been generated. The effects of the eccentricity ratio and of the L/D ratio on performance are studied for different values of bearing preload. Findings are compared with available data.

NOMENCLATURE

- c radial clearance (m)
- e eccentricity (m)
- h film thickness (m)
- $\overline{h}$ h_{min}/c
- m bearing preload $(1 - c_b/c_p)$
- p pressure distribution
- $\overline{p}$ W/BL (pa)
- q volume flow rate/length (m^2/sec)
- B circumferential length of the bearing
- D diameter of the bearing
- H power loss (watt)
- $\overline{H}$ $H.L/D.\dfrac{1}{\omega^2 \eta_s c^3}$

J pressure functional
L length of the bearing (m)
P center-line pressure (pa)
Q total flow rate (m^3/sec)
R journal radius (m)
S Sommerfeld number ($\bar{p}\psi^2/h\omega$)
T temperature (° C)
$\bar{T}$ T/T_s
U surface velocity of the journal (m/sec)
W load (N)
ε eccentricity ratio (e/c)
η absolute viscosity (N/m^2)
ψ c_b/R
ω rotational speed (rad/sec)

Subscripts and superscripts
b bearing
t trailing edge
c cavitation
p pad
i i^{th} pad
l leading pad
z component in axial direction
s supply
θ component in circumferential direction

INTRODUCTION

In contrast to fixed bore cylindrical bearings, tilting-pad journal bearings are inherently stable (see Figs. 1–2). The great stability of these bearings is attributed to the absence of cross-coupled fluid film forces, which are the strongest source of instability in journal bearings. As a result, tilting-pad journal bearings have received considerable attention for applications in modern, high-speed, flexible rotor systems, such as turbines, compressors, and flywheel storage systems.

In the present analysis, circumferentially varying viscosity instead of global viscosity is assumed, and the effect of pad deformation on bearing performance is considered. Not only does elastic deformation depending upon load and speed alter pad clearance but also service clearance under certain pressure gradients may be 50%–100% larger than design clearance. This pad deformation causes distinct changes in bearing performance.

The finite-element method has become an effective tool in the analysis of engineering design problems. Of late the method's application to lubrication problems has gained importance because of its ability to solve a variety of problems. Booker and Huebner [1] applied the finite-element method to the general lubrication problem and provided a full and systematic description of the procedure. Allaire et al. [2] developed a matrix approach to finite elements directly from the variational principle for pressure. This method automatically produces a minimum band width of algebraic equations. A complete elastohydrodynamic, finite, journal-bearing problem is solved in [3]. For thermohydrodynamic (THD) problems, Huebner [4] used the finite-element method. In the present analysis, the finite-element technique is adopted for a thermoelastohydrodynamic (TEHD) solution of a tilting-pad journal bearing problem.

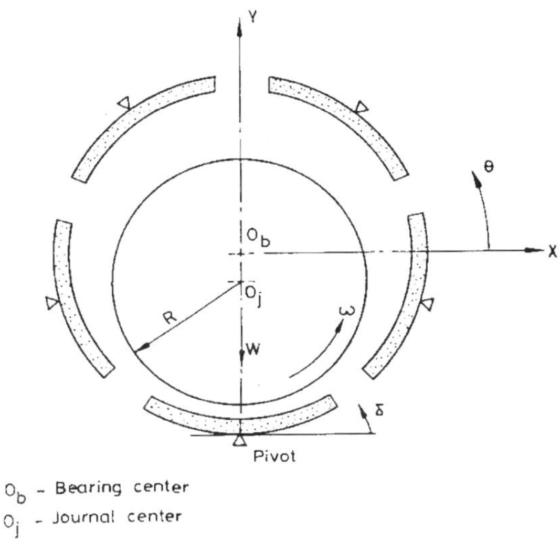

Fig. 1. Tilting-pad bearing geometry.

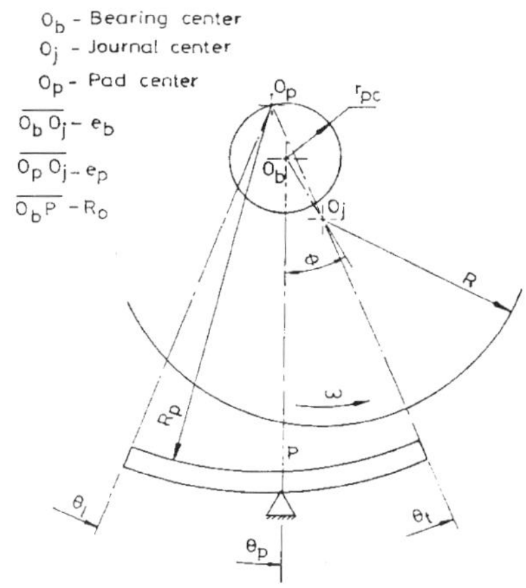

Fig. 2. Single pad details.

ANALYSIS

Instead of the pad assembly approach, which becomes excessively complex for TEHD solutions, the Reynolds equation is solved for the entire bearing. A pressure distribution p in terms of the center-line pressure P is assumed [5], and the bearing is converted into a one-dimensional form:

$$p(\theta,z)=P(\theta,0)\left[1-(2z/L)^n\right]; \quad -L/2\le z\le L/2 \tag{1}$$

The exponent n, which is a function of L/D, is taken as 2 for all practical bearings having $0.25 \le L/D \le 1.0$ [5]. Thus, the Reynolds equation is reduced to an ordinary differential equation.

McEttles [6] has applied beam theory to calculate pad distortion. Recently Zheng et al. [7] compared several deformation models and recommended the curved beam model for priority in application. In the present analysis each pad is considered a curved beam supported at the pivot and loaded by the lubricant film pressures. The elastic deformation of the pads is calculated by the well-known finite-element formulation [2]. The nodal displacements thus obtained are used to modify the film thickness at each iteration.

Calculation of pressure

Variational principles [2,8] can be used to show that the solution of the Reynolds equation also minimizes the hydrodynamic pressure functional after integration with respect to the axial coordinate, by

$$J(p)=\int_{\theta_l}^{\theta_t}\left[\frac{G_\theta h^3}{R^2\eta}\left(\frac{L}{2}-\frac{L}{n+1}+\frac{L}{2(2n+1)}\right)\left(\frac{\partial p(\theta,0)}{\partial\theta}\right)^2\right.$$
$$\left.+\frac{G_z h^3}{\eta}\frac{2n^2}{L(2n-1)}[p(\theta,0)]^2+\left(\frac{\omega}{2}\frac{\partial h}{\partial\theta}+\frac{\partial h}{\partial t}\right)\left(L-\frac{L}{(n+1)}\right)p(\theta,0)\right]d\theta \tag{2}$$

For laminar flow, the turbulence correction factors G_q and G_z are taken as 1/12.

When the pressure functional is approximated by Eq. (1), the functional reduces to a one-dimensional integral and can be written as

$$J(p)=\int_{\theta_l}^{\theta_t}\left[\left(\frac{1}{2}\right)K\left(\frac{\partial p}{\partial\theta}\right)^2-\left(\frac{1}{2}\right)Xp_\theta^2-Yp_\theta\right]d\theta$$

where,

$$K=\frac{G_\theta h^3}{R^2}\left(L-\frac{2L}{n+1}+\frac{L}{2n+1}\right)$$
$$X=-\frac{4G_z h^3}{R^2}\left(\frac{n^2}{L(2n-1)}\right) \tag{3}$$
$$Y=\eta\left(\frac{\omega}{2}\frac{\partial h}{\partial\theta}+\frac{\partial h}{\partial\theta}\right)\left(\frac{L}{n+1}-L\right)$$

The functional can be broken up into integrals over individual elements on a pad, and within each element linear interpolation functions can be used to express pressure in terms of nodal pressures. Nodal pressures are obtained by minimizing J(p) with respect to these pressures as $(\partial/\partial_{pi})J(p) = 0$. This leads to a set of simultaneous equations for each of N elements, and when assembled over all the elements, the system of equations can be written

$$[B]\{p\}=\{c\} \tag{4}$$

Equation (4) can be solved by direct methods for tridiagonal matrices, such as the banded Gaussian elimination technique. Pressure boundary conditions are given in Fig. 3.

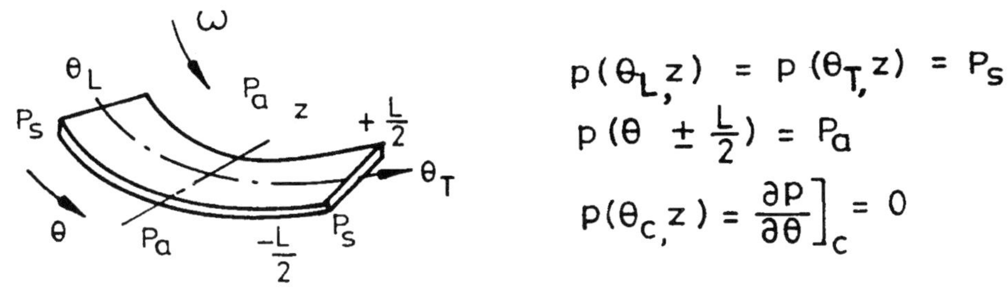

Fig. 3. Pressure boundary conditions.

Calculations of nodal deformations

The well-known finite-element formulation [9] for a curved beam has been used in this analysis to calculate nodal deformations. The displacement model is formulated as

$$Z(\theta) = (1 - 3\xi^2 + 2\xi^3)Z_1 + \ell(\xi - 2\xi^2 + \xi^3)\left(\frac{\partial Z}{\partial \theta}\right)_1$$

$$+(3\xi^2 - 2\xi^3)Z_2 + \ell(-\xi^2 + \xi^3)\left(\frac{\partial Z}{\partial \theta}\right)_2$$

and (4a)

$$\begin{Bmatrix} \dot{\varepsilon}_\theta \\ r_\theta \end{Bmatrix} = \begin{Bmatrix} Z/R_p \\ \dfrac{1}{R_p^2}\left(\dfrac{\partial^2 Z}{\partial \theta^2}\right) \end{Bmatrix}$$

where ξ ranges from 0 to 1, ε = strain, r_θ = change in pad radius R_p, Z_θ = bending displacement, and l = length of the element. These nodal displacements are used to modify film thickness at each iteration (Appendix 1).

Calculation of bearing temperature

Nodal temperatures around the bearing can be found by means of the temperature gradient equation,

$$\frac{\partial T}{\partial \theta} = \frac{\dfrac{h^3}{12R\eta}\left(\dfrac{\partial p}{\partial \theta}\right)^2 + \dfrac{\omega^2 R\eta}{h}}{\rho C_v\left[\dfrac{\omega Rh}{2} - \dfrac{h^3}{12R\eta}\left(\dfrac{\partial p}{\partial \theta}\right)\right]} \tag{5}$$

and by adopting the marching technique with appropriate boundary conditions.

Calculation of lubricant flow rate

Groove mixing theory has been used in the analysis; that is, a part of the outflow from a pad enters the next pad after mixing with the lubricant in the gap between the pads. The circumferential and the axial flows per unit length at any section of a pad are given by

$$q_\theta = -\frac{h^3}{12R\eta}\frac{\partial p}{\partial \theta} + \frac{h}{2}U$$

and

$$q_z = -\frac{h^3}{12\eta}\frac{\partial p}{\partial z} \tag{6}$$

In-flow to a pad Q_{in} is obtained by Eq. (6) at the leading edge, and the side leakage Q_{side} is found by summing q_z components of all elements on a pad. The flow components from each pad are added to obtain total flow Q_{total} (see Fig. 4).

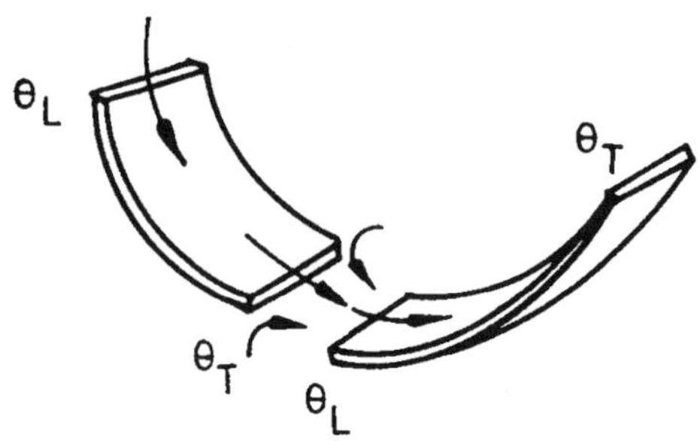

$$Q_{out} = Q_{in} - Q_{side}$$

$$Q_{cool} = Q_{out} + \frac{1}{2}Q_{side}$$

Fig. 4. Flow (groove mixing).

Finite-element model

The finite-element model is formed by dividing each pad into a number of one-dimensional elements of one-degree arc length in the circumferential direction along the center line of the pads. Each element has two nodes. The two nodes on the leading and the trailing edges of a pad are the boundary nodes (see Fig. 5).

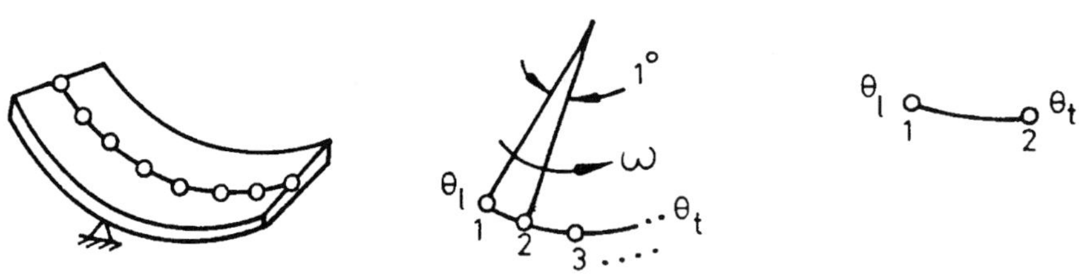

Pad Descretization One element

Fig. 5. Finite-element model.

The unknown nodal pressures on each pad are obtained as a result of minimizing J(P) in Eq. (2). The nodal pressures thus obtained are used to modify film thickness. The load on the pad is obtained by integrating nodal pressures over the surface of the pad. The total bearing load is obtained by vectorial combination of the contributions from each pad.

STEADY-STATE PERFORMANCE PARAMETERS

After obtaining the nodal pressures by solving the Reynolds equation with the finite-element technique and also after modifying the pressures by solving the elastic and energy equations simultaneously, the load on each pad is obtained by integrating the pressures over the surface area of the pad.

The journal equilibrium position is defined in terms of a particular combination of eccentricity ratio and attitude angle, for which the external steady-state load is completely balanced by the generated hydrodynamic force. Once the equilibrium position of the journal has been found, the steady-state performance is evaluated (see Appendix 1).

RESULTS AND DISCUSSION

Performance data have been developed for a 4-pad bearing of a horizontal rotor, with a pad of 75° angular extent, centrally pivoted and load between-the-pads. Findings are compared with available data. The input data required are the same as in Ref. [10]. Each pad is divided into 75 elements, and the CPU time for a run on a Siemens 7580E main frame is 9 seconds.

Results

Tables 1 and 2 give the steady-state performance data for the THD solution for various values of eccentricity ratios and for preload factors of 0 and 0.5. Figure 6 shows the variation of maximum bearing temperature with L/D ratio, and Fig. 7 shows the dependence of the minimum film thickness and radial clearance in THD and TEHD cases. Variation of power loss with speed is shown in Fig. 8, and the effect of load on side flow is brought out in Fig. 9.

Table 1. Steady-state performance data for the THD solution (preload factor = 0).

ε	$\bar{h}$	$\bar{T}$	$\bar{\bar{H}}$	S
0.2	0.8125	1.104	4.27 E13	0.088
0.3	0.7375	1.144	6.27 E13	0.092
0.4	0.6750	1.150	6.65 E13	0.092
0.5	0.5125	1.201	4.93 E14	0.118
0.6	0.5375	1.184	6.87 E14	0.093

Table 2. Steady-state performance data for the THD solution (preload factor = 0.5).

ε	$\bar{h}$	$\bar{T}$	$\bar{H}$	S
0.2	0.8440	1.050	1.14 E13	0.113
0.3	0.7560	1.055	1.29 E13	0.111
0.4	0.6670	1.061	1.03 E13	0.111
0.5	0.5780	1.212	6.08 E13	0.119
0.6	0.4890	1.681	1.94 E14	0.138

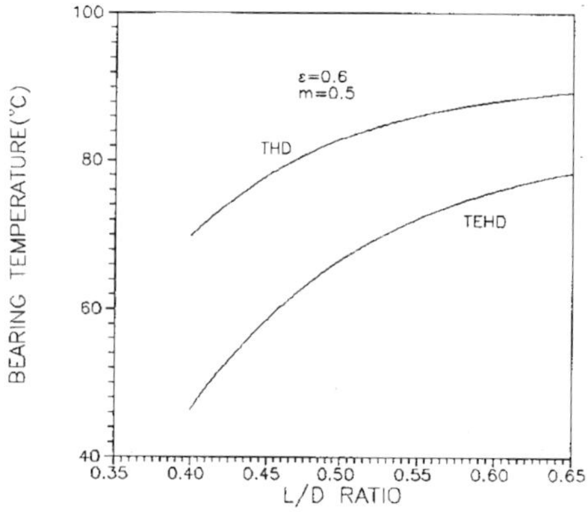

Fig. 6. Variation of maximum temperature with L/D ratio.

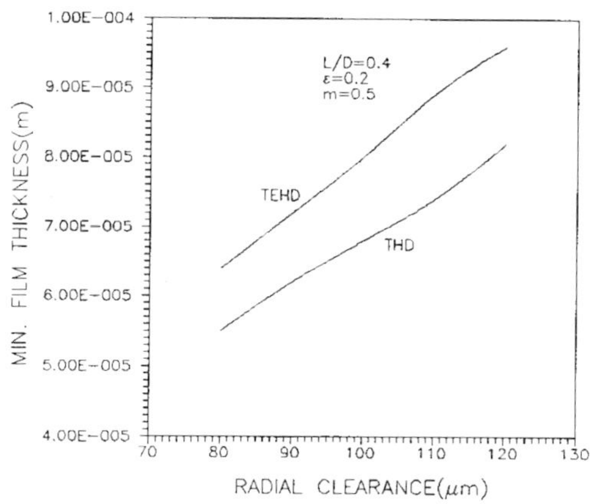

Fig. 7. Effect of radial clearance on min. film thickness.

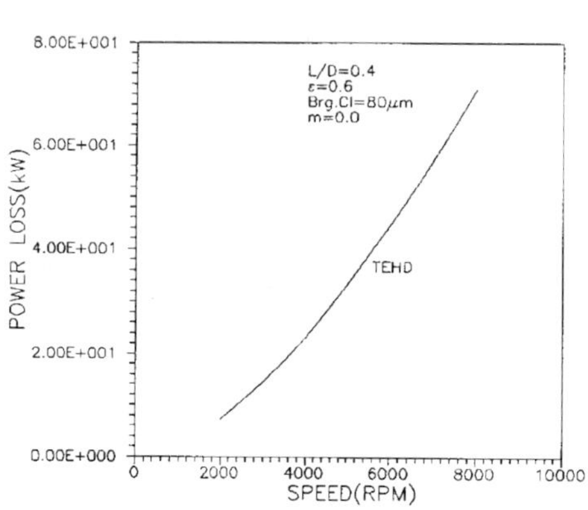

Fig. 8. Variation of power loss with speed.

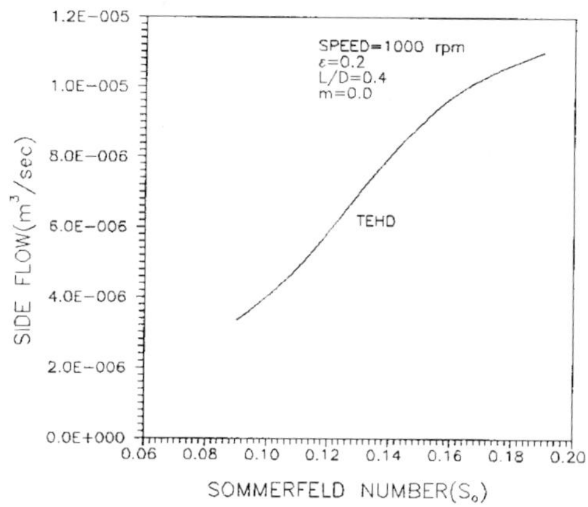

Fig. 9. Effect of Sommerfeld number on side flow.

Discussion

Figure 6 shows that for the same L/D ratio, the TEHD solution predicts lower bearing temperatures than does the THD solution. Likewise the TEHD solution yields a greater minimum film thickness for the same bearing clearance (Fig. 7). As journal speed is increased, power loss is considerably increased, and thermal effects predominate at high speeds (Fig. 8). Side flow continuously increases (Fig. 9) until the load is less than a particular value; then side flow shows a change in the trend [11].

CONCLUSIONS

The finite-element technique is a versatile method for solving complex problems such as THD and TEHD solutions for tilting-pad journal bearing. For rigorous analysis of tilting-pad journal bearings, the thermal effects on both viscosity and elastic deformation of the pads should be considered.

REFERENCES

1 J. F. Booker and K. H. Huebner, J. Lubr. Technol., Trans. ASME (1972) 313–323.

2 P. E. Allaire, J. C. Nicholas and E. J. Gunter, J. Lubr. Technol., Trans. ASME, 99, no. 2 (1977) 187–197.

3 K. P. Oh and K. H. Huebner, J. Lubr. Technol., Trans. ASME, 95, no. 3 (1973) 342.

4 K. H. Huebner, J. Numer. Meth. Eng., 8, no. 1 (1974) 139.

5 J. D. Knight and L. E. Barrett, Trans. ASME, J. Tribol., 110, no. 1 (1988) 128–133.

6 C. M. McEttles, Trans. ASME, J. Lubr. Technol., 102 (1980) 182–192.

7 Z. Zheng, Z. Zhang and Wu, Effect of pad deformation on static performance of journal bearing tilting pad, in: Proc. Internat. Conf. Hydrodynamic Bearings Rotor System Dynamics, Xian, China, 1990, pp. 60–66.

8 J. D. Knight and L. E. Barrett, Trans. ASME, 26, no. 4 (1983) 501–508.

9 O. C. Zienkiewicz, The Finite-Element Method, Tata McGraw-Hill Publishing Co., India, 1987.

10 S. R. Jayaram and B. S. Prabhu, J. Phys. D., 25 (1992) 373.

11 R. Sinhasan and H. N. Chandrawat, Tribol. Inter., 22, no. 5 (1989) 347–353.

Chapter 7
Computing in Design

Issues related to computation, system modeling, and system simulation are presented. This chapter discusses research on optimal sampling procedures, executable CAD/CAM software, a nonlinear optimization technique, polynomial regression, computer simulation, and knowledge-based integrated system approaches for engineering design. Optimal sampling procedures and case studies for power transmission systems, dry cells, and optic fibers are discussed as well. It is shown in this chapter that engineering design is truly a multifaceted subject that requires various computation techniques.

Quality through Engineering Design - W. Kuo (Editor)

A Computational Approach to the Synthesis of an Adjustable Four-Bar, Planar Mechanism for Six-Position, Rigid-Body Guidance

P. M. Padole and J. P. Modak

Department of Mechanical Engineering, Visvesvaraya Regional College of Engineering, Nagpur 440011, India

Abstract

Recently, the authors have suggested graphical approaches for synthesizing an adjustable four-bar, planar mechanism for six-position, rigid-body guidance. Although the graphical approaches give very good insight into the physics of mechanisms, they are time consuming and have poor accuracy. Using the displacement matrix approach, the authors thus present a computer simulation of the graphical approaches suggested for six-position, rigid-body guidance. The numerical results obtained are presented in graphic form with the help of AUTOCAD. Solutions for two case study problems obtained with this technique are presented.

INTRODUCTION

Linkages, cams, gear mechanisms, and their combinations are used in machines to generate predetermined motion programs. The linkage mechanisms offer greater advantages than other mechanisms because they can be readily adjusted to get different kinds of motion from the same mechanism.

Since 1965, many papers have been published about adjustable four-bar mechanisms in relation to various problems. The paper by Ahmed and Waldron [1] gives the graphical technique to synthesize a four-bar mechanism when the coupler is guided through five positions. Along similar lines the authors have suggested graphical approaches for the synthesis of adjustable four-bar mechanisms for six-position, rigid-body guidance [2,3]. Such a synthesis is both time consuming and not very accurate. In the present paper we describe a computer simulation of these graphical approaches and then apply it to a specific case study and present the results in graphical form. In addition, the salient features of the graphical approaches are given [2,3] as a matter of ready reference.

BACKGROUND OF THE PRESENT WORK

Recently, the authors have suggested a graphical approach [2] for synthesizing adjustable four-bar, mechanisms for six-position, rigid-body guidance. The synthesis is possible if there exists at least one moving Burmester point for the given six positions of the coupler. For certain relative orientations of the six positions of the coupler, existence of such a point is possible. Further, two conceptual arguments are put forward suggesting that there could exist a very large number of sets of six positions for which the existence of at least one moving Burmester point could be possible. This high probability of the existence of a moving Burmester point for a very large number of sets of six positions enhances the utility of this contribution to a great extent.

Finally Ref. [2] gives the minimum and necessary condition to ensure the existence of at least one moving Burmester point for the given six positions of the coupler. For three-position, rigid-body guidance all the points on the coupler lie on the circular arcs of corresponding radii; however, in four-position, rigid-body guidance there arc specific points on the coupler plane that only can lie on the circular arcs of different radii for the given four positions. The curve passing through these points is known as a circle point curve. The curve passing through the centers of the corresponding circular arcs is known as the center point curve. For the six positions, existence of a moving Burmester point can be ensured if three such circle-point curves, covering all six positions with three positions common, intersect at one point.

Now suppose the designer has to synthesize a mechanism that has to serve at six different stations (for instance, stations 1–6). If these positions do not satisfy the minimum and necessary condition, there is flexibility to alter any one of the positions (e.g., position 6). Reference [3] suggests a graphical approach to find the region where the position is to be relocated so that the new set of six positions satisfies the minimum and necessary condition and synthesis is possible.

COMPUTER APPROACH

As previously mentioned, synthesizing an adjustable, four-bar mechanism for the given six positions of the coupler is possible if three circle point curves that cover all six positions and have three positions in common intersect at one point. With three positions common, however, many combinations of the three circle point curves are possible, and drawing this large number of curves graphically would be a tedious and time-consuming job. Moreover, the curves drawn would not be very accurate. This problem can be overcome by simulating this graphical approach on the computer. What is useful for the computer simulation is an analytical approach for finding the circle point curves. The following paragraphs briefly describe the important steps involved in this analytical approach.

Let the six positions of the coupler be described by any line PQ on the plane. Then, to describe the six positions, we must specify the coordinates of point P in the respective six positions of the plane and also the inclinations of line Pa with respect to the reference axis. Now, if the plane is displaced from position 1 to position 2 (with combined rotary and translational motion), then in the second position of a plane, coordinates C_{2x} and C_{2y} of any point C can be specified with the help of displacement matrix $[D_{12}]$ if its coordinates C_{1x} and C_{1y} in the first position of the plane are known. This can be given by

$$\overline{C}_2 = [D_{12}]\overline{C}_1$$

where $[D_{12}]$ is a matrix having elements d_{11}, d_{12}, d_{13}, d_{21}, d_{22}, d_{23}, d_{31}, d_{32}, and d_{33}. These elements are the functions of the relative coordinates of point P and the relative orientation. This is applicable for all the positions from position 2 to position 6.

Thus, any point for any position of the plane can be specified in terms of coordinates of that point in the first position of the plane. As previously mentioned, circle point is any point on the plane, four images of which lie on the circular arc for the given four positions of the plane. This means that the distance of that point in all four positions from the center must remain constant, which will be a distance equal to radius of the circle. By applying this constraint and specifying points in the different positions of the plane in terms of coordinates of that point in the first position of the plane, the analytical solution for circle points and corresponding center points results are given by

$$\begin{aligned} &P_{1x}\left[d_{11j}d_{13j} + d_{21j}d_{23j} + (1-d_{11j})P_{ox} - d_{21j}P_{oy}\right] \\ &+P_{1y}\left[d_{12j}d_{13j} + d_{22j}d_{23j} + (1-d_{22j})P_{oy} - d_{12j}P_{ox}\right] \\ &\quad = d_{13j}P_{ox} + d_{23j}P_{oy} - 0.5\left(d_{13j}^2 + d_{23j}^2\right) \end{aligned}$$

for $j = 2$ to 4. Thus, the circle points and center points can be obtained by solving three nonlinear, simultaneous equations. The Newton-Raphson method, which requires initial guesses for the described parameters, is used to solve the nonlinear, simultaneous equations. To ensure the minimum and necessary condition, however, one must draw the various combinations of three circle point curves for the given six positions of the coupler. Therefore, the program is made flexible so that the circle point curves and corresponding center point curves can be determined for any four positions.

If three curves are not intersecting at one point, then the computer program is designed to specify which position is to be altered by minimum deviation and where it is to be located, so that the new set of six positions satisfies the minimum and necessary condition. After ensuring this condition, the synthesis procedure involves splitting the six positions of the coupler in two phases. Phase 1 includes the positions of the coupler plane that are related to the configuration when the fixed pivot is not adjusted, and phase 2 includes those positions when the fixed pivot is adjusted. Table 1 shows the resulting five classes for the synthesis.

Table 1. Classes for six-position, rigid-body guidance.

Class-1	Phase-1 :	Containing 5 positions
	Phase-2 :	Containing 1 position
Class-2	Phase-1 :	Containing 5 positions
	Phase-2 :	Containing 2 positions (one position common)
Class-3	Phase-1 :	Containing 4 positions
	Phase-2 :	Containing 2 positions
Class-4	Phase-1 :	Containing 4 positions
	Phase-2 :	Containing 3 positions (one position common)
Class-5	Phase-1 :	Containing 3 positions
	Phase-2 :	Containing 3 positions

The procedure for getting an input link for all these classes is similar, but the procedure for getting the driving link differs from class to class. The program includes the procedure for all the above mentioned classes. This complete exercise results in a large number of four-bar, adjustable mechanisms for the given six positions of the coupler. For all these mechanisms, however, other parameters, such as transmission angle and mechanical error, may not be within desirable limits [4]. Therefore, another subroutine is added to obtain mechanisms that have a transmission angle within desirable limits. This programming is done in TURBO-BASIC. The program determines the coordinates of the circle points with their corresponding center points for different positions and the coordinates for the fixed and moving pivots of the adjustable, four-bar mechanism. These results are then graphically displayed by using a plotter. The output files of the program are transferred to AUTOCAD with the help of a small program compiled in AUTOLISP.

The graphical display facilitates the view of the synthesized mechanism operation; this is very useful in making the final choice of the mechanism. The flow chart in Fig. 1, shows the important sequence of the computer package.

APPLICATION AND RESULTS

Next the computer package is applied to one of the case studies for which the six positions are given:

$$
\begin{array}{lll}
P_{1x} = 7.00 & P_{1y} = 10.1 & \theta_1 = 0.4861451 \\
P_{2x} = 5.60 & P_{2y} = 7.60 & \theta_2 = 0.7675 \\
P_{3x} = 9.00 & P_{3y} = 8.5 & \theta_3 = 1.431169 \\
P_{4x} = 8.30 & P_{4y} = 7.3 & \theta_4 = 1.570 \\
P_{5x} = 9.5 & P_{5y} = 9.7 & \theta_5 = 1.0001557 \\
P_{6x} = 11.3 & P_{6y} = 10.6 & \theta_6 = 0.93516
\end{array}
$$

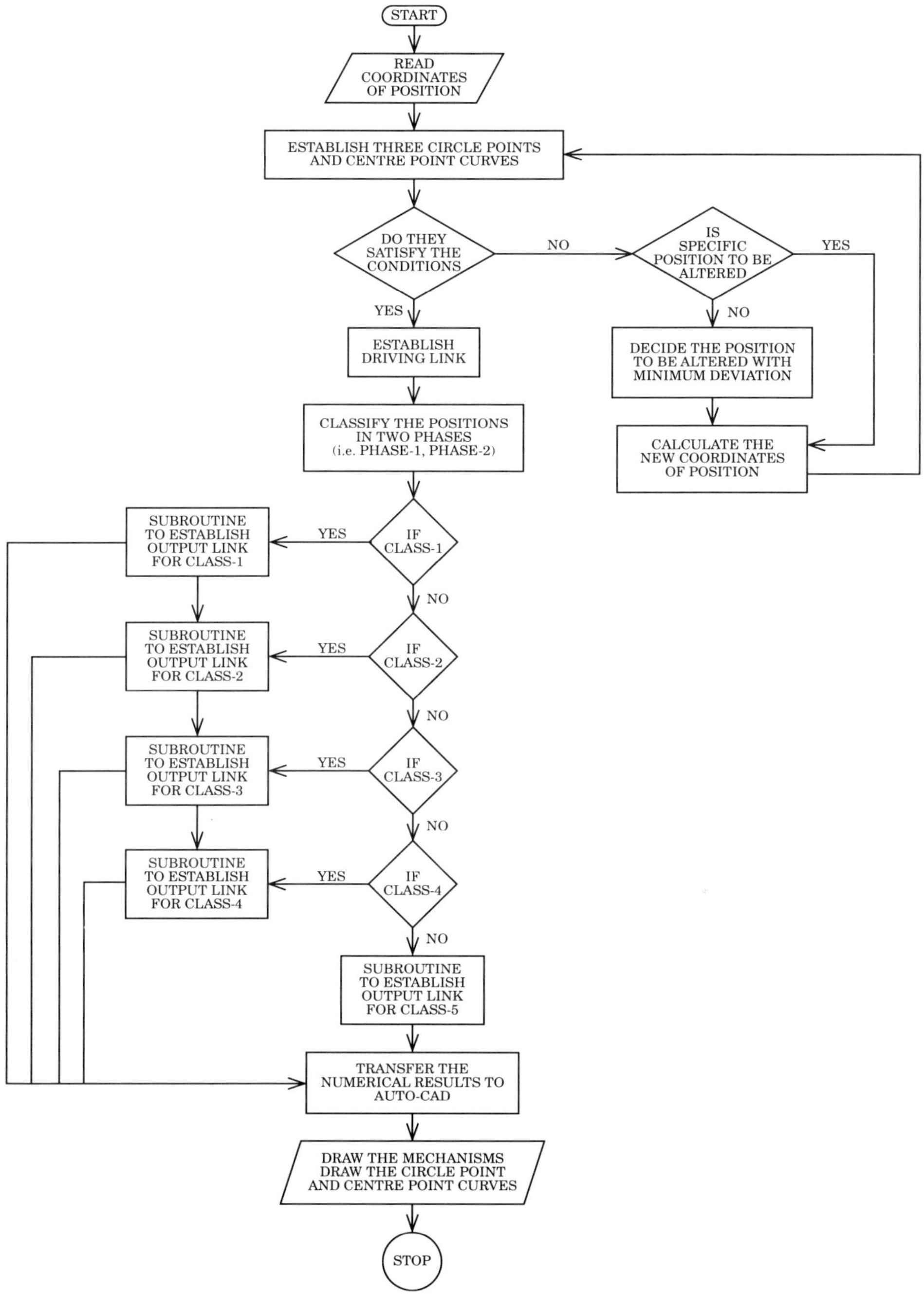

Fig. 1. Flow chart.

For these positions, circle points and center points for positions 1-2-3-4, positions 1-2-3-5, and positions 1-2-3-6 are obtained. To illustrate, Table 2 gives the center and circle points for positions 1-2-3-4. For these positions, the minimum and necessary condition has been satisfied because three circle point curves and the corresponding three center point curves intersect at one point. The graphical presentation of this has been given in Fig. 2; the resulting four-bar, adjustable mechanisms are shown in Fig. 3. The coordinates of the fixed pivots and the moving pivots of these mechanisms are given in Table 3.

Table 2. Center points and circle points for 1-2-3-4.

Positions:	P1X = 7.0	P1Y =10.1	P2X = 5.6	P2Y = 7.6
	P3X = 9.0	P3Y = 8.5	P4X = 8.3	P4Y = 7.3
Angles:	TH1 = 0.4861451	TH2 = 0.7675	TH3 = 1.431169	TH4 = 1.570

SR NO.	DOX	DOY	D1X	D1Y
1	6.600	12.278	9.215	11.620
2	7.200	11.323	8.926	11.606
3	8.000	3.885	4.765	5.342
4	10.000	9.747	9.737	12.138
5	11.000	12.289	12.567	11.714
6	11.500	12.424	13.090	11.505
7	12.000	12.499	13.574	11.254
8	12.500	12.521	14.022	10.964
9	13.900	12.313	15.101	9.940
10	17.000	9.861	16.191	6.135
11	15.000	11.846	15.752	8.896
12	16.000	11.114	16.143	7.709
13	17.500	8.712	15.866	4.988
14	11.000	8.580	9.543	12.378
15	12.000	7.517	9.406	12.772
16	13.000	6.472	9.226	13.499
17	14.100	5.295	8.779	15.630

Table 3. Coordinates of fixed and moving pivots.

Mechanisms for class 3

Input link is same for all mechanisms
Length of input link: 11.624

Coordinates of fixed pivots of input link

POX = 14.100 POY = 5.295

Coordinates of moving pivots of input link

P1X = 8.779	P1Y = 15.630
P2X = 5.774	P2Y = 13.406
P3X = 5.560	P3Y = 13.181
P4X = 4.246	P4Y = 11.460
P5X = 8.331	P5Y = 15.390
P6X = 10.503	P6Y = 16.354

Mechanism: 1

Length of fixed link: 6.613
Length of coupler link: 9.693
Length of output link: 3.044
Transmission angle deviation: 48.57 deg

Coordinates of fixed pivots of output link for first phase

Q0X = 15.000	Q0Y = 11.846

Coordinates of moving pivots of output link for first phase

Q1X = 15.752	Q1Y = 8.896
Q2X = 14.342	Q2Y = 8.874
Q3X = 15.102	Q3Y = 14.889
Q4X = 13.459	Q4Y = 14.472

Length of fixed link: 7.167

Coordinates of fixed pivots of output link for second phase

Q56X = 19.220	Q56Y = 10.310

Coordinates of moving pivots of output link for second phase

Q5X = 17.713	Q5Y = 12.955
Q6X = 19.707	Q6Y = 13.315

Mechanism: 2

Length of fixed link: 4.821
Length of coupler link: 12.785
Length of output link: 4.067
Transmission angle deviation: 39.51 deg

Coordinates of fixed pivots of output link for first phase	
QOX = 17.500	QOY = 8.712
Coordinates of moving pivots of output link for first phase	
Q1X = 15.866	Q1Y = 4.988
Q2X = 15.536	Q2Y = 5.151
Q3X = 18.336	Q3Y = 12.692
Q4X = 16.966	Q4Y = 12.744

Length of fixed link: 7.060

Coordinates of fixed pivots of output link for second phase	
Q56X = 21.142	Q56Y = 5.794
Coordinates of moving pivots of output link for second phase	
Q5X = 19.733	Q5Y = 9.608
Q6X = 21.506	Q6Y = 9.844

In a second example, the six positions of the coupler are considered to be given under

$$
\begin{array}{lll}
P_{1x} = 7.00 & P_{1y} = 10.1 & \theta_1 = 0.4861451 \\
P_{2x} = 5.60 & P_{2y} = 7.60 & \theta_2 = 0.7675 \\
P_{3x} = 9.00 & P_{3y} = 8.5 & \theta_3 = 1.431169 \\
P_{4x} = 8.30 & P_{4y} = 7.3 & \theta_4 = 1.570 \\
P_{5x} = 9.5 & P_{5y} = 9.7 & \theta_5 = 1.0001557 \\
P_{6x} = 14 & P_{6y} = 12 & \theta_6 = 0.93516
\end{array}
$$

For these positions, the minimum and necessary condition cannot be satisfied; therefore, the sixth position is changed from $P_{6x} = 14.00$ to $P_{6xn} = 14.062$ and $P_{6y} = 12.00$ to $P_{6yn} = 11.13$. This result is given in Table 4.

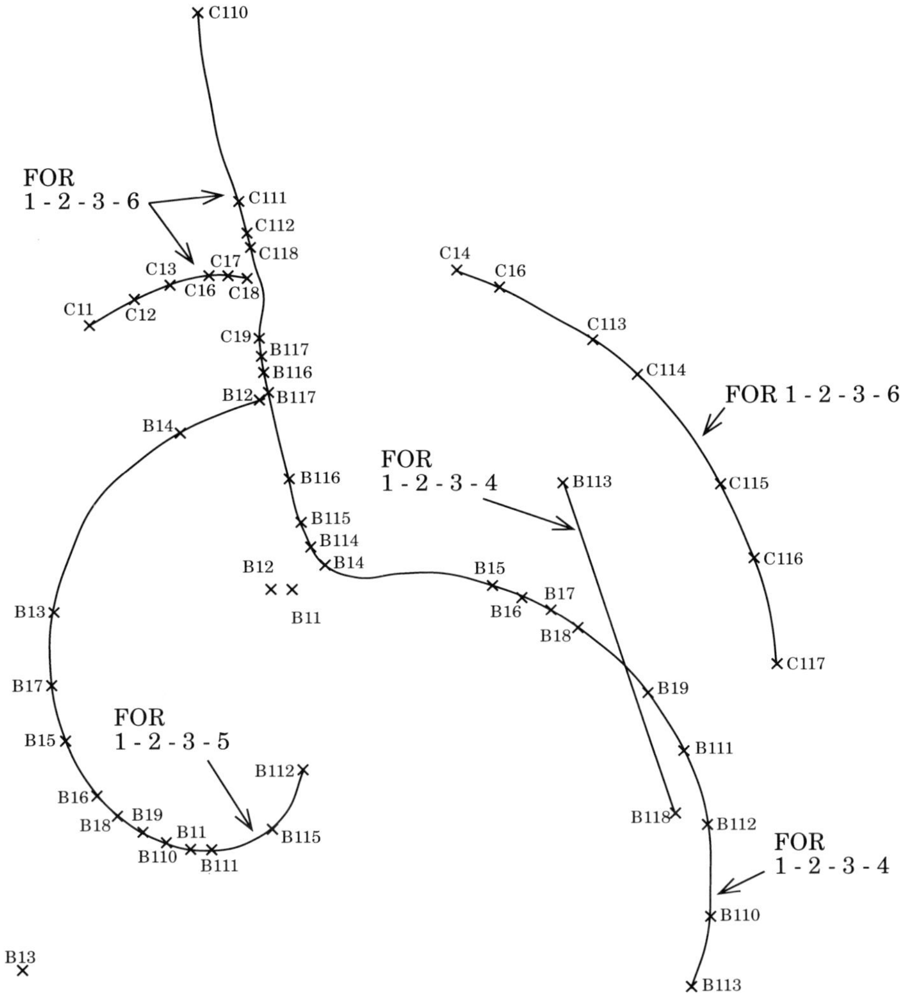

Fig. 2. Intersection of circle point curves.

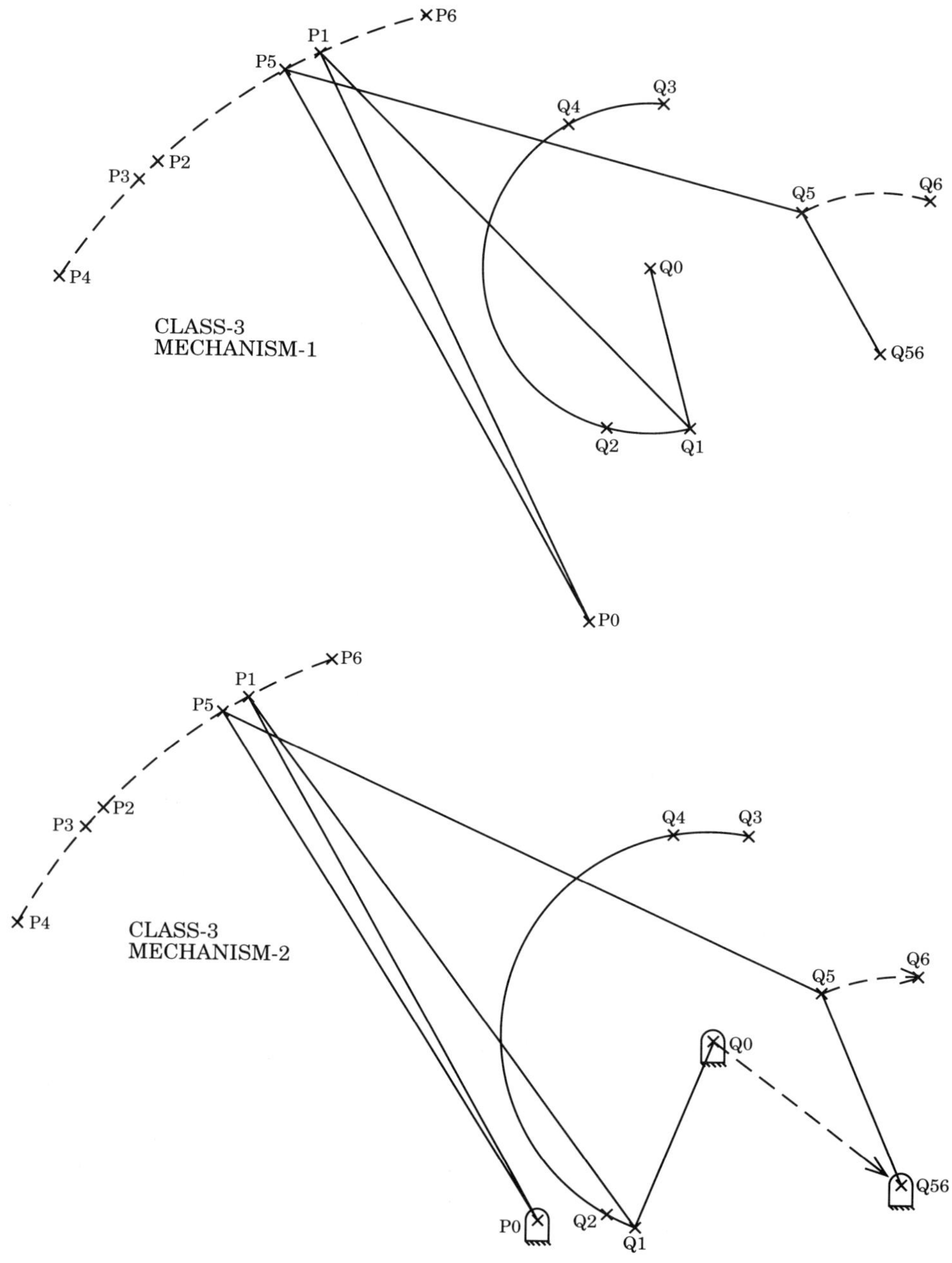

Fig. 3. Adjustable four-bar mechanism.

Table 4. Relocation of an alterable position.

No intersection between three curves
For 1-2-3-4 and 1-2-3-5
Intersection points are
8.779 15.630
sixth position is to be altered with min deviation for the point 8.779 15.630
X coordinate is to be changed from 14.000 to 14.062
Y coordinate is to be changed from 12.000 to 11.135

For 1-2-3-5 and 1-2-3-6
No intersection

For 1-2-3-4 and 1-2-3-6
No intersection

CONCLUSIONS

The graphical approaches we have suggested in the past for synthesizing adjustable, four-bar mechanisms for six-position, rigid-body guidance have some limitations. These approaches are time-consuming and not very accurate. Also, if the mechanism so obtained does not satisfy some other functional requirements, then the complete procedure must be repeated, which is a very tedious job. The computer simulation has made the procedure very fast and accurate. The graphical display facilitates viewing the mechanism and making the final choice of mechanisms.

REFERENCES

1 A. Ahmed and K. Waldron, J. Mech. Mach. Theory, 14 (1979) 405.
2 J. P. Modak and P. M. Padole, Establishing the Minimum and Necessary Condition for Synthesis of Adjustable Four-Bar Planar Mechanism for Six-Position, Rigid-Body Guidance, Modeling, Simulation, Control, Vol. 19, AMSE Press, New York, 1988.
3 J. P. Modak and P. M. Padole, Synthesis of an adjustable four-bar mechanism for six-position, rigid-body guidance with a possibility of some alteration in one of the six positions, in: Proc. Fifth IFTOMM International Conference on Linkages and Computer-aided Design, Bucharest, Romania, July 1989, pp. 593–602.
4 P. M. Padole and J. P. Modak, Type synthesis for function generation problems, in: Abstracts, National Conference on Design Eng., Madras, India, Dec. 1991.

Quality through Engineering Design - W. Kuo (Editor)

An Optimal Sampling Procedure for the Design of a Mixing Conduit for Optic Fibers

M. Jeya Chandra, Brian J. Melloy, Chao Qi, and Allen L. Soyster

Industrial and Management Systems Engineering, The Pennsylvania State University, 207 Hammond Building, University Park, PA 16802, U.S.A.

Abstract

In this paper, we develop a sampling plan to estimate the mean value function of a stochastic process by using polynomial regression. The specific application considered here is the estimation of the mean amount of light energy received in a mixing conduit for optic fibers as a function of the conduit length. The sampling plan variables consist of the sample size and the number of points at which the function is to be estimated. The objective is to determine these variables such that the resulting width of the confidence band obtained for the mean value function is less than or equal to a prespecified criterion. Sampling strategies with both correlated and independent observations are formulated and subsequently compared with respect to Taguchi's loss function.

INTRODUCTION

The general problem that we will study in this paper is the determination of an optimal sampling procedure for estimating certain parameters of a family of probability distributions. More specifically, suppose a family of probability distributions is indexed by ℓ in some interval [a,b]. Let $f(\ell)$ represent the mean of the probability distribution associated with $\ell \in [a,b]$. The objective of this paper is to provide a cost effective and reliable sampling plan for estimating $f(\ell)$.

The particular application considered here involves interconnecting h input fibers with t output fibers via a mixing conduit. For example, consider the case in which there are two input fibers, A and B, and three output fibers C, D, and E, as shown in Fig. 1. Light energy in the form of photons enters the conduit through the input strands and then proceeds across the conduit to the output fibers. A certain energy threshold must be achieved by each output fiber in order for successful activation to occur. Thus, the focus of this investigation is the signal strength of the output fiber receiving the least amount of light energy, where the signal strength is defined as the percentage of photons received by the individual output fibers.

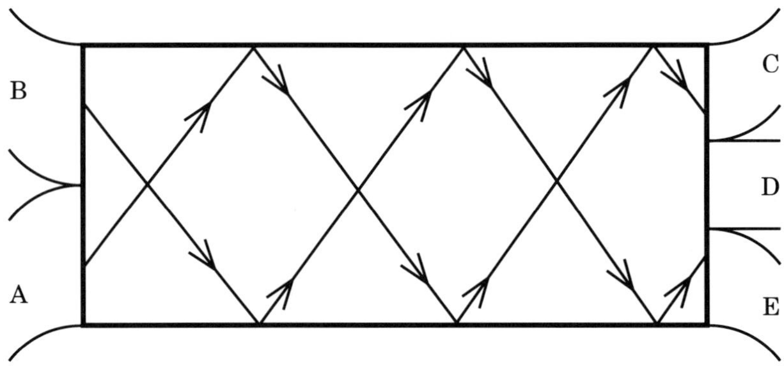

Fig. 1. Mixing conduit for optic fibers.

In order to clarify this concept, suppose that we (somehow) know what the actual proportion is for each possible length ℓ, where $0 < \ell < \ell_n$. The distance of these output fibers from the source is a critical design parameter because these percentages may vary significantly with respect to this factor. For the three output fibers, C, D, and E, there are three corresponding functions, $f_c(\ell)$, $f_d(\ell)$, and $f_e(\ell)$, which represent the true proportion of photons associated with each length. A hypothetical set of these functions is illustrated in Fig. 2, where the darkened lower envelope represents the minimum proportion at each length. The objective is to estimate this lower envelope by sampling at discrete points along the length axis.

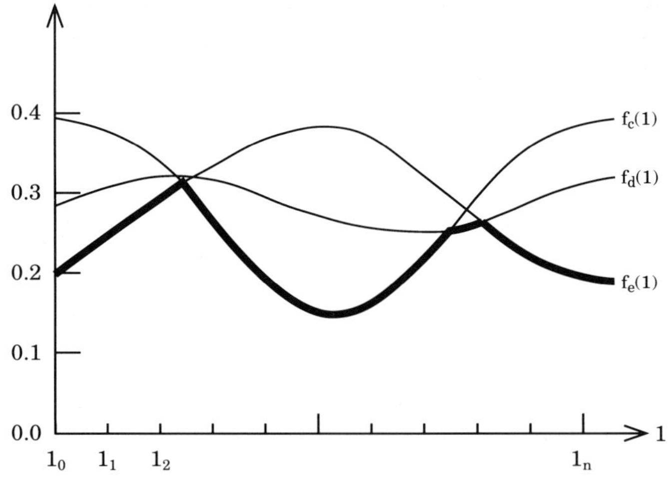

Fig. 2. Minimum proportion.

In order to estimate the smallest proportion of photons received by the output fibers at various lengths of the mixing conduit, we simulate the flow of light from each of the input fibers through the conduit by generating thousands of photon trajectories. The goal is to determine the number of trajectories required to be $(1 - \alpha)\%$ confident that the estimate of the minimum proportion is within δ of its actual value for any length ℓ. One simulation strategy is to generate trajectories for $\ell = \ell_n$ and then utilize these for $\ell = \ell_1, \ell_2, \ldots, \ell_{n-1}$. The problem, of course, is that there may be a high correlation among the estimates. On the other hand, a separate and independent set of trajectories for each $\ell = \ell_1, \ell_2, \ldots, \ell_n$ could be generated, but this would probably be much more expensive.

Thus, the two basic strategies for estimating the minimum proportions are as follows:

1. Launch m photons in each of n independent runs and estimate the minimum proportions. (The total number of photons to be launched is therefore m*n.)
2. Launch r photons in one run and use the same set of trajectories to estimate the minimum proportions at each of the n lengths.

In this paper, plans (1) and (2) will be referred to as independent and dependent processes, respectively. If the dependent process is to be employed, then r must be sufficiently large such that the disadvantage caused by the correlation present among the estimators at different lengths in the dependent process is overcome. Naturally, in either case, too large a sample size results in a waste of resources, and too small a sample diminishes the utility of results.

Thus, the objective of this research is to design a sampling strategy to estimate the function relating the minimum proportion to the length ℓ, which is achieved by using regression analysis. The sampling plan should consist of not only the number of photons to be launched but also the number of discrete lengths at which the sample data is to be collected. The optimal sample size depends on the precision, as defined by the estimation error, that can be tolerated. This tolerance is specified as the width of the confidence interval about the estimate of the minimum proportion. The details of the sampling strategy are presented in the following sections.

ANALYSIS

Preliminary analysis

No existing models in the literature are entirely satisfactory for the application considered in this paper. Although methods are available for determining the requisite sample size for estimation at a given conduit length [1], they are exclusive of regression procedures. On the other hand, the limited number of regression models that incorporate correlated observations are typically restricted to first-order models [2]. Moreover, these models are based on the assumption that the variances of the estimators are homogeneous [3], which is certainly not valid in this case. Therefore, an original formulation is necessary to both capture the complex behavior of the system and provide the necessary performance measures.

As defined earlier, let $f(\ell_i)$ be the mean value function at length ℓ_i:

$$f(\ell_i) = E[G(\ell_i)], \quad \forall i \tag{1}$$

where $G(\ell_i)$ denotes the stochastic process. The objective is to approximate $f(\ell_i)$ by taking m observations of $G(\ell_i)$ at each grid point ℓ_i. Let the mean of these m observations be designated as $\overline{G}(\ell_i)$. Now the problem is reduced to a standard context with observation pairs of the form $[\ell_1, \overline{G}(\ell_i)]$, $[\ell_2, \overline{G}(\ell_2)]$, . . . , $[\ln, \overline{G}(\ell_n)]$.

Next, there are several issues that must be addressed. (1) What should be the functional form of the regression equation? (2) How many grid points should be used, and what should be the sample size at each grid point? (3) What can be done if the samples at grid points ℓ_i and ℓ_j are correlated? We will not treat issue (1) in detail in this paper. We chose a polynomial function here because we knew from the Weierstrass theorem that there exists a polynomial that is as "close" to the given function as desired [4]. Rather, the first step was to determine the number of terms k to be included in the polynomial so that the function represented the true state as closely as possible. The statistical methodology that was developed to determine k is described in [5]. We will assume, therefore, that the value of k is known in the subsequent discussions.

The problem involves estimating the minimum proportion of photons received as a function of the length of the mixing conduit. This is achieved by estimating the proportion at each of the t receivers as a function of the length of the conduit individually and then taking the minimum of these t proportions. Let the true proportion for receiver j at length ℓ_i be $p_j(\ell_i)$, j = 1,2, . . . , t. Then

$$p_j(\ell_i) = B_{0j} + B_{1j}\ell_i + B_{2j}\ell_i^2 + \ldots + B_{kj}\ell_i^k \tag{2}$$

under the assumption of a kth-order polynomial model. The corresponding point estimate of $p_j(l_i)$ obtained from the regression analysis is

$$\tilde{p}_j(\ell_i) = \hat{B}_{0j} + \hat{B}_{1j}\ell_i + \hat{B}_{2j}\ell_i^2 + \ldots + \hat{B}_{kj}\ell_i^k \tag{3}$$

where the estimators of the coefficients obtained through regression are $\hat{B}_{0j}$, $\hat{B}_{1j}$, . . . , $\hat{B}_{kj}$.

It should be emphasized that $\tilde{p}_j(\ell_i)$ is different from the point estimate obtained from the simulations, which will be denoted by $\hat{p}_j(\ell_i)$. This latter estimator is defined as follows. If a total of V photons are launched, of which V_j are received by receiver j at length ℓ_i, then

$$\hat{p}_j(\ell_i) = \frac{V_j}{V}, \quad \forall i, j \tag{4}$$

The values of $\hat{p}_j(\ell_i)$ are the data points used in the regression analysis to obtain the estimators of the coefficients and ultimately $\tilde{p}_j(\ell_i)$.

Because a point estimate does not provide sufficient information, we decided to construct a confidence band to provide an interval estimate of $p_j(\ell_i)$. The half-width of the confidence band for the mean proportion corresponding to receiver j at length ℓ_i is

$$\delta_j(\ell_i) = Z_{\frac{\alpha}{2}} \sqrt{Var\left[\tilde{p}_j(\ell_i)\right]}, \quad \forall i, j \tag{5}$$

where $\delta_j(\ell_i)$ is the precision of the estimator, $(1 - \alpha)$ is the confidence level, and $Z_{\alpha/2}$ is the value of the standard normal variable Z such that $P(\,|z| > Z_{\alpha/2}\,) = \alpha$. The variance of $\tilde{p}_j(\ell)$ is

$$Var\left(\tilde{p}_j\right) = \ell^t\left(\mathbf{L}^t\mathbf{L}\right)^{-1}\mathbf{L}^t CoV\left(\hat{p}_j\right)\mathbf{L}\left(\mathbf{L}^t\mathbf{L}\right)^{-1}\ell, \quad \forall j \tag{6}$$

where

$$\ell = \left[1 \;\; \ell \;\; \ell^2, \; . \; . \; . \; , \; \ell^k\right] \tag{7}$$

ℓ^t denotes the transpose of ℓ,

$$\mathbf{L} = \begin{bmatrix} 1 & \ell_1 & \ell_1^2 & \cdots & \ell_1^k \\ 1 & \ell_2 & \ell_2^2 & \cdots & \ell_1^k \\ \vdots & & & & \\ 1 & \ell_n & \ell_n^2 & \cdots & \ell_n^k \end{bmatrix} \tag{8}$$

and $\mathbf{L}^t$ is the transpose of $\mathbf{L}$.

In Eq. (7), the covariance matrix is expressed as

$$CoV\left(\hat{p}_j\right) = \begin{bmatrix} Var\left[\hat{p}_j(\ell_1)\right] & CoV\left[\hat{p}_j(\ell_1), \hat{p}_j(\ell_2)\right] & \cdots & CoV\left[\hat{p}_j(\ell_1), \hat{p}_j(\ell_n)\right] \\ CoV\left[\hat{p}_j(\ell_2), \hat{p}_j(\ell_1)\right] & Var\left[\hat{p}_j(\ell_2)\right] & \cdots & CoV\left[\hat{p}_j(\ell_2), \hat{p}_j(\ell_n)\right] \\ \vdots & & & \\ CoV\left[\hat{p}_j(\ell_n), \hat{p}_j(\ell_1)\right] & \cdots & \cdots & Var\left[\hat{p}_j(\ell_n)\right] \end{bmatrix}, \forall j \tag{9}$$

where

$$Var\left[\hat{p}_j(\ell_i)\right] \approx \frac{\hat{p}_j(\ell_i)\left[1 - \hat{p}_j(\ell_i)\right]}{V}, \quad \forall i, j \tag{10}$$

However, no similar expression exists for the covariance terms, and consequently, they are estimated directly from the simulation data. Naturally, if the runs for the different lengths are performed independently, then all of the covariance terms are equal to zero.

In this work, we develop a sampling plan and methodology to achieve specified values of $(1 - \alpha)$ and δ such that the desired precision is

$$\delta \geq \underset{j}{Max}\left[\delta_j(\ell_i)\right], \quad \forall i \tag{11}$$

It can be seen from Eqs. (5–10) that the precision depends on n, the number of grid points at which data is collected, and V, the total number of photons launched in a simulation for given values of $(1 - \alpha)$ and k. The development of the sampling plan that yields the values of n and V is discussed in the next subsection.

Development of the sampling plan

Determining a sampling strategy requires the resolution of Eq. (5). However, it is not possible to obtain closed-form expressions for n and V from this equality. Thus, a numerical procedure is sought to obtain these parameter values. In particular, the objective is to find a value of r for which the properties of the dependent process will be better than those of the independent process. Unfortunately, incorporating the dependent process makes the development of this procedure more complex. Nevertheless, the use of the dependent process is motivated by its potentially greater efficiency.

Rather than proceeding with the dependent approach directly, though, let us first assume that the independent process is used for estimating the proportions and that the number of grid lengths is fixed at n. This means that a certain number of photons m are launched in every simulation run for estimating the t proportions at each of the n lengths. In this case, a total of nm observations are collected from the photon trajectories. Clearly, the m observations at any given length are independent of the m observations at any other lengths, and hence the estimates of the mean proportions at these lengths are also mutually independent.

Now, if X_{ij} photons are received by the jth receiver at length ℓ_i, then the estimate of the mean proportion of photons is

$$\bar{X}_{ij} = \frac{X_{ij}}{m}, \text{ for } i = 1, 2, \ldots, n, \text{ and } j = 1, 2, \ldots, t \tag{12}$$

It is well known that $\bar{X}_{ij}$, as defined here, is an unbiased estimator of p_{ij} for all i and j.

Next, for the dependent process, a total of r photons are launched in a single simulation and the estimates of the p_{ij}s are obtained at each of the n lengths by using only those photons. Let the number of photons received by the jth receiver at length ℓ_i be Y_{ij}. Then the estimate of p_{ij} is

$$\bar{Y}_{ij} = \frac{Y_{ij}}{r}, \text{ for } i = 1, 2, \ldots, n, \text{ and } j = 1, 2, \ldots, t \tag{13}$$

This estimator of p_{ij} is also unbiased, as was the case with the independent process. In contrast, however, the $\overline{Y}_{ij}$'s are correlated because the Y_{ij}'s are not mutually independent (for any given j).

Now we need a measure to compare the "goodness" of the independent estimator in Eq. (12) and the dependent estimator in Eq. (13). Clearly, this measure must take into consideration the correlation inherent in the dependent process. One such possible measure is the average sum of squared errors, which for the independent and the dependent processes, are

$$SSE_x = \sum_{i=1}^{n} \frac{\left(\overline{X}_i - p_i\right)^2}{n} \tag{14}$$

and

$$SSE_y = \sum_{i=1}^{n} \frac{\left(\overline{Y}_i - p_i\right)^2}{n} \tag{15}$$

respectively. For convenience, the subscript j has been dropped because the analysis is identical for each of the t fibers.

It is evident that any criterion used to compare the two methods must be a function of the expected values as well as the variances of SSE_x and SSE_y. Moreover, since these statistics are measures of the error of estimation, the expected values of these statistics must be equal to zero and their variances should be as small as possible. The expected loss function developed by Taguchi [6] has similar properties and thus will be used here as a basis to compare the two methods.

Taguchi's loss function for a random variable W (assuming that the coefficient of the quadratic function is equal to 1) [6] is defined as

$$L(W) = (W - W_0)^2 \tag{16}$$

where W_0 is the target value of W. The expected value of the loss function is therefore

$$E[L(W)] = E(W - W_0)^2 = [E(W) - W_0]^2 + Var(W) \tag{17}$$

Specifically, in this context $W_0 = 0$, since the random variable is a measure of the estimation error; hence Eq. (17) reduces to

$$E\,[L(W)] = E(W)^2 + Var(W) \tag{18}$$

(Observe that the expected loss is directly proportional to both the variance and the difference between the mean and the target). Thus, a comparison of the two processes requires that $E(SSE_x)$, $E(SSE_y)$, $Var(SSE_x)$, and $Var(SSE_y)$ be found.

First, the equations for $E(SSE_x)$ and $E(SSE_y)$ are given by

$$E(SSE_x)=\frac{1}{n}\sum_{i=1}^{n}Var(\overline{X}_i)=\frac{1}{mn}\sum_{i=1}^{n}p_i(1-p_i) \tag{19}$$

and

$$E(SSE_y)=\frac{1}{n}\sum_{i=1}^{n}Var(\overline{Y}_i)=\frac{1}{rn}\sum_{i=1}^{n}p_i(1-p_i) \tag{20}$$

respectively. Observe that the two quantities, $E(SSE_x)$ and $E(SSE_y)$, are not the same because $\overline{X}_i$ and $\overline{Y}_i$ are based on sample sizes of m and r, respectively.

Next, the variance of SSE_x is given by

$$Var(SSE_x)=Var\sum_{i=1}^{n}\frac{(\overline{X}_i-p_i)^2}{n} \tag{21}$$

Now, since X_i is a binomially distributed random variable with parameter p_i [7],

$$Var\left[(\overline{X}_i-p_i)^2\right]=\frac{1}{m^3}\left\{2mp_i^2(1-p_i)^2+p_i(1-p_i)\left[1-6p_i(1-p_i)\right]\right\},\ \forall i \tag{22}$$

Therefore it follows that

$$Var(SSE_x)=\frac{1}{m^3n^2}\sum_{i=1}^{n}\left\{2mp_i^2(1-p_i)^2+p_i(1-p_i)\left[1-6p_i(1-p_i)\right]\right\} \tag{23}$$

Similarly, the variance of SSE_y [7] is

$$\begin{aligned}Var(SSE_y)=&\frac{1}{r^3n^2}\sum_{i=1}^{n}\left\{2rp_i^2(1-p_i)^2+p_i(1-p_i)\left[1-6p_i(1-p_i)\right]\right\}\\&+\frac{2}{n^2}\sum_{i=1}^{n-1}\sum_{j=i+1}^{n}CoV\left[(\overline{Y}_i-p_i)^2,(\overline{Y}_j-p_j)^2\right]\end{aligned} \tag{24}$$

Observe that the variance of SSE_y includes additional covariance terms since the $\overline{Y}_i$'s are not independent. Since the covariance expression in Eq. (24) is intractable, we will replace it by an upper bound.

An individual covariance term in Eq. (24) is bounded [8] by

$$CoV\left[\left(\overline{Y}_i - p_i\right)^2, \left(\overline{Y}_j - p_j\right)^2\right] \le \left[Var\left(\overline{Y}_i - p_i\right)^2 * Var\left(\overline{Y}_j - p_j\right)^2\right]^{1/2}, \forall i, j \quad (25)$$

Clearly, the right-hand side will be maximized when each of the variance terms attains its maximum value. By employing Eq. (22), we can show that this occurs when the product $p_i(1 - p_i)$ is maximized; that is, when

$$G = Max\left[p_i(1 - p_i)\right] = \frac{1}{t}\left(1 - \frac{1}{t}\right) = \frac{(t-1)}{t^2}, \forall i \quad (26)$$

and r > 3. (This latter restriction is extraneous since r will certainly be much greater than 3.) The derivation of Eq. (26) is straightforward since

$$\underset{j}{Min}\left(p_{ij}\right) \le \frac{1}{t}, \forall i \quad (27)$$

and the product $p_i(1 - p_i)$ is a nondecreasing function on [0,1/t]. Hence, from Eqs. (22), (25), and (26),

$$CoV\left[\left(\overline{Y}_i - p_i\right)^2, \left(\overline{Y}_j - p_j\right)^2\right] \le \frac{1}{r^3}\left[2rG^2 + G(1-6G)\right], \forall i, j \quad (28)$$

Finally, because of Eqs. (24), (26), and (28), the variance of SSE_y can be bounded by

$$\begin{aligned} Var\left(SSE_y\right) &\le \frac{1}{r^3 n^2}\sum_{i=1}^{n}\left[2rG^2 + G(1-6G)\right] + \frac{2}{n^2}\sum_{i=1}^{n-1}\sum_{j=1+1}^{n}\frac{1}{r^3}\left[2rG^2 + G(1-6G)\right] \\ &= \frac{1}{nr^3}\left[2rG^2 + G(1-6G)\right] + \frac{(n-1)}{nr^3}\left[2rG^2 + G(1-6G)\right] \\ &= \left[2rG^2 + G(1-6G)\right]\frac{1}{r^3} \end{aligned} \quad (29)$$

Now, recall that the objective is to find the value of r for which the expected loss of the dependent process will be less than that of the independent process. Hence, using the results of Eqs. (19-24), (26), and (29), we define the following inequality:

$$\begin{gathered} E\left(SSE_y\right)^2 + Var\left(SSE_y\right) \le E\left(SSE_x\right)^2 + V\left(SSE_x\right) \\ \frac{G^2}{r^2} + \frac{1}{r^3}\left[2rG^2 + G(1-6G)\right] \le \frac{G^2}{m^2} + \frac{1}{nm^3}\left[2mG^2 + G(1-6G)\right] \end{gathered} \quad (30)$$

Equation (30) can be used to find the number of observations r required in the dependent process for given values of m, the number of observations in each replication in the independent process, and the number of lengths n. (The initial value of n and the number of polynomial terms to be included in the regression model have been determined by using a statistical procedure [5].) For example, Table 1 contains values of m, n, and r that satisfy Eq. (3) when k = 7. Note that the value of r is significantly less than the product mn, which reflects the superior efficiency of the dependent process. Moreover, since an upper bound was used for the covariance term, the value obtained for r is conservative.

Table 1. A comparison of dependent and independent process results.

n	m	m*n	r
5	200	1000	293
10	50	500	80
10	100	1000	159
10	200	2000	293

The number of observations necessary (for each replication of the independent process) to obtain the desired half-width of the confidence interval at the specified confidence level [9–10] is

$$m = \left(\frac{Z_{\frac{\alpha}{2}}}{\delta} \right)^2 p_i(1 - p_i) \tag{31}$$

Unfortunately, this formula requires the value of p_i, which is not known. However, since the minimum proportion is of interest in our problem, p_i can be replaced by 1/t to obtain

$$m \leq \left(\frac{Z_{\frac{\alpha}{2}}}{\delta} \right)^2 \frac{1}{t}\left(1 - \frac{1}{t}\right) \tag{32}$$

as the equation for the upper bound of m.

The half-width of the confidence band can be reduced by increasing the number of grid points, because the variance of p_j in Eq. (6) is indirectly proportional to this parameter. This relationship is used to determine the final number of grid points in the following manner. First, the half-widths of the confidence bands at each of the n lengths are computed. Then, the value of the largest half-width is compared with the predetermined precision, δ. If this value is greater than δ, then the value of n is (arbitrarily) doubled. This process is repeated until the maximum half-width of the confidence band is less than or equal to δ. Finally, this procedure is summarized in Table 2.

Table 2. Algorithm summary.

1. Determine the order of the regression function k and the initial value of n, the number of lengths at which the mean values are to be estimated [2].
2. On the basis of the specified value of the precision δ, compute the sample size m required by the independent process by using Eq. (42).
3. Using Eq. (40), obtain the equivalent number of observations r for the dependent process.
4. Using the simulation model, obtain r trajectories from the dependent process and estimate the mean $f(\ell i)$ at $\ell_i, \ell_2, \ldots, \ell_n$.
5. Obtain the estimates of the regression coefficients by using standard regression formulas, and construct a confidence interval for the mean $f(\ell)$.
6. Compare the half-width of the widest confidence interval with the prespecified precision value δ. If the width obtained is $\leq \delta$, stop. Otherwise, go to step (7).
7. Set $n = 2n$ and go to step (4).

CONCLUSIONS

In this paper, a sampling strategy was developed to estimate the mean value (polynomial) function of a stochastic process. The particular application considered in this paper involved interconnecting input and output fibers via a mixing conduit. The sampling strategy defines the final number of observations. A method for determining the initial number of observations and the order of the polynomial is described in Ref. [5]. The plan may be used with either independent replicated experiments or dependent nonreplicated experiments. (The formulation compensates for any correlation that may exist across time or space within a given experiment.) Experimental evidence suggests, however, that the latter approach is significantly more efficient and optimal than the former because it yields the same or better precision while requiring a smaller sample size. The particular application considered in this paper involved interconnecting input and output fibers via a mixing conduit.

REFERENCES

1 S. K. Thompson, The American Statistician, 41, no. 1 (1987) 42–46.
2 J. D. Tew and J. R. Wilson, Estimating simulation metamodels using integrated variance reduction techniques, in: Proceedings of the Joint Statistical Meetings of the American Statistical Association and the Biometric Society, 1988.
3 J. D. Tew, IE News: Operations Research, 24, no. 2 (1990) 1–9.

4 R. L. Burden, J. D. Faires and A. C. Reynolds, Numerical Analysis, 2nd ed., PWS Publishers, Boston, Mass., 1981.

5 M. J. Chandra, B. J. Melloy, C. Qi and A. L. Soyster, An Iterative Algorithm for Determining Both the Order and Sample Size for a Polynomial Regression Model, Working Paper, Department of Industrial and Management Systems Engineering, The Pennsylvania State University, 1991.

6 G. Taguchi, A. Elsayed and T. Hsiang, Quality Engineering in Production Systems, McGraw Hill Book Company, New York, 1989.

7 S. M. Ross, Introduction to Probability Models, 4th ed., Academic Press, New York, 1990.

8 V. K. Rohatgi, An Introduction to Probability Theory and Mathematical Statistics, John Wiley and Sons, Inc., New York, 1976.

9 W. G. Cochran, Sampling Techniques, 3rd ed., John Wiley and Sons, Inc., New York, 1977.

10 J. L. Devore, Probability and Statistics for Engineers and the Scientists, 2nd ed., Brooks/Cole Publishing Company, Monterey, Calif., 1989.

Quality through Engineering Design - W. Kuo (Editor)

Optimization of Flashlight Performance of Dry Cells

Deepankar Ray Chaudhury and Tridib K. Dutta

Quality Management and Operations Research Division, Indian Statistical Institute, 203, B. T. Road, Calcutta 700035, India

Abstract

This study deals with improving the flashlight performance of a dry cell. A mixture experiment was conducted with five ingredients used in the bobbin of a cell. A second-order, rotatable, central-composite design in four factors was used to construct the layout of the experiment. A second-degree polynomial was adequate to represent the flashlight performance data over the zone of exploration. A nonlinear programming technique was employed to arrive at an optimum mix formulation that resulted in about a 20% improvement in flashlight performance.

INTRODUCTION

The most important quality characteristics that determine the functional performance of a battery are its open-circuit voltage (VOC), short-circuit amperage (SCA), performance of flashlight test (LIF), and performance of radio test (RADIO). Depending upon the intended end use, the batteries are stamped as flashlight, transistor, or multipurpose. For flashlight batteries, the performance on the LIF test is the major deciding factor, whereas performance on the RADIO test is more important for transistor batteries. Multipurpose batteries are not meant for any specific end use and are expected to perform satisfactorily on both the LIF and RADIO tests.

This study was conducted in a premier battery-manufacturing organization, but the organization was facing tough competition in the battery market. This necessitated continuous improvement in all the products to sustain a competitive advantage in the market. The flashlight batteries, being the flagship product of the company, accounted for more than 65% of their total volume of production. Thus, the management was keen to improve the performance of flashlight batteries. The management also imposed an additional constraint of achieving an improvement of at least 10% over the existing LIF performance of about 320 minutes.

BRIEF DESCRIPTION OF THE PROCESS

The battery cans are impact-extruded from calots, which are small hexagonal or circular zinc plates. These cans function as the negative electrode as well as the container for the dry cells. Bobbins with a small carbon rod are inserted into these cans. The carbon rod works as the positive electrode.

The bobbins are manufactured out of a mixture of carbon (acetylene black), electrolyzed manganese dioxide, manganese ore, zinc-chloride solution, sal ammoniac (ammonium chloride), water, hydrophil, and other proprietary ingredients. The blending of the mix is carried out in jaeggars, which are rotary blenders placed at an angle to the horizontal floor and made to rotate for proper blending. The mix is then fed to stampers, in which the mixture is compressed into a cylindrical shape and a carbon electrode is inserted in the middle.

Bobbins are then inserted into the zinc cans in the Cooker lines. A white electrolytic paste separates the bobbin from the can. The paste acts as a buffer between the anode and the cathode. After assembly, the cells are sealed with asphalt. The top collar is then placed over the container, and a metal cap is placed over the carbon electrode. Finally, jackets are wrapped around the cells.

OBJECTIVE

Discussions with the technical personnel led to the suspicion that LIF might be functionally related to the proportion of ingredients in the battery mix and that by varying the composition of the mix through changing ingredient proportions, the LIF also might be changed. Hence, a study was undertaken with the following objectives:

1. Determine an optimum mix of the ingredients that will enable an LIF of at least 350 minutes.
2. Estimate the functional relationship between LIF and the proportion of ingredients in the mix.

APPROACH

The objectives of the study suggested that the functional relationship of LIF with the mixture ingredients be explored for empirical model building through statistically designed experiments. Further, LIF was suspected to be a function of the ingredient proportions rather than the corresponding absolute amounts present in the mix. Thus, this problem belonged to the class of mixture experiments. Consequently, we decided to adopt the following approach for conducting the study:

1. Conduct a statistically designed mixture experiment with the mix ingredients (see Ref. [1]).
2. Employ response surface methodology (RSM) for exploring the functional relationship of LIF with the ingredient proportions (see Ref. [2]).

3. Determine an optimal mix of the ingredients by using suitable mathematical optimization techniques and perform a sensitivity analysis to study the effect of deviations from the optimum (see Ref. [3]).

CONDUCTING THE MIXTURE EXPERIMENT

Selection of factors and the region of exploration

The factors and the corresponding region of exploration were decided with active involvement of the technical personnel. Considerations for selecting a factor were based on the assumptions that

1. the factor affected the LIF of a cell
2. the factor was controllable
3. the effect of such a factor needed to be investigated

Table 1 shows the selected factors, along with their respective region of exploration.

Table 1. Factors and their region of exploration.

Sl. no.	Factor (Mix ingredient)	Factor code	Region of exploration (%)		
			Minimum	Principal level*	Maximum
1	Ore	τ_1	50.0	55.0	60.0
2	Carbon	τ_2	6.0	7.0	8.0
3	Sal	τ_3	10.0	14.0	18.0
4	Zinc Chloride	τ_4	1.0	5.5	10.0
5	Water	τ_5	17.0	18.5	20.0

*denotes the existing level

Model specification and selection of mixture design

Discussions with the research and development (R&D) personnel and a survey of the technical literature indicated that the effects of the factors would be nonlinear over the range of interest. Further, the actual optimum mix was expected to be in the vicinity of the existing mix composition. Thus, a second-order response surface model in the variables τ_1 to τ_5, with the residual variation ascribed principally to uncorrelated random error and with mean zero and variance σ^2, was thought to be adequate for the purpose of optimization. The model selected was

$$E(y|\tau) = \sum_{i} \beta_i \tau_i + \sum_{i<j} \beta_{ij} \tau_i \tau_j \tag{1}$$

where the model coefficients β_i and β_{ij} were estimated from the data of the experiment.

Because the original mixture variables τ_1, τ_2, τ_3, τ_4, and τ_5 were dependent (for a mix to be feasible, the sum must be 100.0), these variables were transformed to a set of four independent design variables — x_1, x_2, x_3, and x_4 — for the purpose of selecting a mixture design. The selected design in four independent design variables was a uniform, precision-rotatable, central-composite design. The functional relation of the original mixture components (τ) and the design variables (X) is given in Eq. (2).

$$\tau = c[X|0]T'H + Jdiag(\tau_0) \tag{2}$$

where

$\tau(31 \times 5)$ is the design with the mixture ingredients,
$diag(\tau_0)$ is a 5×5 diagonal matrix of center points of τ,
c is the radius multiplier,
$X(31 \times 4)$ is the design in independent variables,
$T(5 \times 5)$ is the transformation matrix,
$H(5 \times 5)$ is the diagonal matrix of half ranges, and
$J(31 \times 5)$ is a matrix with all the elements as 1.

The layout or master plan of the experiment is given in Table 2.

Control factors

To prevent the experimental units from being affected by wide variations in the extraneous factors, researchers controlled the sequence of mixing, blending time, mix aging period, quantity of hydrophil, bobbin height, cooking-bath temperature, source of the mixture ingredients, quantity of electrolyzed manganese dioxide, and the amount of electrolytic paste at their existing levels.

Response

The response considered was the LIF test performance in minutes at 1.1 V, which was the time taken for initial voltage of a dry cell (after four weeks of aging) to drop to a level of 1.1 V, when it was drained for 4 min/hr, 8 hr/day, through a circuit having 4Ω resistance.

Conduct of the experiment

The entire experiment was conducted at the laboratory of the Research and Development Center according to the master plan given in Table 2. Thus 2.5 kg of mix were used for each of the trials. The order of experimentation was randomized and is shown in the master plan of the experiment as the Trial no. For any trial, the ingredients were preweighed accurately, and cells were manufactured out of the mix prepared. After four weeks of aging, one cell from each of the trials was subjected to the LIF test and the response was noted. The entire experiment thus gave 31 readings on LIF; results are shown in Table 2.

Table 2. Master plan and data on LIF values at 1.1 V.

Sl. no.	Trial no.	Mix ingredients (%) Ore	Carbon	Sal	$ZnCl_2$	Water	LIF (minute)
1	14	58.43	6.62	13.31	3.88	17.76	238
2	2	57.44	7.61	13.31	3.88	17.76	325
3	27	55.37	6.50	16.49	3.88	17.76	320
4	2	54.38	7.49	16.49	3.88	17.76	317
5	15	56.21	6.53	11.90	7.60	17.76	269
6	22	55.22	7.52	11.90	7.60	17.76	325
7	20	53.16	6.41	15.07	7.60	17.76	339
8	9	52.17	7.40	15.07	7.60	17.76	352
9	30	57.83	6.60	12.93	3.40	19.24	267
10	5	56.84	7.59	12.93	3.40	19.24	310
11	26	54.78	6.48	16.10	3.40	19.24	288
12	21	53.79	7.47	16.10	3.40	19.24	315
13	10	55.62	6.51	11.51	7.12	19.24	296
14	8	54.63	7.50	11.51	7.12	19.24	299
15	23	52.56	6.39	14.69	7.12	19.24	317
16	13	51.57	7.38	14.69	7.12	19.24	309
17	1	55.99	6.01	14.00	5.50	18.50	291
18	24	54.01	7.99	14.00	5.50	18.50	288
19	4	58.05	7.12	10.83	5.50	18.50	320
20	29	51.95	6.88	17.17	5.50	18.50	344
21	18	57.22	7.09	15.42	1.77	18.50	246
22	19	52.78	6.91	12.58	9.23	18.50	325
23	3	55.60	7.02	14.38	5.98	17.02	338
24	16	54.40	6.98	13.62	5.02	19.98	309
25	7	55.00	7.00	14.00	5.50	18.50	325
26	28	55.00	7.00	14.00	5.50	18.50	308
27	12	55.00	7.00	14.00	5.50	18.50	311
28	11	55.00	7.00	14.00	5.50	18.50	323
29	31	55.00	7.00	14.00	5.50	18.50	320
30	17	55.00	7.00	14.00	5.50	18.50	311
31	25	55.00	7.00	14.00	5.50	18.50	320

ANALYSIS AND RESULTS

The collected data were analyzed by using response-surface methodology to develop an empirical model. The stationary point was calculated and characterized as a saddle point. Thus the regression equation of LIF on design variables was not optimized by using the canonical form. Rather nonlinear programming techniques were used to find the optimum proportion of ingredients satisfying the requirement of a minimum LIF value of 350 minutes on the basis of the empirical model. The relevant details of the analysis and results are described in subsequent sections.

Estimation of the response surface

A second-degree polynomial in four independent variables — x_1, x_2, x_3, and x_4 — was fitted to all the 31 observations by the method of least squares. A stepwise regression-analysis technique was employed. At each step of the regression analysis, the adjusted R^2 (R^2_A) and the mean-square-error (MSE) values were compared with those obtained in the previous step. The stepwise procedure was terminated when it showed a decrease in R^2_A and an increase in MSE comparable to that of the previous step. The equation so obtained was considered to be an adequate representation of the data over the region of exploration. Residual analysis was employed to check for the validity of the model assumptions.

Fitted surface

The fitted second-order surface for LIF values in terms of the independent variables is as follows:

$$\begin{aligned}\hat{y}(x) = 318.17 \ & +8.83x_1 + 11.50x_2 + 11.83x_3 - 5.92x_4 \\ & -7.35x_1^2 + 3.27x_2^2 - 8.35x_3^2 - 10.00x_1x_2 \\ & -5.63x_1x_3 - 5.50x_1x_4 - 7.13x_2x_4\end{aligned} \tag{3}$$

Regression summary

The coefficient of determination (R^2) was 0.806. Thus, about 81% of the variability in the data was explained by the fitted model. This was considered to be a satisfactory fit. Analysis of variance was performed to test the significance of the regression. The residual, with 19 degrees of freedom (DOF), was partitioned into two components, one due to lack of fit (13-DOF) and the other due to pure error (6-DOF). The lack of fit was insignificant and was subsequently pooled with pure error to obtain a pooled estimate of MSE of 14.4739. The analysis of variance is given in Table 3.

Table 3. Analysis of variance on LIF.

Source	DOF	S.S.	M.S.	F value
Regression	11	16,555.300	1,505.0273	7.1841
Residual	19	3,980.3777	209.4936	
Total	30	20,535.677		

Probability of (F > F value) = 0.000

Checking model adequacy

Residual analysis was employed to verify the underlying assumptions of the model. The normal probability plot of the residuals, the plot of the residuals against the fitted values, and the plot of the residuals against the independent variables did not reveal any notable departure from the assumptions.

Characterization of the surface

To gain an understanding of the fitted response surface, we performed a canonical analysis. The canonical equation is as follows:

$$\hat{y} - 300.98 = -12.2885w_1^2 - 7.0622w_2^2 + 0.6344w_3^2 + 6.2881w_4^2 \qquad (4)$$

Center of the system (coded variables)	
$x_{1s} = -1.2970$	$x_{3s} = 1.1453$
$x_{2s} = 0.1707$	$x_{4s} = 3.5912$

We observed that two of the coefficients (namely, eigenvalues) in the preceding equation were positive, whereas the remaining coefficients were negative. Thus, the canonical equation represented a hypersaddle surface. An examination of the coordinates of the center of the system revealed that the center was outside the region of interest, which was indicative of the surface being an inclined ridge or an inclined trough.

Location of the optimum of the fitted surface

The fitted equation was that of a hypersaddle surface with the center lying outside the zone of exploration. Thus, the optimum of the surface could have been discovered only through extensive search or by using techniques like ridge analysis. Such techniques, however, either were inefficient or did not consider the variance of the predicted response. Thus, we obtained the optimum of the surface by using nonlinear-programming techniques with due consideration to the variance.

Because a rotatable design was used for experimentation, the variance of the predicted response for any given $x = (x_1, x_2, x_3, x_4)'$ was a function of the distance of x from the center of the design (0,0,0,0). Therefore, we decided to maximize the 95% lower prediction limit for the LIF of a cell in the zone of exploration.

Problem formulation

The predicted value of LIF corresponding to a mix composition for which x is the associated design variable was given by $\hat{y}(x)$, that is, the estimate obtained from the regression equation. We can easily show that the 95% lower prediction limit for the predicted value of LIF of a single future cell is given by

$$L(x) = \hat{y}(x) - t_{19,0.05}\sigma\sqrt{\left(1.1429 - 0.0298r^2 + 0.0350r^4\right)} \qquad (5)$$

where

$t_{19,0.05}$ = 5% upper tail value of the t-distribution with 19-DOF = 1.729
s = standard error of estimate = 14.4739

r = distance of x from the design center = $\sqrt{x_1^2 + x_2^2 + x_3^2 + x_4^2}$

Thus, the problem was formulated as

$$\begin{aligned} &\text{Maximize } L(x_1, x_2, x_3, x_4) \\ &\text{subject to } -2.0 \le x_i \le 2.0 \text{ for each } i = 1, 2, 3, 4 \end{aligned} \tag{6}$$

This is a general nonlinear-programming problem in four variables with bounds on each variable.

Solution

The nonlinear programming problem in Eq. (6) was solved by using IMSL library routines called from a FORTRAN program. The program was executed under the VAX/VMS environment. The routine DBCONF from the IMSL library was used to maximize the objective function. A summary of the results of optimization is given in Table 4. Table 5 shows a comparison of LIF at the existing and optimum mix compositions.

Table 4. Results of optimization of LIF.

Optimal mix values		Predicted LIF $[\hat{y}]$	Standard error $[se(\hat{y})]$	95% lower prediction limit $[y-1.729\ se(\hat{y})]$
Design variables	Mixture variable			
$x_1 = 0.1386$ $x_2 = 2.0000$ $x_3 = 0.5195$ $x_4 = 2.0000$	$\tau_1 = 52.04$ $\tau_2 = 6.81$ $\tau_3 = 17.19$ $\tau_4 = 6.95$ $\tau_5 = 17.01$	398.78	26.29	353.32

Sensitivity analysis

To assess the impact of deviations from the suggested optimum on LIF performance, we conducted a simulation study. Discussions with the production personnel revealed that it would be possible to control the proportion of the ingredients within 1% of the suggested value. Thus, it was assumed that the proportion of the individual ingredients would have a normal distribution during actual production, with the mean as the suggested optimum value and

Table 5. Comparison of LIF (existing versus optimum).

Ingredients		Range of exploration		Mix formulation		Predicted LIF	
		Min.	Max.	Existing	optimum	Existing	optimum
Ore	(τ_1)	50.0	60.0	55.0	52.04		
Carbon	(τ_2)	6.0	8.0	7.0	6.81		
Sal	(τ_3)	10.0	18.0	14.0	17.19	318.17	398.78
$ZnCl_2$	(τ_4)	1.0	10.0	4.5	6.95		
Water	(τ_5)	17.0	20.0	18.5	17.01		

the variance calculated on the basis of ±1.0% error in the proportion of the mix. On the basis of this assumption of the distribution of the ingredients, a total of 500 observations were simulated around the optimum composition of the mix. The corresponding values of the expected LIF and lower prediction limit on LIF of a single cell were computed from the derived expression of the response surface.

We note that with 1% variation around the suggested optimum, the average LIF performance did not deteriorate beyond 1.7% of the corresponding predicted value. Further, for LIF corresponding to individual cells, no serious violations of the minimum LIF requirement of 350 minutes occurred; the simulated minimum was 349.932 minutes.

CONCLUSIONS

In the region of exploration, the LIF value can be adequately predicted by using a second-degree equation. Around 81% of the variability in LIF values can be explained by this regression equation. The standard error of prediction was less than 7% of the predicted value.

Substantial improvement in LIF performance is possible without changing the proportions of the ingredients in the mix to a great extent.

There was no local maximum of the fitted surface in the region of exploration, as judged from the canonical analysis. The stationary point was a saddle point. Thus there might exist a mix composition, outside the region of experimentation and in the direction suggested by the fitted optimum, that might yield a better LIF performance compared with the suggested optimum.

A decrease in the ore-carbon ratio of the mix increases the cost of mix preparation. The suggested optimum mix has a slightly lower ore-carbon ratio (7.64) compared with the existing mix (7.86). Thus, use of the optimum mix will not result in an appreciable increase in the cost of production per cell.

CONFIRMATORY TRIAL

Three batches of cells were prepared according to the suggested optimum formulation to confirm the findings of the study. Ten cells from each batch were tested for LIF performance. The results complied with the expected performance predicted in the study. The average LIF value of these cells was 394.27 minutes and the minimum was 361 minutes.

IMPLEMENTATION

The findings of the study were discussed with the factory personnel in a joint meeting, and they were encouraged to note the results. The factory personnel decided to implement the suggested optimum mix formulation in the routine production of flashlight batteries immediately.

REFERENCES

1 J. A. Cornell, Experiments with Mixtures, John Wiley and Sons, Inc., New York, 1981.
2 R. H. Myers, Response Surface Methodology, Allyn and Bacon, Boston, 1971.
3 W. E. Biles and J. J. Swain, Optimization and Industrial Experimentation, John Wiley and Sons, Inc., New York, 1980.

Quality through Engineering Design - W. Kuo (Editor)

Application of a Knowledge Based, Integrated-System Approach to the Conceptual Design of a Power Transmission System

Daizhong Su

Department of Mechanical Engineering, University of Surrey, Guildford, Surrey, U.K. GU2 5XH

Abstract

This paper presents an approach for performing engineering design tasks by integrating expert systems, numerical calculations, data processing, and computer-aided design drawings to provide a computer-based implementation. With this approach, a prototype conceptual design for a power transmission system has been developed. The design has proved feasible, and the results produced are believable.

INTRODUCTION

In addition to decision making, engineering design usually involves the following three types of activities—data processing, numerical calculations and analyses, and engineering drawing—for which the techniques of computer-aided design have been successfully applied. However, in most cases, they are isolated. How to link them through the whole design procedure is still a challenging research topic.

The conceptual design of power transmission systems can be categorized as routine because the elements and structures involved are all well defined and widely used. However, it is still a complex task because the searching space containing design alternatives is open ended. The design quality depends highly on the designer's knowledge and expertise in decision making, handling a wide variety of design data, numerical analysis, and layout drawing. This paper applies a knowledge-based, integrated system (KBIS) approach to the conceptual design of a power transmission system.

A KBIS APPROACH FOR AN ENGINEERING DESIGN PROCEDURE

A concept is constructed first by assembling basic units (such as gear, chain, etc.) into a pattern that is a set of empty positions for basic units. Then when each position is occupied by a particular basic unit, a concept is formed.

The conceptual design procedure, shown in Fig. 1, includes four steps: identifying product design specification (PDS), constructing concepts according to the identified PDS, generating rating marks and creating layout drawings for the constructed concepts, and evaluating concepts based on the rating marks and layout.

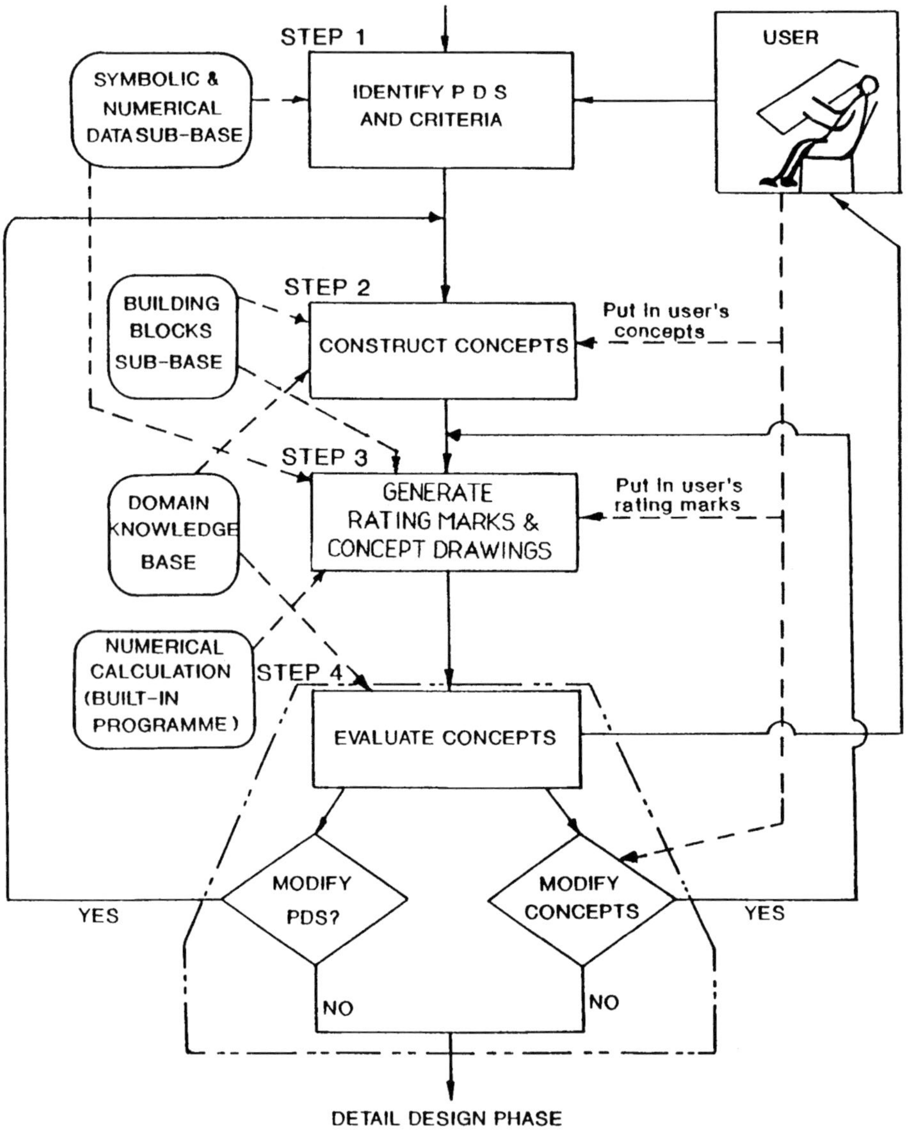

Fig. 1. Conceptual design process.

The main structure of the facilities employed by this approach (shown in Fig. 2) includes five basic elements: the inference engine, knowledge base (KB), numerical calculation facilities (NC), engineering drawings (ED), and data bases (DB). The inference engine controls the others. For the design procedure mentioned above, design expertise is stored in KB; NC is used to calculate the concept rating marks; ED is used to create concept drawings; and DB is used to store all data in the design process: symbolic data (PDS and criteria items), numerical data (rating marks of basic units, arrangement patterns, and concepts), symbolic-numerical mixed data (information of all constructed concepts), and graphic data (drawing files of basic units and concepts).

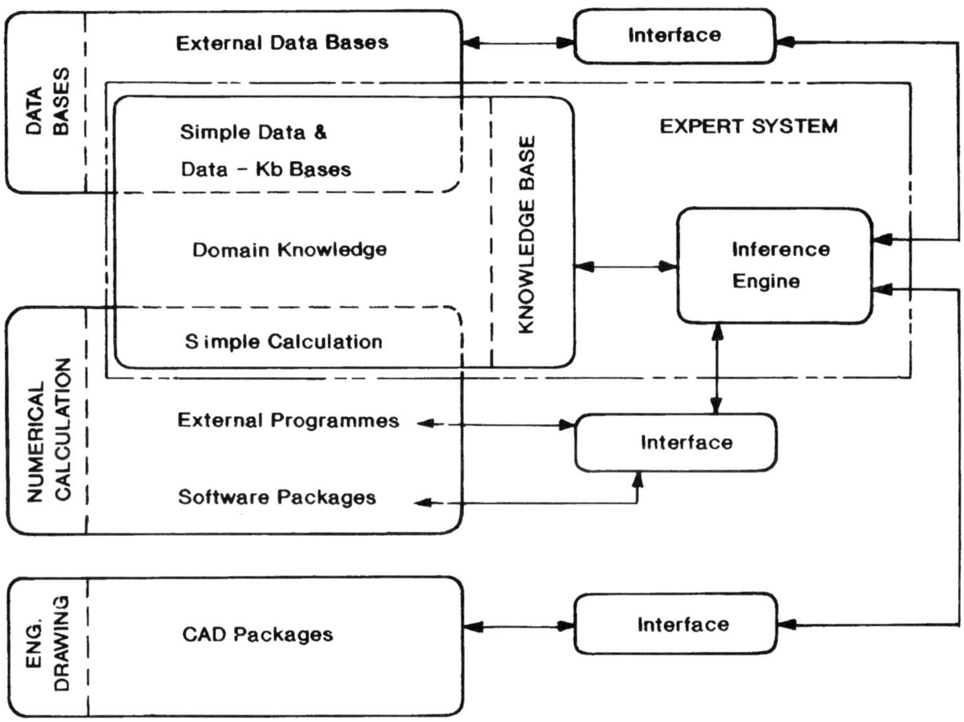

Fig. 2. Main structure of the approach.

A PROTOTYPE OF KBIS FOR CONCEPTUAL DESIGN OF A POWER TRANSMISSION SYSTEM

Based on the KBIS approach, we developed a prototype system by using the XI-Plus expert system shell [1], the AutoCAD drawing package [2], and the Quick BASIC 4 programming language. Currently, the prototype can construct up to 75 concepts by using seven types of basic units (see Fig. 3) and ten types of arrangement patterns (see Fig. 4). The design process is described in the following sections.

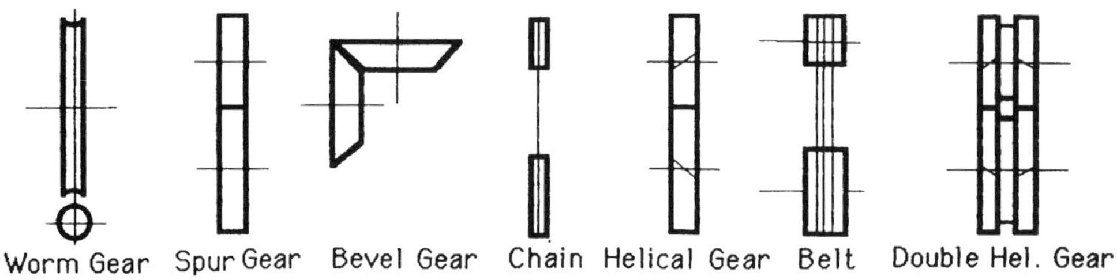

Fig. 3. Basic units.

Identify PDS

When the design process begins, the system lists all PDS. Then the designer identifies the PDS for his particular design task. When the PDS is identified, the criteria are automatically determined because each PDS item associates with one or more criteria inside the KB. So far, the prototype system contains PDS items and their associated criteria as follows:

PDS ITEMS	CORRESPONDING CRITERIA
A. system's speed ratio	ranges of overall speed ratios
B. input and output axes orientation	parallel orientation cross orientation perpendicular orientation
C. central distance of input and output axes	large distance middle or small distance changeable distance
D. accuracy of system speed ratio	ratio accuracy
E. operation noise	transmission smoothness manufacturing accuracy
F. system's transmission efficiency	transmission efficiency
G. system's maintenance and assembly	ease of maintenance
H. easy to manufacture	ease of manufacturing
I. manufacturing cost	cost
J. system's weight	weight
K. system's size	size
L. load distribution on shafts	load distribution

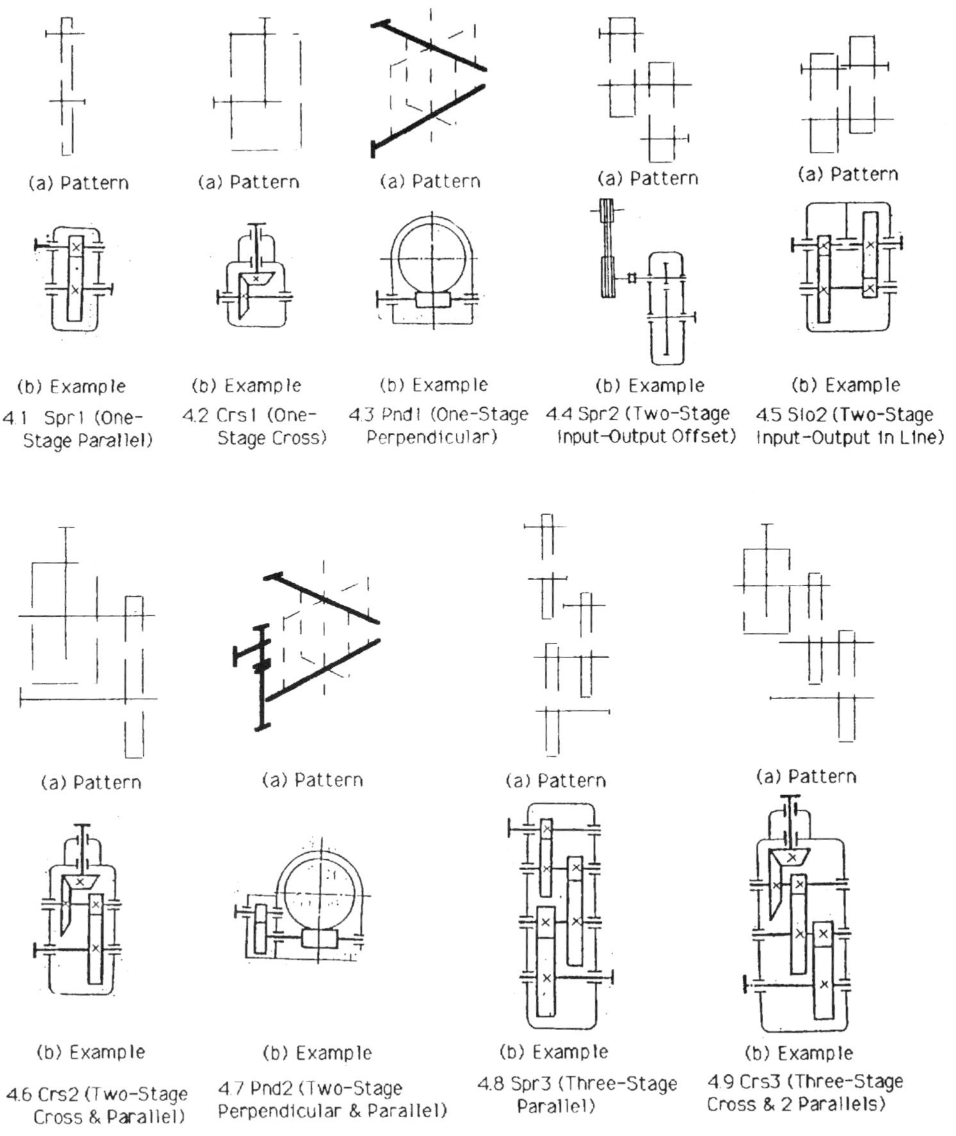

Fig. 4. Arrangement patterns.

PDS and corresponding criteria are classified into two types: critical and flexible. In the above PDS list A, B, and C are critical, while PDS C to L are flexible. Critical criteria are rated by yes or no. If the rating result of a critical criterion is no, then the rated object (e.g., arrangement pattern or basic unit) should be refused. Flexible criteria are rated by numerical rating marks. The higher the rating mark, the better the rating result.

Inside the system's knowledge bases, the critical PDS items are used to decide which concepts should be constructed, while flexible PDS items are used to select rating criteria to evaluate the constructed concepts.

Construct concepts

The construction process involves calling in the relevant subknowledge base(s) first and then constructing the concepts within each subknowledge base.

Calling in the relevant knowledge base depends on the decision made on the basis of the result of PDS, B specified by the user. The following rules are used to determine which subknowledge base should be called:

if *parallel orientation*
then call KB parallel system with criteria of an *overall speed ratio* and a type of *system central distance*

if *perpendicular orientation*
then call KB perpendicular system with criteria of an *overall speed ratio* and a type of *system central distance*

if *cross orientation*
then call KB cross system with criteria of an *overall speed ratio* and a type of *system central distance*

In the above rules, the italicized items are critical criteria specified by the user already. The criteria in the "then" clauses of the rules are passed into relevant subknowledge bases.

Constructing concepts inside the subknowledge bases involves first determining the number of stages of the concept on the basis of the overall speed ratio. Then, depending on the number of stages and the ***system center distance***, the relevant arrangement patterns and basic units are chosen to construct concepts. Let's take the concept construction inside the subknowledge base ***parallel system*** as an example.

On the basis of the overall speed ratio, the possible stages of the concepts are determined first by the following rules:

if overall speed ratio < 4
then one-stage parallel concepts are to be constructed

if overall speed ratio ≥ 4
and overall speed ratio ≤ 7
then two-stage parallel concepts are to be constructed
and one-stage parallel concepts are to be constructed

if overall speed ratio > 7
and overall speed ratio ≤ 100
then two-stage parallel concepts are to be constructed

if overall speed ratio > 30
and overall speed ratio ≤ 100
then two-stage parallel concepts are to be constructed

Next the system constructs the concepts with a determined number of stages (one-stage, two-stage, or three-stage). To do so, the system first selects an arrangement pattern(s) and then assembles relevant basic units into the arrangement pattern. For example, to construct the concept hssio2 having parallel input-output orientation and a small value of system central distance, use the following rules:

when two-stage parallel concepts are to be constructed
and system central distance is small
then two-stage parallel input-output in line is selected

when two-stage parallel input-output in line is selected
then helical-gear pair is in stage 1
and spur-gear pair is in stage 2
and concept hssio2 is formed

The above choices were made because the rating results for the critical criteria small distance and parallel orientation were all yes.

In the same way, up to 48 concepts can be constructed inside the parallel system subknowledge base; up to seven concepts can be constructed inside the perpendicular system subknowledge base; and up to 20 concepts can be constructed inside the cross system subknowledge base.

Generating concept rating marks and drawing description files

When a concept is constructed, its relevant information is passed into an external program written in Quick Basic 4 that has been integrated into the expert system.

Using the passed information about the concept's basic units, arrangement pattern, and coordinates, the external program picks up relevant data from the graphic database to create a concept-drawing description file in AutoCAD format and calculates from the numerical database the concept rating marks, CRMark, using the following formula [3]

$$\text{CRMark} = \sum_{i=1}^{n}\left[\left(\sum_{j=1}^{m} U_j\right)_i + A_i + B_i\right]$$

where

$$B_i = 0 \text{ if } k = m$$

$$B_i = P_i\left(\sum_{j=m+1}^{k} U_j\right)_i \text{ if } k > m$$

$$B_i = Q_i(m-k)U_i^* \text{ if } k < m$$

where m is the number of positions in the data arrangement pattern; n is the total number of rating criteria; U_j is the rating mark of basic unit j; U^* is the rating mark of the datum basic unit; A is the rating mark of an arrangement pattern; k is the number of positions in a concept's arrangement pattern; P is the more-positions modification coefficient; and Q is the less-positions modification coefficient. Values of U, U^*, A, P, and Q are given in Table 1.

Table 1. Values of rating marks.

	Ctr1	Ctr2	Ctr3	Ctr4	Ctr5	Ctr6	Ctr7	Ctr8	Ctr9	Ctr10
BASIC UNITS		**Rating**		**Marks**		**U**				
belt	6	3	4	3.5	1	2	4	1	1.5	2
bevel gear	2.5	2	3	2	2.5	2.5	3	3	2	3
chain	2	3	4	3	1	2.5	4	2	1.5	2
d. hel. gear	4.5	2.5	3	2	2.5	2.5	2.5	3	3	3
helical gear	4	3	3	2.5	3	3	3	3	2	3
spur gear	*3*	*3*	*3*	*3*	*3*	*3*	*3*	*3*	*3*	*3*
worm gear	5	2.5	2	1.5	8	6	2	3	2	1
ARRANGEMENT PATTERNS		**Rating**		**Marks**		**A**				
spr1	0	4	3.5	4.5	4	4	3.5	0	3	0
crs1	0	3.5	3.5	3.5	3	3	3.5	0	3	0
pnd1	0	3	2.5	2.5	3	3	3	0	3	0
spr2	*0*	*3*	*3*	*3*	*3*	*3*	*3*	*0*	*3*	*0*
sio2	0	2.5	3	2.5	3	3	2.5	0	3	0
crs2	0	2.5	2.5	2.5	2.5	3	3	0	3	0
pnd2	0	2	2	2	3	3	3	0	3	0
nbx2	0	5	5	5	5	5	5	0	5	0
spr3	0	2.5	3	2.5	2	2.5	3	0	3	0
crs3	0	1.5	1	1	2	2	2.5	0	3	0
More position coefficient P	−0.2	0	−0.2	−0.5	−0.2	−0.5	0	0	0	−0.2
Less position coefficient Q	1.2	1	1.2	2	1.2	1.5	1	1	1	1.2

NOTE:

- The values are for the following criteria (from Ctr. 1 to Ctr. 10), respectively; transmission smoothness, manufacturing accuracy, ease of manufacturing, cost, size, weight, ease of maintenance and assembly, ratio accuracy, load distribution, and efficiency.
- The italic values are for U^* and A^*.
- For the meaning of arrangement pattern names, please see Fig. 4.

Evaluating the concepts

After each CRMark is calculated, the program selects the concept having the highest CRMark to be the best concept, and presents the results on the screen with the following information:

- the selected concept's arrangement pattern, basic units, and total rating mark
- listing of all the arrangement patterns, basic units, and total rating marks of the constructed concepts.
- each concept's layout drawing, presented by using AutoCAD to access the created drawing files.

At this stage, the system asks the user, "Do you wish to accept the above result?" (yes/no) If the user's response is no, then the following information appears on the screen: "Which concept should be the best one? Please give its concept number (ConNo) in the listing of constructed concepts." In this way the user is allowed to re-select the best concept if the user is not satisfied with the result selected by the system. The re-selection cycle repeats until the user types in yes when responding to the first question above.

EXAMPLE

The problem is to design a power transmission system with the following specifications: overall speed ratio of 10; all types of central distances (large, middle, small, and changeable); output and input orientations including perpendicular, parallel, and cross; concerned flexible PDS including accuracy of speed ratio, transmission efficiency, system size, ease of manufacture, system maintenance, and manufacturing cost. Table 2 and Figure 5 show the design results produced by the prototype system with the above specifications.

CONCLUSIONS

The approach developed here performs engineering design tasks, covering the whole conceptual-design procedure. It integrates expert systems, numerical calculations, data processing, and CAD drawings to provide a computer-based implementation.

The approach has been applied to the conceptual design of a power transmission system, and a prototype system has been developed. All aspects of the approach have been implemented and shown to work, giving believable results. The approach is applicable not only to real problems but also to development of useful design packages.

Two areas are still being refined at present. Sometimes not much differentiation occurs between concept-rating marks. Thus it may be hard to say that the concept with the highest CRMark is the best one. The reason for this is that the relative importance factors of the criteria have not been considered; in other words, all criteria are treated equally. To avoid this, importance factors should be used.

Table 2. Design results of the example.

BEST CONCEPT

Arrangement Pattern:	nb × 2 (two stage parallel, input-output offset, no gearbox case)
Basic Unit in First Stage:	belt
Basic Unit in Second Stage:	chain
Basic Unit in Third Stage:	none
Concept Name:	bcno
Concept Number:	4
Concept Total Rating Mark:	52.5

LIST OF CONCEPTS CONSIDERED (EVALUATED)

ConNo	Concept	B.Unit1	B.Unit2	B.Unit3	Arran.Pattern	CRMark
1	bsspr1	belt	spurg	none	spr2	49
2	bhspr1	belt	helg	none	spr2	48.5
3	bdspr1	belt	dhelg	none	spr2	47
4	bcno	belt	chain	none	nbx2	52.5
5	ssspr2	spurg	spurg	none	spr2	48
6	scspr1	spurg	chain	none	spr2	49.5
7	hsspr2	helg	spurg	none	spr2	47.5
8	hhspr2	helg	helg	none	spr2	47
9	hcspr1	helg	chain	none	spr2	49
10	dsspr2	dhelg	spurg	none	spr2	46
11	dhspr2	dhelg	helg	none	spr2	45.5
12	ddspr2	dhelg	dhelg	none	spr2	44
13	dcspr1	dhelg	chain	none	spr2	47.5
14	sssio2	spurg	spurg	none	sio2	47
15	hssio2	helg	spurg	none	sio2	46.5
16	hhsio2	helg	helg	none	sio2	46
17	dssio2	dhelg	spurg	none	sio2	45
18	dhsio2	dhelg	helg	none	sio2	44.5
19	ddsio2	dhelg	dhelg	none	sio2	43
20	bdcrs2	bevelg	dhelg	none	crs2	43
21	bhcrs2	bevelg	helg	none	crs2	44.5
22	bscrs2	bevelg	spurg	none	crs2	45
23	bccrs1	bevelg	chain	none	crs2	46
24	bbcrs1	belt	bevelg	none	crs2	45.5
25	wpnd1	wormg	none	none	pnd2	51.3

NOTE:

- For the meanings of concept names and arrangement pattern names, please refer to Figs. 4 and 5.
- The higher the CRMark, the better the concept is.

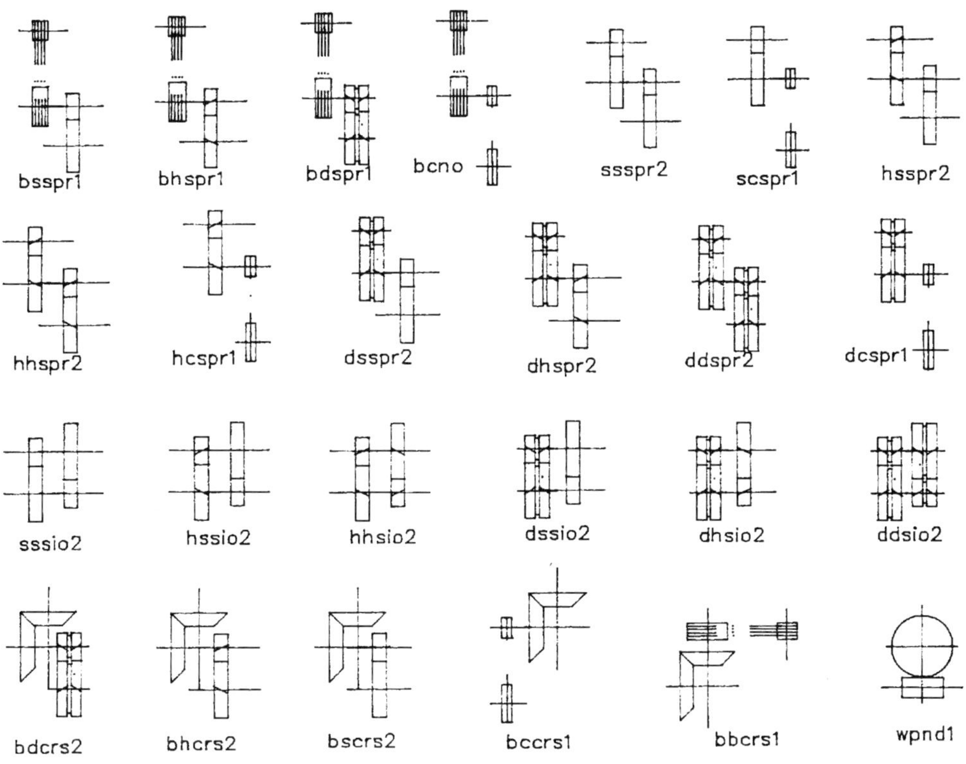

Fig. 5. Constructed concepts in the example.

The values in current databases were given on the basis of an experienced designer's expertise and some design cases. To make the values more accurate, they need to be further studied on the basis of more design cases. When performing a particular design task, the user should be given guidance to check the values used in the databases against previously completed designs.

REFERENCES

1 Xi Plus User Manual, Version 3.00, Expertech Ltd., Slough, England, Aug. 31, 1988.
2 Software package: AutoCAD, Release 10, Autodesk Ltd., London, 1989.
3 D. Su, Knowledge-Based Integrated System Approach for an Engineering Design Procedure, Ph.D. diss., University of Strathclyde, Glasgow, U.K., Nov. 1990.

Chapter 8
Materials Processing and Metrological Design

Since modern products influence large manufacturing markets, we address quality and reliability from a life-cycle approach that can best provide a global view of design. Materials processing plays a very important role in this aspect of design. The subprocesses involved in the manufacture for advanced ceramic materials are discussed by highlighting major parameters that affect properties of the ceramics. A case for the application of statistics in the development of robust processes and its role in experimentation is also described. At Japan's National Research Laboratory of Metrology, a computer-aided measurement and process design and control system (CAMPS) is introduced for plastic and ceramic injection molding processes as a system applied for securing production quality prior to starting mass production. It emphasizes the important role of product design and production process design.

Quality through Engineering Design - W. Kuo (Editor)
1993 Elsevier Science Publishers B.V.

Process Design for Ceramic Materials

Subhas G. Malghan[1]

Ceramics Division, National Institute of Standards and Technology, Technology Administration, U.S. Department of Commerce, Gaithersburg, MD 20899, U.S.A.

Abstract

Process design for advanced ceramic materials such as silicon nitride constitutes consideration of the entire process of manufacture from synthesis of powders to the densification of ceramics. The subprocesses involved in the manufacture are discussed by highlighting major parameters that affect properties of the ceramics. One major subprocess for each stage of the manufacturing of silicon-nitride ceramics is described to provide more details. Because the task of scale-up for complex processes of this kind requires a systematic engineering approach, some possibilities are identified. Finally, a case for the application of statistics in the development of robust processes and its role in experimentation is described.

INTRODUCTION

The synthesis, processing, testing, and commercial applications of advanced ceramic materials have accelerated in the past two decades because of the excellent properties of these materials. The advanced ceramic materials are incorporated in the modern designs mainly because of their superior mechanical strength, hardness, and thermal and chemical stability, and their enhanced magnetic, electrical, optical, and thermal properties. Although a number of these materials have demonstrated superior properties in structural and functional applications, their ultimate and widespread commercialization is yet to be exploited [1]. Two applications that have used some of the significant benefits are electronics and optics, in which the ceramic materials are used for the overall superiority of their electrical, optical, thermal, and magnetic properties. On the other hand, the structural ceramic materials (oxides and nonoxides) are gradually finding their way into commercial applications. Although the oxide ceramics such as alumina and zirconia have been exploited extensively, nonoxide ceramics such as silicon nitride and silicon carbides are yet to be used to their full potential.

Introducing advanced technology for the manufacture of these materials has created new problems and risks that must be minimized by careful control of processing and design variables. The properties of the ceramics vary because of processing effects during manufacturing. In addition, ceramics are brittle, which can cause catastrophic failure from the presence of structural defects. In the application of these technologically superior materials, the primary obstacles are expense, poor repeatability in manufacturing, and lack of design data [2]. Because the process engineering of each of these materials is very specific, we will focus on silicon nitride.

The silicon-nitride ceramics have attracted the attention of designers and applications engineers in the automotive industry and for other applications which require superior mechanical and chemical properties [3]. In recent years, the processing of silicon-nitride ceramic materials has been actively pursued to address some of the issues just described [4–8]. As a result, significant progress has occurred in the processing and properties of these materials. In this paper, process design, as related to the development and application of silicon-nitride ceramics, is described. In each major area of powder processing, an example of a unit operation is described to demonstrate the complexity of parameters involved in the manufacture of silicon-nitride ceramics.

PROPERTIES AND PROCESSING OF SILICON NITRIDE

Properties of silicon nitride ceramics

Silicon nitride is one of the leading materials for use in high-temperature structural components, such as rotors in advanced gas-turbine engines, because of such outstanding properties as high strength, low thermal expansion coefficient, low specific gravity, and high heat resistance. Components include rotor blades, valves, stator vanes, combustion chambers, turbo chargers, piston caps, and cylinder liners [3]. Such use of ceramics offers a number of advantages, including the ability to operate at higher temperatures that leads to increased efficiency and decreased specific fuel consumption, decreased weight, less potential life-cycle cost, and decreased complexity through the use of noncooled components. Like steel, silicon-nitride materials of different properties can be produced by using different composition and processing conditions. Pure silicon nitride shows remarkable resistance to sintering. For example, about 85% of theoretical density was achieved by using reactive-decomposition hot pressing [4] or under high-pressure forming conditions at 1750° C and 600 MPa pressure. In densifying silicon nitride to greater than 98% theoretical density, sintering aids (oxides of yttrium, magnesium, aluminum, and others) from 1% to 20% by weight are used.

Manufacture of ceramics

Current methods of processing ceramic materials have not changed very radically over the years, although many improvements have been made in the equipment [9]. Currently, the processing is batch-mode and labor intensive. Time-dependent processes are not well understood; therefore, defects occurring because of heat and mass transfer in the three-dimensional shapes and from chemical reactions are difficult to identify and eliminate. Hence, process and quality control by conventional techniques is rather difficult and at times is not achievable. All these factors have been responsible for high rejection rates, which range from 5% to 50%, depending on the unit operation. Most often the scrap value of ceramics is extremely low, unlike that of metals. One of the current goals of the advanced ceramic manufacturing industry is to consistently reproduce properties of the materials so that the rejection rate is kept low. To accomplish this, one must correlate considerable physical and chemical data with performance so that the production process can be controlled at a set level to achieve the desired properties in the final products.

The manufacturing process for silicon-nitride ceramics consists of a number of steps, including powder synthesis, powder preparation, consolidation to green body, densification, and finishing. A schematic of dense ceramic manufacture is shown in Fig. 1, in which the consolidation to green body is by injection molding. The unit operations are interactive in the sense that depending on the route selected for consolidation, the entire powder-processing route, often including the powder-synthesis route, can change.

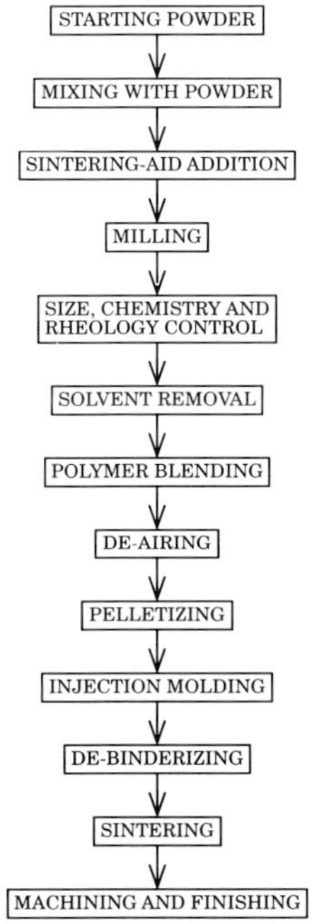

Fig. 1. Generalized flowsheet for ceramic powder processing in the manufacture of silicon nitride ceramics. Shape forming is done by injection molding.

Powder synthesis

Silicon-nitride powders are produced by five major processes, depending on the composition of the starting raw materials. The processes and properties of the resulting powders are described elsewhere in detail; only a brief description is given here [10]:

1. direct nitridation methods, in which the silicon metal is heated at high temperatures in the presence of nitrogen or ammonia
2. carbothermal reduction and nitridation, in which a mixture of silicon oxide and carbon are heated simultaneously at a high temperature in the presence of nitrogen
3. vapor-phase reactions, in which mixtures of silicon tetrachloride and gaseous ammonia diluted with nitrogen are reacted at high temperatures
4. thermal decomposition of amides or imides, in which silicon tetrachloride and liquid ammonia are reacted at ambient or lower temperature, and the resulting metallic amides or imides are heated at high temperatures in an atmosphere of ammonia or nitrogen after they are separated from the reaction mixture
5. pyrolysis of silane, an emerging method, in which high-purity silane gas is heated in a reactor with nitrogen or ammonia; the resulting product could be either an amorphous or crystalline silicon nitride

With such a large number of possibilities for powder synthesis, the resulting powders can possess different degrees of oxidation resistance and sinterability. In silicon-nitride powders, the major chemical impurities are considered to be oxygen, carbon, metals (Al, Fe, Ca, etc.), and halides (F, Cl, etc.). Of these, oxygen is the single most abundant impurity in the starting powders, often ranging from 0.3% to 2% by weight of commercial powders. The oxygen content of the starting powders and that added via sintering-aids constitutes the total oxygen that participates in glass formation at sintering temperatures. Therefore, data on the oxidation of silicon nitride powders and composition of the oxidized layer are some of the necessary pieces of information required to evaluate powder sinterability. The processing, sinterability, and oxygen content of powders produced by amide decomposition has been studied extensively in our laboratory and elsewhere.

The thermal decomposition of imides produces a high-quality silicon-nitride powder by liquid-phase reaction of silicon tetrachloride and ammonia [11]. A schematic of the process is described in Fig. 2. Because the nitridation of silicon begins from the liquid phase, the main feature of this process is a more complete conversion to silicon nitride in the bulk phase. In addition, the final powder is very pure because the starting raw materials are very pure. Another notable advantage of this process is its ability to produce powders of different and narrow size distribution without resorting to milling. Some disadvantages of the powders produced by this process are a high degree of agglomeration, blocky morphology of particles, and a relatively high concentration of chlorine.

Powder consolidation

Depending on the form of the powder, we have a large number of powder consolidation or shape-forming processes available, such as dry, molded with polymer, or slurry. Some of these options are summarized in Table 1.

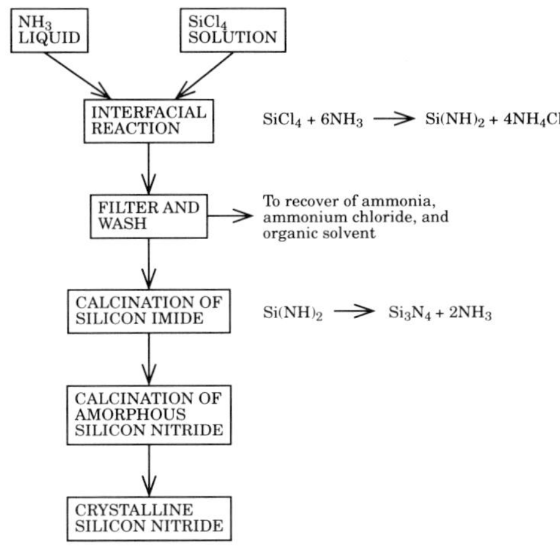

Fig. 2. Schematic of silicon-nitride synthesis by interfacial reaction.

These processes are required because the powders must be prepared in various ways that are compatible with the forming processes in question [12]. Although a large number of possibilities exist, only a selected few forming processes can actually meet the requirements of high volume, cost effectiveness, and shape and net-shape forming capability during production. As a result, dry pressing, injection molding, extrusion, tape casting, and slip casting have been widely used. Of all these possibilities, injection molding has been the one most studied by the ceramic industry. In injection molding, the blended powder is mixed with a heated polymer, and the resulting mixture is injected under pressure into a mold and cooled to room temperature. A schematic of the injection-molding process is shown in Fig. 3. Size-matched and well-mixed powders are homogenized with 10% to 30% weight of polymer blend. The powder-polymer mixture is homogenized thoroughly and pelletized, and the resulting pellets are molded. The major parameters of molding are the mold and die design, injection speed and pressure, barrel and the mold temperature, holding time, and most of all, the properties of the mixture [13]. In addition, the most important step after molding is the debinderizing in which the polymer is removed in a slow and controlled process. Severe defects such as cracking, blistering, and warping can occur in this step [13,14].

Powder densification

Densifying silicon-nitride powders is one of the key factors that determines the final properties of the ceramic. The selection of an appropriate chemical composition is necessary to achieve the desired design criteria for this material. The chemical composition, in turn, is determined by the raw materials (powder

Table 1. Powder consolidation or shape-forming processes for silicon-nitride powders.

- Primary Consolidation Processes
  - Dry Processes
    - Dry pressing
    - Isostatic pressing
    - Hot isostatic pressing
    - Hot pressing
  - Plastic Forming Processes
    - Injection molding
    - Extrusion
    - Compressing molding
    - Ram pressing
    - Jiggering
  - Fluid Forming or Slurry Casting Processes
    - Slip casting
    - Pressure casting
    - Wet pressing
    - Tape casting
    - Soluble mold casting
- Secondary Forming Processes
  - Machining Processes
    - Fired part machining
    - Bisque part machining
    - Green part machining

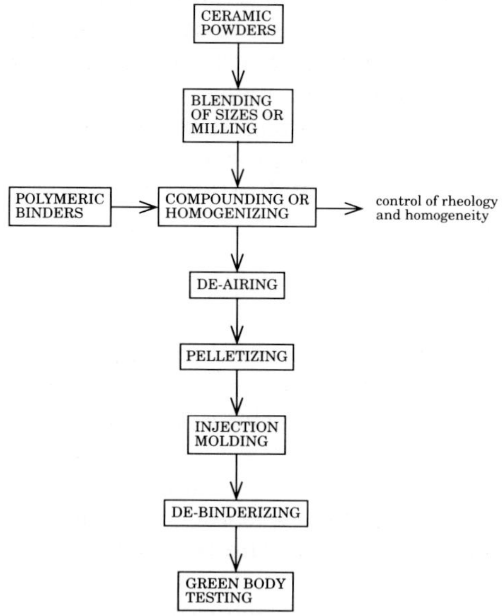

Fig. 3. Schematic of injection-molding process showing major steps.

and sintering-aid) used in the powder preparation and by the densification route. Some examples of the densification processes and types of silicon nitride [7] are

- gas-pressure sintering (GPS)
- sintered silicon nitride (SSN)
- hot-pressed silicon nitride (HPSN)
- conventional clad, hot-isostatic pressing (HIP)
- sintered reaction-bonded silicon nitride (SRBSN)

Each of these processes has specific merits to produce silicon nitrides with different properties. The sintering aids added to cause densifying, react at high temperatures with a layer of silica present around each particle to form an oxynitride liquid. This liquid allows for shrinkage of the green body, and the alpha-phase dissolves in the liquid and transforms by precipitating as beta-silicon nitride. After sintering, during cooling, the liquid remains as a metal-silicon-oxynitride grain-boundary phase, which is either a glass or a crystalline phase depending on the original concentration of sintering aids and their composition. The nature of these phases has a marked effect on subsequent thermo-mechanical behavior of the ceramic. Among several possibilities previously mentioned to carry out densification, the GPS process is the most promising sintering technique because it can produce pore-free, dense ceramics even with small amounts of sintering aids.

In GPS, green bodies of complex shapes can be densified, while decreasing the overall cost. Two-step GPS is a modification of the primary GPS process in which sintering at low pressures in the range of 0.1 MPa is followed by a final densification by raising the pressure to 2.0 MPa at 1780°C. Fully dense silicon nitrides have been developed in a single-step GPS process by using nitrogen overpressures in the range of 1 to 10 MPa and temperatures between 1900°C and 2100°C. The nitrogen pressures and temperatures are determined by the equilibrium nitrogen pressures required to retard the following reaction: $Si_3N_4(s) \rightarrow 3Si(l) + 2N_2(g)$. For example, in the temperature range of 1900°C to 2100°C, the respective minimum N_2 pressures of approximately 1 and 5 MPa (~10 and 50 atm) permit good sinterability and <2% weight loss [8].

PROCESS DESIGN

For new technologies to become a reality, many questions must be answered. Two of these questions are (1) Has the technical feasibility of the process been established? and (2) What technical and cost data are required for the design of a production plant? The overall sequence of steps required in the process development are shown in Fig. 4. The approach to developing processing technologies and ultimate commercialization of advanced ceramic materials continues to be a concern because of a smaller production scale, a fast-changing materials knowledge base, and large capital requirements. The laboratory- and bench-scale studies have a unique place in the process development for advanced materials because this is where the concept is fully tested, cause-effect relationships are developed, and a preliminary process design is generated.

Mathematical models and their applications for computerized design alter the strategies for process development by permitting forecasting of process results through preliminary design and by feeding back conclusions to guide the experimental program [15]. Because most of the experiments can now be conducted on laboratory- or bench-scale units, the purpose of the pilot plant is becoming one of proving predictions and calculating results rather than generating data. However, pilot plants do have a key role in the process design because they are the last opportunity to commit blunders on a small scale so that profits can be made on a large scale.

As shown in Fig. 4, there is no unique sequence of operations. A number of feedback loops are possible, depending on the complexity of the process and the amount of effort spent at each step. Occasionally, the data from laboratory tests can be analyzed, and the design of a production plant can be developed. A significant feedback loop is between bench-scale development and the pilot plant, where most of the cause-effect relationships are tested and a set of quantitative correlations is established.

The concept of a miniplant has proven to be a powerful technique for developing new processes in the chemical-process industry. According to Robins [16], some basic guidelines for miniplant design and operation include:

1. developing a conceptual design of the full-scale plant from fundamental thermodynamic and kinetic data

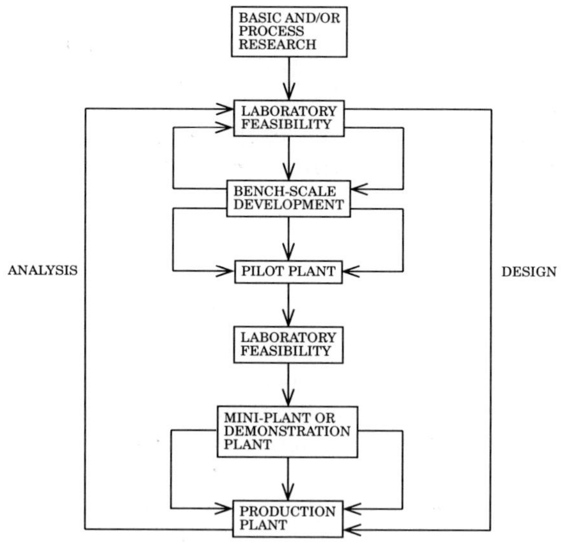

Fig. 4. Schematic of sequential steps involved in the process development and design of a manufacturing process.

2. scaling-down the plant concepts into a miniplant design to demonstrate the feasibility and performance of the steps in question, while maintaining good engineering design and plant constraints
3. using the type and size of miniplant equipment where credible scale-up relationships exist
4. operating the miniplant on actual raw materials to be used in the plant

Because many of the unit operations in processing ceramic powders are well-proven in other applications such as minerals processing and chemicals production, the miniplant approach seems to have merit for advanced materials-process design.

APPLICATION OF STATISTICAL ANALYSIS

The design of experiments and analysis of the resulting data by statistical techniques is invaluable because the amount of experimentation can be reduced and complex interactions between the variables can be obtained. One additional, highly significant benefit of the statistically designed experiments is the ability to obtain interactions of process parameters across the boundaries of a single-unit operation. One example is the degree of aging or degradation due to the environment of submicron powders in the bench-scale development versus the pilot plant. Unless the experiments are designed carefully, such variations can cause significant differences between the results. Design and planning of all experiments aimed at evaluating the interaction effects can account for such variations.

One area that will gain prominence in the future in the manufacture of advanced ceramic materials is the application of mathematical models for simulating bench-scale and pilot-plant data. This will significantly reduce the amount of experimentation. The primary limitation at this time is the unavailability of such models. One of the first steps toward the development of such models is to develop statistical correlations, which require elaborate experimentation using statistical experimental design. The ultimate goal is to develop deterministic models on the basis of first principles.

CONCLUSIONS

This paper addresses one of the major areas of concern in quality through design as applied to the advanced ceramic materials manufacturing. This concern is related to the design of processes by which improvements in manufacturing repeatability can be increased for advanced materials. Because advanced ceramic materials offer certain unique properties, their manufacture has been actively pursued, but success has been limited because of the lack of sufficient process design data. The problems identified by the industry have been largely an outcome of numerous empirical studies. The scope of these studies has been increasingly complex and costly because of a lack of data on physical and chemical interrelationships between the process parameters and the quality of the resulting ceramic. A large number of unit operations and process parameters are involved in the process development and scale-up of manufacturing processes. Statistical design of the experiments and evaluation of data is one answer to overcoming the complexity, cost, and quality issues. In the long run, the development of process models will play an important role in the design of processes of pilot- and plant-scale.

REFERENCES

1 Ceramic Application and Design, Ceramic Industry (Feb. 1988) s3–s18.
2 E. J. Kubel, Jr., Adv. Mater. and Processes (Sept. 1989) 55–60.
3 K. Takahara, Techno Japan, 20, no. 6 (1987) 8–14.
4 P. E. D. Morgan, Research on Densification, Character and Properties of Dense Silicon Nitride, NTIS AD-757-748, March 1973.
5 D. W. Richerson, Ceramic Bull., 7 (1973) 560–69.
6 G. Schwier, On the preparation of fine silicon nitride powders, in: Progress in Nitrogen Ceramics, F. L. Riley (ed.), Martinus Nijhoff Publishers, Boston, 1983, pp. 157–166.
7 R. N. Katz, Nitrogen Ceramics 1976–1981, in: Progress in Nitrogen Ceramics, F. L. Riley (ed.), Martinus Nijhoff Publishers, Boston, 1983, pp. 3–20.
8 C. Greskovich, A gas pressure sintering process for producing dense Si_3N_4, in: Progress in Nitrogen Ceramics, F. L. Riley (ed.), Martinus Nijhoff Publishers, Boston, 1983, pp. 283–293.
9 J. C. Agarwal, Adv. Ceram. Mater., 1, no. 4 (1986) 332–334.
10 G. Schwier, On the preparation of fine silicon nitride powders, in: Progress in Nitrogen Ceramics, F. L. Riley (ed.), Martinus Nijhoff Publishers, Boston, 1983, 157–166.
11 Advanced Ceramic Materials Technical Report, 3.6000, Ube Industries, Tokyo, Japan, 1987.
12 R. L. Lehman, Overview of Ceramic Design and Process Engineering, Engineered Materials Handbook-Ceramics and Glasses, 4 (1991) 29–37.
13 M. J. Edirisinghe and J. R. G. Evans, Mater. Sci. Eng., A109 (1989) 17–26.
14 F. L. Kennard, Ceramic Eng. Sci. Proc. (1980) pp. 1095–1111.
15 J. M. Berty, CEP (Sept. 1979) 48–51.
16 L. A. Robins, CEP (Sept. 1979) 45–47.

Quality through Engineering Design - W. Kuo (Editor)

CAMPS and Metrological Design

Masayoshi Koike[a] and Hiroshi Yano[b]

[a]Instrumentation Mathematics Section, Measurement System Department, National Research Laboratory of Metrology, 1-1-4, Umezono, Tsukuba, IBARAKI 305, Japan

[b]Department of Technology, Miyagi University of Education, Aoba, Aramaki, Sendai, MIYAGI 980, Japan

Abstract

At Japan's National Research Laboratory of Metrology (NRLM), a systematic approach to improving the measurement system and utilizing measurement results based on quality engineering has been applied to metrological design in a quality design system CAMPS (computer-aided measurement and process design and control system). CAMPS was developed for plastic and ceramic injection molding processes as a system that is applied to secure production quality prior to starting mass production. It emphasizes the important role of product design and production process design to improve product quality and of metrological design to perform them effectively. NRLM proposed a theme of developing intelligent CAMPS to the Ministry of International Trade and to industry's Intelligent Manufacturing System (IMS) project, an international research cooperation project.

INTRODUCTION

The National Research Laboratory of Metrology, NRLM, has a responsibility for research and development of measurement standards and technologies in Japan similar to the National Institute of Standards and Technology, NIST, formerly the National Bureau of Standards, in the United States of America. The Instrumentation Mathematics Section of NRLM is studying software aspects of measurement technology and management of measurement systems, such as error evaluation methods; experimental design for measurement; analysis of measured data; and utilization of measurement results, calibration systems, and measurement control. In the course of these studies, we applied quality engineering concepts to metrological fields and found the methods to be advantageous. Especially, error evaluation by signal-to-noise (S/N) ratio, measurement system improvement by robust design, and calibration interval determination by on-line loss function are the key concepts of metrological design. Metrological design can be said to be the quality design system in the metrological fields. It has been systematized in the development of the quality engineering system CAMPS, computer-aided measurement and process design and control system.

This paper presents the concepts of CAMPS, quality engineering, metrological design, and the application of the metrological design system, using Japanese Industrial Standard for calibration systems.

QUALITY ENGINEERING AND QUALITY DESIGN SYSTEM

Quality design system: CAMPS

Quality engineering includes various kinds of statistical methods and design strategies based on Taguchi's methods and concepts and applied to development of a quality design system. The goal of this system is to secure product quality in the design stages of products and production processes prior to mass production and to achieve high-quality products from the first stage of mass production. It is also a system that uses the results of measurement effectively in the production system.

CAMPS, developed by NRLM, is a personal computer software which facilitates easy implementation of a quality design system in a specific production system. As shown in Fig. 1., CAMPS consists of measurement design (origin of metrological design), parameter design (robust design in the United States), on-line control design and on-line control [1].

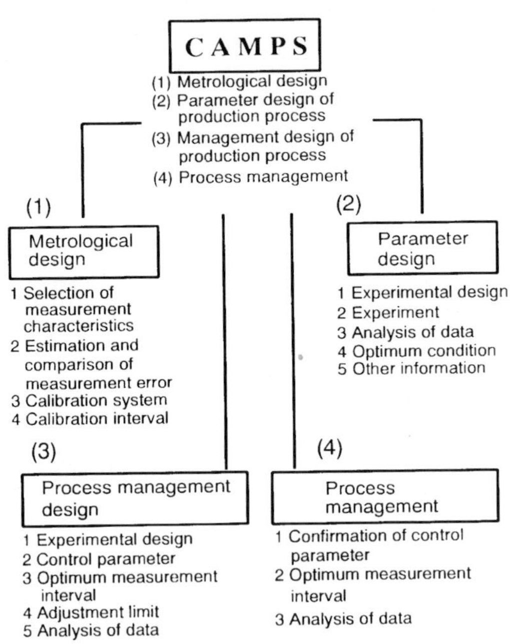

Fig. 1. Composition of CAMPS.

Quality engineering

Quality engineering, the technical base of CAMPS, offers two important means for improving product quality:

1. an evaluation method to determine the quality of products by S/N ratio
2. experimental methods for improving the S/N ratio

The S/N ratio also indicates the performance of a system, that is, the stability or controllability of the system.

There are two phases of quality engineering to improve quality: off-line and on-line. Off-line quality engineering comprises methods used in product design and production process design steps. On-line quality engineering refers to methods used in mass production.

In the off-line phase, products and production process should be designed as a system that is robust to noise sources and is well controlled with a signal factor. To achieve a high-performance system, in general, involves the following four production stages:

1. system selection
2. parameter design (robust design)
3. tolerance design
4. tolerance determination

The system selection stage involves designing system components that achieve intrinsic functions, which are determined on the basis of the knowledge of the specific technology. The other three stages form the objects of the quality design system [2]. Parameter design, using carefully planned experimentation including simulation, can determine the optimum system parameters. Tolerance design can estimate the realistic error of the system and determine the rational allowance of the parameters. Allowance of the quality characteristics can be determined in the tolerance determination step. Off-line quality engineering offers an up-stream means for quality improvement.

In the on-line phase, management and control systems for the mass production process are designed and operated with technical and economical efficiency [3]. For the mass production process a feedback control that is based on on-line measurement is applied to get products close to the designed value. Although the process is already optimized by robust design, the process is controlled to decrease process fluctuation caused by such noise sources as environmental and internal changes. In on-line control, products are periodically sampled and measured. If a large process deviation is detected, the signal factor is changed to bring the product's characteristic to its target value. On-line control is achieved through Taguchi's β coefficient method, which is a statistical and predictive control method [4].

Application of CAMPS to the plastic molding process

Applying the procedure of CAMPS to designing a production process is shown in Fig. 2. We assume that a product's design and process hardware are given. That means the design of the product and system selection of the production process are already completed prior to applying CAMPS.

In the trial production stage, process parameters are optimized by the robust

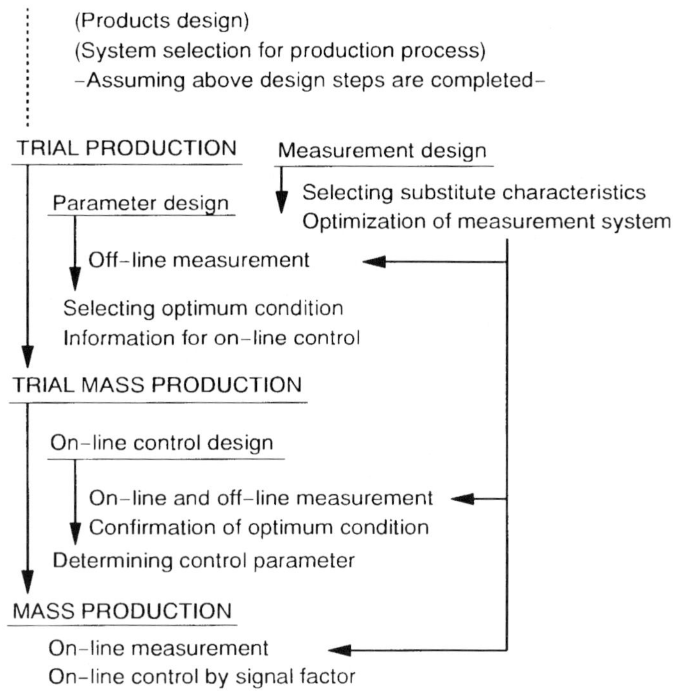

Fig. 2. Applying procedure of CAMPS.

design from the viewpoint of variability. Also, an adjustment process parameter that is controlled to get a target value at the start of mass production and a signal factor for on-line control are selected, on the basis of the results of robust design experiment, from the parameters which change the value of the product's characteristics linearly but do not change the S/N ratio.

In the trial mass production stage, a predicting equation for process fluctuation and a relational equation between the measured product's characteristic and the signal factor for on-line control are derived as an on-line control design. Conditions for mass production are determined in these procedures.

In the mass production stage, periodical sampling and on-line measurements are performed to detect process fluctuation. When the process fluctuation exceeds the control limit, process is corrected by operating the signal factor to adjust the product's value of the characteristics to the target value.

In the measurement design stage, measurement methods used in parameter design, on-line control design, and on-line control are optimized. Characteristics to be measured in off-line and on-line steps are selected on the basis of their purposes. The calibration system for the measurement instruments used in the production system is also determined in this stage. The measurement design concept is the prototype of the metrological design system.

CAMPS was first applied to a plastic molding process. The variability of the plastic model products decreases to one-third in standard deviation of their dimension by using robust design. On-line measurement detects the process deviation from the designed value and corrects the process by changing molding pressure, a signal factor. The on-line control system of two characteristics was also designed with two signal factors [5].

Research on CAMPS was an epoch-making project in the history of the Instrumentation Mathematics Section of NRLM. We applied individual methods such as S/N ratio, design for experiments, and loss function to particular cases before we applied CAMPS to the plastic molding process. The S/N ratio method has been applied to the optimization of dimension measurement for plastic parts from the late 1970s. Dimension measurement of plastic parts is strongly influenced by environmental factors such as temperature and humidity and measuring conditions such as measuring force and operators when compared to metal parts measurement. The research on plastic measurement was connected to the research on plastic production process because it is essential to improve the plastic production process rather than to detect errors in the plastic products. The plastic CAMPS was developed in parallel with systematizing quality engineering.

Fine-ceramic CAMPS

The plastic molding process was a rather simple and independent process. For the purpose of applying CAMPS to more complex processes, we studied a fine-ceramics production process with molding. As shown in Fig. 3, it is composed of four basic processes: kneading, injection molding, degreasing, and sintering. Each process except molding is operated in batch. Measurement characteristics in and between the basic processes are studied to explore the possibility of feedback or feedforward on-line control. Optimization was performed by robust design for the whole process. The variability in dimension of sintered products decreases one-half in standard deviation [6].

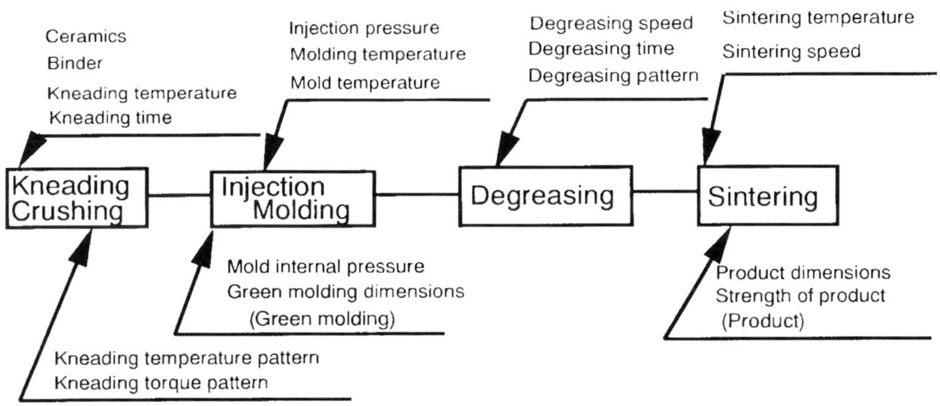

Fig. 3. Measurement characteristics and injection molding process of fine ceramics.

In evaluating performance of the production process, the concept of transferability is applied. Transferability means the ability of the production process to copy the form of the mold to the form of ceramic products without deformation and variation. In plastic CAMPS, controllability and stability were the main targets of the optimization, but in the fine-ceramic CAMPS, we emphasized transferability as a basic function of the production process. The process was optimized with regards to less deformation. Transferability was originated in the proposal of technology development by Taguchi and developed in the fine-ceramic CAMPS.

In the fine-ceramic CAMPS project, we learned the importance of the basic function of the process or the system concerned. The basic function of the system can be defined as the ideal relational equation between the signal factor and the output. The choice of signal factor and characteristics to be measured varies according to the definition of the basic function. The choice of transferability in fine-ceramic CAMPS made progress in the basic technological approach to the molding production process. How applicable the results of experiments are also varies by the experimental design. The choice of noise factors is especially important for technology development.

METROLOGICAL DESIGN SYSTEM

Outline of metrological design

Metrological design is a more general concept than measurement design. In Japanese, the word measurement is used with rather limited meaning for action in getting data. Measurement design was developed in relation to the more general concept of metrological design.

The measurement system that we are studying is composed of instruments, operator, environment, and situation. The steps of measurement include the setting of the purpose, devising a data acquisition plan, analyzing the data, and using the measurement results. From the viewpoint of measurement system performance and loss function, whole steps should be optimized. The role of measurement activities in the production system should be emphasized more. For this purpose, we used quality engineering in the design of the measurement system and in this context called it the metrological design system.

Metrological design is the key concept in the quality design system because measurements give information on the production system and product quality, and correct information makes high quality production. Metrological design consists of the following two parts:

1. designing measurement system and instruments
2. designing management system of measurement systems

Designing measurement systems or instruments is similar to designing products in quality engineering. The four stages of design mentioned above are applied. Developing new measurement technology corresponds to system selection. Robust designs for instrument parameters are performed by taking environmental conditions in using the instrument as noise factors to optimize in the design of instruments as shown in Fig. 4. For the production process of instruments, tolerances of instrument parameters are determined in the tolerance design stage. Measurement error is estimated in the tolerance design step and compared with requirements in the tolerance determination step.

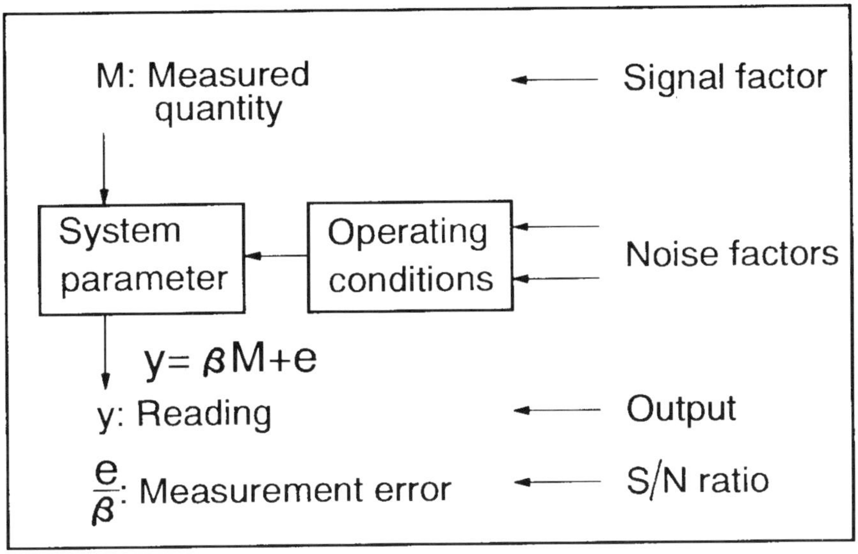

Fig. 4. Robust design of measurement system.

The variability of a measuring instrument and its parameters must be within the tolerances determined. The production process of the instrument should be optimized to keep the production errors within the tolerances. This part of an application of off-line quality engineering to the measurement system or instrument is rather straightforward. Four stages in the design of the production process are applied to this part.

The management system is a software system for effective measurements. It includes

- selecting measurement characteristics
- defining the measurement system
- evaluating measurement errors
- designing a data acquisition plan
- designing the calibration system of measurement system
- extracting information from measurement data
- utilizing measurement results

The management system is optimized to secure measurement quality and economical effectiveness. Various methods of quality engineering and statistical methods can be applied to the design of the management system. The loss function of quality engineering is especially useful for optimizing the system for economical efficiency. The calibration system will be discussed in the next chapter.

Objective system of metrological design

Traditionally measurement error was thought to arise only from the measurement device used, but research on error evaluation in NRLM revealed that this idea is not true. Measurement error can be estimated when the measurement system is defined. For example, the measurement object, the operator, the measurement condition such as temperature and moisture, and the management system should be defined as parts of the measurement system to evaluate the error. Accordingly the objective system of the metrological design should be the measurement system that includes measurement instruments such as a hardware system and the management system as a software system as shown in Fig. 5. The measurement system is evaluated and optimized through the metrological design.

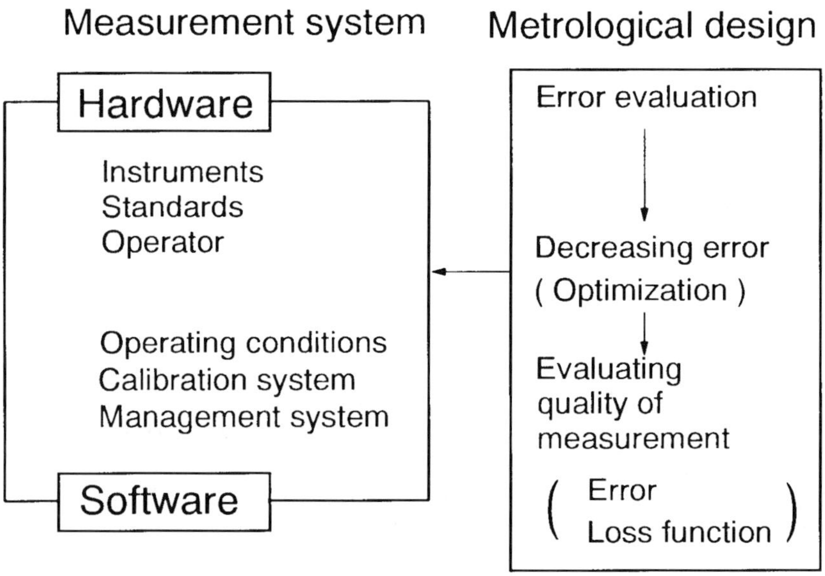

Fig. 5. Metrological design of measurement system.

The NRLM research project looking at a measurement system as a man-machine system clarified the importance of the definition of the measurement system in the mid 1970s. Taguchi's S/N ratio method was used in the project as well as traditional statistical methods including analysis of variance and experimental design. In the project, personal difference in measurement results was studied, and the measurement error varied depending on the definition of the objective measurement system.

Error evaluation for metrological design

The S/N ratio method is the key to the error evaluation method. Error is defined as the difference between measured value and true value. Following this definition of error, we cannot know the error exactly as long as true value is unknown. Commonly, however, the true value is unknowable. Nevertheless, a method of performance evaluation based on the definition of error is proposed. The size of error can be estimated even in the situation where the true value is unknowable. The S/N ratio method can be applied to estimate the size of the error by using the proportionality of the reading y_i to measured quantity M_i. Using regression analysis, we can formulate the S/N ratio as

$$\eta = \frac{\beta^2}{\sigma^2} = \frac{\frac{1}{r}\left(S_\beta - V_e\right)}{V_e} \tag{1}$$

where

$$\begin{aligned} S_T &= \sum_{i=1}^{n} y_i^2 \\ S_\beta &= \frac{\sum\left[\left(M_i - \overline{M}\right)y_i\right]^2}{r} \\ r &= \sum\left(M_i - \overline{M}\right)^2 \\ V_e &= \frac{S_T - S_\beta}{n-1} \end{aligned} \tag{2}$$

The advantages of the S/N ratio method are as follows:

- It allows the possibility of error evaluation without the true values of measured quantity because it is based on the evaluation of linearity.
- It suggests that noise factors should be taken into account in experiments to evaluate realistic measurement error.
- The S/N ratio is the index of variability after the system is calibrated properly. It can evaluate the performance of the system managed in the best way.

In quality engineering, the S/N ratio method is applied to the evaluation of the performance of systems such as products and production processes. The main purpose of improving a system is to minimize the system variation, that is, to maximize the S/N ratio of the system. S/N ratio is a good index for system variation which has not been investigated directly with traditional statistical analysis.

The S/N ratio method is especially important for measurement or instrumentation. Historically the first application of the S/N ratio was in the metrological field. Since the 1970s, the S/N ratio and experimental design have been applied to error evaluation experiments for hardness standards and other measurement systems in NRLM.

METROLOGICAL DESIGN FOR CALIBRATION SYSTEM

JIS Z 9090 general rules for calibration systems

Metrological design emphasizes the importance of the calibration system as a management system for measurement systems. Calibration is essential to establishing a relation between the quantities to be measured and the readings at the time of measurement. Under the assumption that all the actions we take will be based on this relation, the major target of quality engineering is to evaluate the error variance that remains after calibration and to minimize it.

The standard JIS Z9090 [7] was established in 1991 for designing calibration systems. The drafting committee organized in the Japanese Standards Association was led by Taguchi and by Yano of NRLM. It shows an error evaluation method and management system for measurement systems.

The scope of JIS Z9090 is regulated as follows: "This Japanese industrial standard specifies the common items for the definition of a calibration system for measuring instruments to be used for the measurement of physical quantities and the determination of the size of errors related to measurement" [7].

There is no technical reason why the scope of JIS Z9090 has to be limited to physical quantities. However, there already exists a JIS standard for permissible tolerance, including error evaluations, in chemical analysis. There are some differences between the chemical and physical fields in the definition of terms and in the approach to the standards and measurements. Even JIS standards have some inconsistencies. The standard for calibration systems can also be applied to chemical measurements, but to avoid fruitless arguments about terminology, we have limited the scope of JIS Z9090 to physical measurements.

The contents of JIS Z9090 are as follows:

1. Scope
2. Definition (Glossary of terms)
3. Basic concept of calibration system
4. Items to be determined for calibration system
5. Procedures to define feature of calibration system

Appendix 1. Calibration interval and correction limit
Appendix 2. Error estimation in measuring the products
Appendix 3. Error estimation in calibration operations using standards
Appendix 4. Error estimation in indicated value of standards

The JIS specifies the procedures to determine the following items.

- classification of calibration
- standard and levels of standard
- calibration interval and correction limit
- procedures for calibration
- action after calibration
- conditions for use of measuring instrument
- procedures for measurement

One of the most important items is the classification of calibration, which is determined on the basis of the relational formula between the readings y and the values of standards M. The JIS deals with calibrations by linear-type formulas.

For the error variance to be a realistic one, we should include in it both the nonlinearity and variability of the measurement system. The size of errors, for example, depends on the assumed relation between true values and the readings. Also the variance associated with the measurement of standards can be quite different from the variance associated with the measurement of products. In many cases, calibration of instruments is conducted in a good environment, but product measurement is conducted on the shop floor. Differences in environments and measuring objects can cause large errors.

Suppose that a measurement instrument is used after it is properly calibrated with standards, the true values of which are known. Then the total error variance is composed of the following three components:

1. error in the indicated value of the standards, σ_0^2
2. error in calibration operation with the standards, σ_c^2
3. error in measuring the products, σ^2

These error components can be estimated by the experimental methods described in the appendixes. The variance of the measurement on objective products is estimated as a sum of these components. That means the total variance can be formulated as

$$\sigma_T = \sqrt{\sigma_0^2 + \sigma_c^2 + \sigma^2} \tag{3}$$

The result of measurement can be expressed as

$$\hat{M} \pm 2\sigma_T \tag{4}$$

where $\hat{M}$ denotes the estimated value of the quantity to be measured. There are many arguments concerning the confidence interval for the true value. The JIS does not specify a confidence interval, but it is important to have rules that the components of variance are added to form the total variance and that 2σ is taken as the expression of uncertainty without a probability expression. The more important thing is to estimate realistically the size of the error for each component.

When the total variance is estimated, the following loss function is tested to judge the validity of the calibration system:

$$L = \frac{A}{\Delta^2}\sigma_T^2 + b \tag{5}$$

where A and Δ denote the rejection loss and the tolerance of the measured products, respectively, and b denotes the measurement cost per product.

When the first and the second terms on the right side of Eq. (5) are almost equal, the calibration system is valid and balanced with respect to cost and loss.

The appendixes in the JIS give the technical background of the calibration to simplify the main body of the standard. Appendix 1 gives a rational determination of the calibration interval. Appendixes 2, 3, and 4 give procedures in experimental design and the calculations for estimating the components of measurement error. These procedures are applied in defining the features of the calibration system.

Calibration interval

Appendix 1 gives a method of rationally determining the calibration interval on the basis of loss function. This method is an application of optimizing the feedback control system in on-line quality engineering. Total loss of the calibration system is expressed as a sum of management cost and quality loss as follows:

$$L = \frac{B}{n} + \frac{C}{u} + \frac{A}{\Delta^2}\left[\frac{D^2}{3} + \left(\frac{n+1}{2} + 1\right)\frac{D^2}{u^2}\right] \tag{6}$$

where n is the check interval,
u is the correction interval,
Δ is the tolerance of products,
A is the rejection loss of products,
B is the check cost,
C is the correction cost,
D is the correction limit,
1 is the time lag of correction.

The management cost consists of the check cost of the first term and the correction cost of the second term. Quality loss of the third term comes from two sources of variability: the variability of apparatus and the uncertainty of the standards.

Optimum calibration interval n and correction limit Δ can be determined by the following equations to minimize the total loss:

$$n = \sqrt{\frac{2u_0 B}{A}\frac{\Delta}{D_0}}$$
$$D = \left[\frac{3C}{A}\frac{D_0^2}{u_0}\Delta^2\right]^{1/4} \tag{7}$$

where the subscript 0 indicates that the value used in the present calibration system is substituted.

This procedure is similar to the optimization of production process control in on-line quality engineering. The variability of the measurement instrument is similar to the variability of the production process, and the measurement error is effectively the same for products as for production processes as shown in the loss function.

Experimental design for error estimation

Appendix 2 shows the method for experimentally determining the size of errors generated by the measuring instrument or errors caused by environmental conditions within the interval of calibration that cannot be removed by calibration. A signal factor and noise factors are assigned in an orthogonal array L_{18} in Table 1, and the error variance can be obtained from 18 data points taken according to the L_{18} design. The second level is set as the average condition during the calibration interval. Interval h of the noise factor is set as $(3/2)^{\frac{1}{2}}\sigma_E$ where σ_E denotes the standard deviation of the noise factor's change during the calibration interval. Then the calculated variance by L_{18} data corresponds to the actual size of variance in the calibration interval. If the levels of error factors are changed by control, a new standard deviation or error variance is estimated from the results of an L_{18} experiment. The calculation procedure is an application of tolerance design in quality engineering.

Table 1. Orthogonal array L_{18} having been assigned with signal and error factors.

Line	A	M	B	C	D	E	F	G	Data
No.	1	2	3	4	5	6	7	8	
1	1	1	1	1	1	1	1	1	y_1
2	1	1	2	2	2	2	2	2	y_2
3	1	1	3	3	3	3	3	3	y_3
4	1	2	1	1	2	2	3	3	y_4
5	1	2	2	2	3	3	1	1	y_5
6	1	2	3	3	1	1	2	2	y_6
7	1	3	1	2	1	3	2	3	y_7
8	1	3	2	3	2	1	3	1	y_8
9	1	3	3	1	3	2	1	2	y_9
10	2	1	1	3	3	2	2	1	y_{10}
11	2	1	2	1	1	3	3	2	y_{11}
12	2	1	3	2	2	1	1	2	y_{12}
13	2	2	1	2	3	1	3	2	y_{13}
14	2	2	2	3	1	2	1	3	y_{14}
15	2	2	3	1	2	3	2	1	y_{15}
16	2	3	1	3	2	3	1	2	y_{16}
17	2	3	2	1	3	1	2	3	y_{17}
18	2	3	3	2	1	2	3	1	y_{18}

Appendix 3 shows the method for determining the size of the error generated by the calibration operation. An example is a two-way experiment with a signal factor and error factors. Regression analysis is applied to calculate the error from the data.

Appendix 4 shows the method for determining the size of the error of indicated values of standards used for calibration. Historically a way of evaluating the error of standards was not clarified in public research laboratories or private sectors. Basically the method of Appendix 2 is applied in Appendix 4. A systematic approach to the error of standards will be conducted following the JIS in the future.

Although error evaluation for measuring instruments has not been defined clearly, the JIS Z9090 proposes a basic concept and technical background of error evaluation. In addition, it gives common rules for error evaluation, and the procedures are outlined with detailed examples along with the technical concepts. Moreover, it will be the basis for discussions with researchers in different fields, and it will be applied to various measurement fields.

Historically, Japan has imported standards from foreign countries for establishing JIS standards and has put forth much effort to make JIS consistent with the international standard. Japan can do much by proposing new and reliable concepts for international standards. For example, the concept of quality-cost performance based on a loss function is new for JIS and internationally. This aspect of JIS Z7109 has attracted much attention in the United States. Quality engineering concepts can benefit metrology and its application.

CONCLUSIONS

Research based on quality engineering at NRLM has been developed with the systematization of quality engineering. First, we applied experimental design as a statistical concept and the S/N ratio method to error evaluation in individual measurement areas. Statistical methods were not accepted easily. There were many arguments on the methods of error evaluation. Research on man-machine measurement system proved the importance of defining the measurement system. These individual applications were integrated into a quality design system, CAMPS. In the project, the concepts of metrological design were clarified.

NRLM proposed a theme of constructing an intelligent CAMPS in the MITI-proposed international cooperation project, IMS. In the project, we intend to develop a combined system of CAMPS, the database of specific technology, and the knowledge base of quality engineering to design various manufacturing systems and measurement systems efficiently. Real data for the database will be effectively collected by methods of quality engineering such as experiments based on a robust design-type of configuration.

REFERENCES

1 M. Koike, K. Tanaka, T. Kamoshita, J. Matsuda and H. Yano, Bull. NRLM, 38 (1989) 443.

2 G. Taguchi, Introduction to Quality Engineering, Asian Productivity Organization, Tokyo, 1986.

3 G. Taguchi, E. A. Elsayed and T. Hsiang, Quality Engineering in Production Systems, McGraw-Hill, New York, 1989.

4 G. Taguchi, System of Experimental Design Vol. 2, UNIPUB/Kraus International, New York, 1987.

5 K. Tanaka, M. Koike, T. Kamoshita, J. Matsuda and H. Yano, CAMPS, a system for designing a production process, in: Proc. Design Productivity, International Conference, Hawaii, 1991.
6 T. Kamoshita, M. Koike, J. Matsuda and H. Yano, Quality engineering approach of injection molding system for fine ceramics, Ninth Symposium on Taguchi Methods, Detroit, 1991.
7 Japanese Industrial Standard, JIS Z9090 Measurement—General Rules for Calibration Systems, Japanese Standard Association, Tokyo, 1991.

Author Index

Key Word Index